EVOLUTION OF FOSSIL ECOSYSTEMS

SECOND EDITION

Paul A. Selden
University of Kansas, Lawrence, Kansas, USA

John R. Nudds
University of Manchester, Manchester, UK

CRC Press
Taylor & Francis Group
Boca Raton London New York

CRC Press is an imprint of the
Taylor & Francis Group, an **informa** business

DEDICATION

We dedicate this book to the memory of Fred Broadhurst, a dear teacher, colleague, and friend.

First published 2012 by Manson Publishing Ltd

CRC Press
Taylor & Francis Group
6000 Broken Sound Parkway NW, Suite 300
Boca Raton, FL 33487-2742

First issued in hardback 2018

ISBN 13: 978-1-138-42405-0 (hbk)
ISBN 13: 978-1-84076-160-3 (pbk)

Visit the Taylor & Francis Web site at
http://www.taylorandfrancis.com

and the CRC Press Web site at
http://www.crcpress.com

Cover design: Cathy Martin
Book design and layout: Cathy Martin

A CIP catalogue record for this book is available from the British Library.

CONTENTS

ACKNOWLEDGEMENTS
Photography and Illustrations

American Museum of Natural History Library: 233, 236.

Cristoph Bartels, Deutsches Bergbau-Museum, Bochum: 87, 88, 101, 106, 109.

Günter Bechly, Museum für Naturkunde, Stuttgart: 358.

Fred Broadhurst, University of Manchester: 138, 139, 141, 142, 143, 144, 145, 146, 147, 148, 149, 150, 151, 152, 153, 154, 155, 156, 157.

P-J. Cheng, Nanjing Institute of Geology and Paleontology: 282.

Simon Conway Morris, University of Cambridge: 25, 27, 28, 36, 58.

Angela Delgado, Universidad Autonoma de Madrid: 322, 327, 332, 338.

Jason Dunlop, Leibniz Institute for Research on Evolution and Biodiversity at the Humboldt University, Berlin:132.

M. Ellison, American Museum of Natural History, New York: 287.

Andres Estrada, Fundidora Park, Monterrey: 239.

Susan Evans, University College London: 335.

Richard Fortey, Natural History Museum, London: 67.

Dino Frey, Staatliches Museum für Naturkunde, Karlsruhe: 365.

Sarah Gabbott, University of Leicester: 65, 68, 69, 70.

Jean-Claude Gall, Université Louis Pasteur de Strasbourg: 196, 197, 198, 199, 200, 202, 203, 204, 205, 206, 207.

David Green, University of Manchester: 30, 43, 46, 49, 53, 57, 99, 277, 361.

Richard Hartley, University of Manchester: 1, 4, 6, 12, 18, 19, 22, 38, 40, 59, 61, 86, 91, 117, 118, 123, 124, 125, 126, 127, 128, 131, 133, 134, 135, 137, 140, 158, 193, 195, 208, 211, 228, 230, 249, 255, 278, 279, 344, 345, 392, 393, 419, 420, 436, 439.

Rolf Hauff, Urwelt-Museum Hauff: 212, 214, 215, 217, 219, 220, 222, 225, 226, 227.

Sam Heads, University of Illinois: 301, 302.

Ken Higgs, University College, Cork: 440.

Cindy Howells, National Museum of Wales, Cardiff: 89, 98, 252, 254, 259, 267, 275, 284, 292, 293, 299, 300, 373, 374, 375, 394, 395, 396, 397, 399, 401, 402, 403, 405, 407, 408, 409, 410, 411, 412, 413, 414, 415, 416.

Jørn Hurum, Natural History Museum, University of Oslo: 390.

Antonio Lacasa-Ruiz, Institut d'Estudis Ilerdencs: 303, 306, 315, 316, 317, 318, 319, 320, 326, 328, 331, 336, 340, 341.

Robert Loveridge, University of Portsmouth: 353, 354, 355, 356, 357, 362, 363, 366, 367, 368.

David Martill, University of Portsmouth: 352, 370, 372, 378.

Federica Menon, University of Manchester: 359.

Natural History Museum of Los Angeles County: 441, 444, 446, 453.

John Nudds, University of Manchester: 2, 3, 5, 8, 11, 15, 20, 21, 90, 92, 94, 96, 97, 100, 102, 104, 209, 210, 229, 231, 232, 240, 243, 244, 246, 250, 251, 253, 261, 263, 264, 265, 266, 268, 269, 271, 272, 273, 274, 280, 281, 291, 294, 295, 296, 297, 298, 346, 347, 348, 349, 350, 351, 369, 376, 437, 438, 449, 451.

Burkhard Pohl, Wyoming Dinosaur Center: 242.

Graham Rosewarne, Avening, Gloucestershire: 23, 24, 26, 29, 31, 32, 33, 35, 37, 42, 45, 47, 51, 54, 93, 95, 103, 105, 107, 108, 213, 216, 218, 221, 223, 224, 235, 238, 241, 245, 248, 256, 258, 260, 262, 270, 276, 283, 285, 288, 290, 364, 371, 377, 398, 400, 404, 406, 442, 443, 445, 447, 448, 450, 452.

Sauriermuseum Aathal, Switzerland: 234, 237, 247.

Paul Selden, University of Kansas: 7, 9, 10, 13, 14, 16, 17, 60, 62, 63, 64, 111, 112, 113, 114, 115, 116, 119, 120, 121, 122, 129, 130, 136, 159, 160, 161, 162, 163, 164, 165, 166, 167, 168, 171, 172, 173, 174, 175, 176, 177, 178, 179, 180, 181, 183, 189, 190, 191, 194, 201, 304, 305, 307, 308, 309, 310, 311, 312, 313, 314, 329, 330, 333, 334, 337, 339, 342, 343, 360, 379, 380, 417, 418.

Forschungsinstitut und Naturmuseum Senckenberg, Messel Research Department: 381, 382, 383, 384, 385, 386, 387, 388, 389, 391.

David Siveter, University of Leicester: 39, 41.

Derek Siveter, Oxford University Museum: 44, 48, 50, 52, 55, 56 (all from Hou, X-G., Aldridge, R. J., Bergström, J., Siveter, D. J., Siveter, D. J. and Feng, X-H. 2004. The Cambrian Fossils of Chengjiang, China: the flowering of early animal life. Wiley-Blackwells); 72, 75, 76, 77, 78, 79, 80, 81, 82, 83, 84, 85.

Roger Smith, Iziko South African Museum: 169, 170, 182, 184, 185, 186, 187, 188, 192.

Carmen Soriano, Universitat de Barcelona: 321, 323, 324, 325.

Wouter Südkamp, Bundenbach, Germany: 110.

Geoff Thompson, University of Manchester: 257.

Wolfgang Weitschat, University of Hamburg, Germany: 421, 422, 423, 424, 425, 426, 427, 428, 429, 430, 431, 432, 433, 434, 435.

H. B. Whittington, University of Cambridge: 34.

Rowan Whittle, British Antarctic Survey: 66.

Xing Xu, Institute of Vertebrate Paleontology and Paleoanthropology, Beijing: 286, 289.

Access to sites and other help

Raimund Albersdörfer, Artur Andrade, Marion Bamford, Brent Breithaupt, Paulo Brito, Des Collins, John Dalingwater, Angela Delgado, Jason Dunlop, Bob Farrar, Mike Flynn, Jim Gehling, Zhouping Guo, David Green, Rolf Hauff, Sam Heads, Andre Herzog, Ken Higgs, Cindy Howells, Mary Howie, Jørn Hurum, James Jepson, Antonio Lacasa-Ruiz, Neal Larson, Robert Loveridge, Terry Manning, David Martill, Cathy McNassor, Federica Menon, Urs Möckli, Chris Moore, Robert Morris, Sam Morris, Robert J. Nudds, Burkhard Pohl, Annesuse Raquet-Schwickert, Helen Read, Glenn Rockers, Martin Röper, Hans-Peter Schultze, Chris Shaw, Bill Shear, David Siveter, Derek Siveter, Roger Smith, Carmen Soriano, Wouter Südkamp, Kent Sundell, Edie Taylor, Tom Taylor, Hannes Theron, Brian Turner, Rene Vandervelde, Jane Washington-Evans, Xing Xu.

ABBREVIATIONS

Specimen repositories are abbreviated in the figure captions as follows:

AMNH	American Museum of Natural History, New York, USA
BHIGR	Black Hills Institute of Geological Research, Hill City, South Dakota, USA
BKM	Bad Kreuznach Museum, Germany
BM	Bundenbach Museum, Germany
BMH	Berger Museum, Harthof, Germany
BMM	Bürgermeister-Müller Museum, Solnhofen, Germany
BPI	Bernard Price Institute for Palaeontological Research, University of the Witwatersrand, Johannesburg, South Africa
BSPGM	Bayerische Staatssammlung für Paläontologie und Historische Geologie, München, Germany
CAGS	Chinese Academy of Geological Sciences, Beijing, PRC
CFM	Field Museum, Chicago, USA
CNU	Capital Normal University, Beijing, PRC
DBMB	Deutsches Bergbau-Museum, Bochum, Germany
FPM	Fundidora Park, Monterrey, Mexico
GCPM	George C. Page Museum, Los Angeles, USA
GGUS	Grauvogel–Gall Collection, Université Louis Pasteur, Strasbourg, France
GMC	Geological Museum, Copenhagen, Denmark
GPMH	Geologische-Paläontologisches Museum, Universität Hamburg, Germany
GSSA	Council for Geosciences of South Africa, Bellville, South Africa
HMB	Humboldt Museum, Berlin, Germany
IEI	Institut d'Estudis Ilerdencs, Lleida, Spain
IVPP	Institute of Vertebrate Paleontology and Paleoanthropology, Beijing, PRC
JM	Jura Museum, Eichstätt, Germany
KMNH	Kitakyushu Museum of Natural History and Human History, Kitakyushu, Japan
MCCM	Museo de las Ciencias de Castilla-La Mancha, Cuenca, Spain
MM	Manchester University Museum, UK
MU	Manchester University, School of Earth, Atmospheric and Environmental Sciences, UK
NBI	National Botanical Institute, Pretoria, South Africa
NGMC	National Geological Museum of China, Beijing, PRC
NHM	Natural History Museum, London, UK
NIGP	Nanjing Institute of Geology and Paleontology, PRC
NMW	National Museum of Wales, Cardiff, UK
PBM	Paläobotanik Museum, Westfälische Wilhelms-Universität, Münster, Germany
PC	Private Collection
RC	Rubidge Collection, Wellwood, Graaff-Reinet, Eastern Cape, South Africa
RCCBYU	Research Center for the Chengjiang Biota, Yunnan University, PRC
SAM	Iziko South African Museum, Cape Town, South Africa
SI	Smithsonian Institute, Washington DC, USA
SM	Simmern Museum, Germany
SMA	Sauriermuseum Aathal, Switzerland
SMFM	Senckenberg Museum, Frankfurt-am-Main, Germany
SMNK	Staatliches Museum für Naturkunde, Karlsruhe, Germany
SMNS	Staatliches Museum für Naturkunde, Stuttgart, Germany
STMN	Shandong Tianyu Museum of Nature, Pingyi, PRC
TMC	Tate Museum, Casper, Wyoming, USA
UMH	Urwelt-Museum Hauff, Holzmaden, Germany
UOM	Natural History Museum, University of Oslo, Oslo, Norway
UWGM	University of Wyoming Geological Museum, Laramie, Wyoming, USA
WAM	Western Australian Museum, Perth, Western Australia
WDC	Wyoming Dinosaur Center, Thermopolis, Wyoming, USA

FOREWORD

Most major advances in understanding the history of life on Earth in recent years have been through the study of exceptionally well preserved biotas (Fossil-Lagerstätten). Indeed, particular Fossil-Lagerstätten, such as the Burgess Shale of British Columbia and the Solnhofen Limestone of Bavaria, have gained exceptional fame through popular science writings. Study of a selection of such sites scattered throughout the geological record – windows on the history of life on Earth – can provide a fairly complete picture of the evolution of ecosystems through time.

This book arose from the realization that there was an obvious void in the range of palaeontology texts available at present for a book which brings together succinct summaries of most of the better-known Fossil-Lagerstätten, primarily for a student and interested amateur readership. The authors teach an undergraduate course at third-year level which is based on case studies of a number of Fossil-Lagerstätten; we have also collaborated on the design of a new fossil gallery at Manchester University Museum based around this theme.

Following an introduction to Fossil-Lagerstätten and their distribution through geological time, each chapter deals with a single Fossil-Lagerstätte, or a number of related sites. Each chapter follows the same format: after a brief introduction placing the fossil occurrence in an evolutionary context, there then follows a history of study of the locality; the background sedimentology, stratigraphy and palaeoenvironment; a description of the biota; discussion of the palaeoecology; and a comparison with other Lagerstätten of a similar age and/or environment. At the end of each chapter is a list of museums to visit which display fossils of each locality and suggestions for visiting the sites.

In this Second Edition we have taken the opportunity to revise the original chapters in the light of many recent advances in knowledge, and also to add six new topics. The new chapters follow the same format as the others and, in some cases, more than one Lagerstätte is included for comparative or other purposes. For example, in Chapter 15, El Montsec and Las Hoyas, we compare two similar localities in the Cretaceous of Spain. In Chapter 9, Karoo, several important localities are included in order to tell the story of the evolutionary events which span the Permo–Triassic boundary, so wonderfully exposed in South Africa. There will, of course, always be exceptional localities worthy of inclusion which we are unable to cover. In this Second Edition we have added some Lagerstätten which are now world famous but were less well-known at the time of writing the First Edition. Chengjiang, the Herefordshire Nodules, and the Jehol Group are three such which are included for the first time. A welcome addition to the list of important localities of Cenozoic age is the White River Group, which preserves the finest examples of mammals around the Eocene–Oligocene boundary, including many now-extinct groups.

INTRODUCTION

The fossil record is very incomplete. Only a tiny percentage of plants and animals alive at any one time in the past get preserved as fossils, so that the palaeontologist attempting to reconstruct ancient ecosystems is, in effect, trying to complete a jigsaw puzzle without the picture on the box lid, and for which the majority of pieces are missing. Under normal fossilization conditions probably only around 15% of organisms are preserved. Moreover, the fossil record is biased in favour of those animals and plants with hard, mineralized shells, skeletons or cuticle, and towards those living in marine environments. Thus, the *preservation potential* of a particular organism depends on two main factors: its constitution (better if it contains hard parts), and its habitat (better if it lives in an environment where sedimentary deposition occurs).

Occasionally, however, the fossil record presents us with surprises. Very rarely, exceptional circumstances of one sort or another allow unusual preservation of soft parts of organisms, or in environments where fossilization rarely happens. Rock strata within the geological record which contain a much more completely preserved record than is normally the case are windows on the history of life on Earth. They have been termed *Fossil-Lagerstätten* (Seilacher *et al.*, 1985), a name derived from German mining tradition to denote a particularly rich seam of ore, to which Seilacher compared a bed rich in fossil remains. A Fossil-Lagerstätte can be translated into English as a fossil bonanza!

There are two main types of Fossil-Lagerstätten. *Concentration Lagerstätten* (Konzentrat-Lagerstätten), as the name suggests, are deposits in which vast numbers of fossils are preserved, such as coquinas (shell accumulations), bone beds, cave deposits, and natural animal traps. The quality of individual preservation may not be exceptional, but the sheer numbers are informative. *Conservation Lagerstätten* (Konservat-Lagerstätten), on the other hand, preserve quality rather than (often as well as) quantity, and this term is restricted to those rare instances where peculiar preservation conditions have allowed even the soft tissue of animals and plants to be preserved, often in incredible detail. It may also be used for deposits that yield articulated skeletons without soft tissue. Most of the Lagerstätten described in this book are examples of Conservation Lagerstätten, and some fit into both categories. Some chapters (e.g. Chapter 9, Karoo, and Chapter 16, The Santana and Crato Formations) deal with more than one Lagerstätte, some (e.g. Chapter 12, The Morrison Formation) bear numerous horizons which are rich in fossils, while Chapter 5 (The Herefordshire Nodules) deals with just one horizon at a single locality.

There are various types of Conservation Lagerstätten including conservation traps such as entombment in amber, deep-freezing in permafrost, pickling in oil swamps, and mummification by desiccation. On a larger scale are obrution deposits, where episodic smothering ensures rapid burial of mainly benthonic (sea-floor) communities, and stagnation deposits, where anoxic (low oxygen) conditions in stagnant or hypersaline (high salinity) bottom waters ensure reduced microbial decay, in predominantly pelagic (open-sea) communities. In fact, most Conservation Lagerstätten combine obrution and stagnation in the preservation of soft tissue.

Taphonomy is the name given to the process of

preservation of a plant or animal as a fossil. It actually consists of three main processes: *necrosis*, which refers to the changes which occur at or shortly after death, such as rigor mortis; *biostratinomy*, which covers the course of events from the time after death to burial in sediment (or entombment in amber, cave deposits, and so on). The time taken by this process can vary from a few minutes (e.g. insects trapped in amber, or mammals in tar) to many years for an accumulation of bones or shells. Ideally, for exceptional preservation of soft tissues, the time between death and isolation from oxygen and decaying organisms should be short. Following burial, the process of *diagenesis* begins: the conversion of soft sediment or other deposits to rock. Further destruction of organic molecules can occur during diagenesis; the action of heat can turn organic molecules into oil and gas, for example, and crushing in coarse sand can fragment plant and animal cuticles.

Soft-tissue preservation has three important implications. First, the study of soft-part morphology alongside the morphology of the shell or skeleton allows better comparison with living forms and provides additional phylogenetic information. Second, it enables the preservation of animals and plants which are entirely soft bodied and which would normally stand no chance of fossilization. For example, it has been estimated that 85% of the animals in the Burgess Shale (Chapter 2) are entirely soft bodied and are therefore absent from Cambrian biotas preserved under normal taphonomic conditions. The third implication follows – such Conservation Lagerstätten therefore preserve for the palaeontologist a complete (or much more nearly complete) ecosystem. Comparison of such horizons in a chronological framework gives us an insight into the evolution of ecosystems over geological time.

The fossil occurrences described in this book are arranged in chronological order, from the late Precambrian Ediacara biota to the Pleistocene Rancho La Brea (*Table 1*), so it is possible to follow the development of the Earth's ecosystems through a series of snapshots of life at a number of points in time. While this does not give a complete picture of the evolving biosphere, Lagerstätten are important because they preserve far more of the biota than occurs under normal preservation conditions, so the palaeontologist can see as completely as possible the ecological interactions of the organisms in that particular habitat. In the late Precambrian Ediacara biota, for example, it is possible that we are seeing a different grade of organization of organisms and modes of life than we see in later, Phanerozoic time. This biota existed before hard parts of animals evolved and predation became widespread. By the Cambrian Period, the time of the Chengjiang and Burgess Shale biotas, almost all animal phyla had developed, and it is possible to reconstruct the ecological dynamics of the sea floor, complete with predators, scavengers, and deposit feeders. Similarly diverse assemblages are seen in the Silurian Herefordshire Nodules and the Devonian Hunsrück Slate. Most of these biotas preserve mainly benthos (sea-floor dwellers). Benthos can be divided into infauna (animals living within the sediment) and epifauna (animals living on the sediment surface). Some of the biota in these Lagerstätten belong to the nekton (swimmers), and in the Ordovician Soom Shale the nekton is dominant because there were only rare occasions when the sea floor was conducive to benthonic life. To complete the picture of marine life-styles, organisms which float are called plankton.

A major evolutionary advance which occurred in mid-Palaeozoic times was the colonization of land by plants and animals. The Devonian Rhynie Chert was one of the first, and still the best-known, biota preserving some of the earliest land plants and animals. By late Carboniferous times, the land in tropical regions had become colonized by forest, with its accompaniment of insects and their predators. The Mazon Creek biota preserves a forest ecosystem mingled with non-marine aquatic organisms in a deltaic setting, so common in this period of major coal formation. The Permian Period saw the rise of mammal-like reptiles on land, which ultimately gave rise to the true mammals. But before that could come about, the end-Permian extinction event, by far the greatest of all time, saw the demise of some 80% of living things. The Karoo Supergroup of southern Africa records these tremendous events in minute detail.

Life recovered from the mass extinction, and when we examine the biota of the Triassic Grès à Voltzia delta, we find many similarities to that of Mazon Creek. Three Lagerstätten are represented from the Jurassic Period, two marine and one terrestrial. The Holzmaden *Posidonia* shales

Million years before present	Era	Period			Lagerstätte
2.6	CENOZOIC	Quaternary		Holocene	Rancho La Brea
				Pleistocene	
			Neogene	Pliocene	
				Miocene	
23		Tertiary		Oligocene	Baltic Amber
					White River Group
			Palaeogene	Eocene	Grube Messel
65				Palaeocene	
	MESOZOIC	Cretaceous			Santana and Crato
					El Montsec and Las Hoyas
					Jehol Group
146					Solnhofen Limestone
		Jurassic			Morrison Formation
200					Holzmaden Shale
		Triassic			Grès à Voltzia
251					Karoo
	PALAEOZOIC	Permian			
299		Upper Carboniferous (Pennsylvanian)			Mazon Creek
320		Lower Carboniferous (Mississippian)			
359		Devonian			Hunsrück Slate
					Rhynie Chert
416		Silurian			Herefordshire Nodules
444		Ordovician			Soom Shale
488		Cambrian			Burgess Shale
					Chengjiang
542	PRECAMBRIAN	Ediacaran			Ediacara
~635					
4600					

Table 1 Stratigraphic table of Fossil-Lagerstätten described in this book.

represent a snapshot of pelagic marine life in the Jurassic Period, in which large marine vertebrates such as plesiosaurs, ichthyosaurs, and crocodiles are found together with their prey: cephalopods and fish. In contrast, the Solnhofen Plattenkalk (lithographic limestone) preserves marine plankton, nekton and benthos, such as ammonites, horseshoe crabs, crustaceans, as well as rare flying animals, e.g. *Archaeopteryx* – the first bird, all swept together into a lagoon by severe storms. On land in the Jurassic Period dinosaurs dominated the scene, and the Morrison Formation of western USA is the best-known Lagerstätte preserving these giants.

The Jehol biota of north-east China has now become very famous for its preservation of early birds, alongside feathered dinosaurs, as well as many other interesting plants and animals. It is dated to the early Cretaceous Period. Also early in the Cretaceous are two Lagerstätten in close proximity to each other in north-east Spain: El Montsec and Las Hoyas. Together, these two Lagerstätten provide a detailed insight on the palaeoecology of life in and around lakes during the early Cretaceous. A region of what is now Brazil also witnessed two Fossil-Lagerstätten: the Santana and Crato Formations. The former is best known for its fish and pterosaurs in nodules, while the latter is well known for its insects and plants in a Plattenkalk.

The dinosaurs died out together with ammonites, marine reptiles, and some other plant and animal groups at the end of the Cretaceous Period. In the following Cenozoic Era mammals became the dominant vertebrate group, and some of the finest fossils of these occur in the Grube Messel locality in Germany, which has preserved terrestrial plants and animals in a peculiar setting: a crater lake. Grube Messel has revealed some of the earliest diversity of mammals, including horses, bats, and lemurs. Not long after Grube Messel was being deposited, the White River Group of North America was preserving some of the largest land mammals ever seen, along with the first occurrence of several modern groups of carnivores. Land animals and plants are generally much rarer as fossils than those which live in places where sediments are being laid down, such as lakes and the sea, so Lagerstätten which preserve terrestrial biotas are especially prized. Grube Messel and White River are two good examples, and the amazing fauna (especially insects) of Baltic amber is another. Amber (fossilized tree resin) acts as a sticky trap for insects and their predators and, in a similar manner, the tar pits of Rancho La Brea attracted mammals and birds in search of a drink, trapping them and their predators and scavengers in sticky tar. Rancho La Brea thus preserves a snapshot of land life in southern California over the last 40,000 years.

EDIACARA

BACKGROUND: FIRST LIFE ON EARTH

Life on Earth arose some 3.5 billion years ago. There is some debate concerning what actually constitutes 'life' and, indeed, whether life actually arose on this planet or arrived here from outer space in a simple form and further evolved here. Nevertheless, the earliest fossil evidence of microbial prokaryotes akin to modern cyano-bacteria (blue-green algae) comes from cherts in Western Australia. For some 2.5 billion years after its origin, life evolved slowly. Eukaryotes (cells which contain a nucleus and organelles) evolved from prokaryotes, but it was not until about 1,000 million years ago that multicellular forms developed. These first multicellular organisms form the subject of this chapter. Whether they are plants (metaphytes), animals (metazoans), both, or neither is hotly debated, but they were typically flattish organisms, with a high surface area/body mass ratio. The development of multicellularity was a major step in the evolution of life: it enabled organisms to grow in size, to develop organ systems through tissue differentiation, and led to the plants and animals with which we are familiar today.

Until the middle of the last century it was thought that rocks older than Cambrian in age, collectively called Precambrian, were devoid of fossils of multicellular creatures. Fossils of shelled animals, brachiopods, trilobites and sponges, for example, appear apparently suddenly in Cambrian-age rocks. The discovery of soft-bodied organisms similar in appearance to jellyfish and worms in rocks of late Precambrian age was therefore a major surprise, and led to a complete reappraisal not only of the fossil record of multicellular organisms but also of the evolution of life and its relationship to the Earth's physical systems (atmosphere, oceans). The question changed from 'why did multicellular life suddenly appear at the base of the Cambrian?' to 'why did multicellular organisms suddenly develop hard parts at the start of the Cambrian?' (see Chapter 2, The Burgess Shale).

HISTORY OF DISCOVERY OF THE EDIACARA BIOTA

In 1946 Reginald C. Sprigg, a government geologist, was exploring an area of the Flinders Ranges some 300 km (c. 190 miles) north of Adelaide, Australia, known as the Ediacara Hills (**1**). In these hills he found fossilized imprints of what were apparently soft-bodied organisms, preserved mostly on the undersides of slabs of quartzite and sandstone (**2, 3**). Most were round, disc-shaped forms that Sprigg called 'medusoids' from their seeming similarity to jellyfish (Sprigg, 1947, 1949). Others resembled worms and arthropods, and some could not be classified.

Initially, Sprigg thought that these rocks were Cambrian in age because they contained fossils, but later work established that their age was, in fact, late Precambrian. While scattered reports of soft-bodied organisms had appeared in the scientific literature as far back as the mid-nineteenth century, this was the first diverse assemblage of well-preserved Precambrian fossils to be discovered and was studied in detail by Martin Glaessner and Mary Wade of the University of Adelaide (Glaessner & Wade, 1966). Not long after Sprigg's discovery, assemblages of soft-bodied organisms were discovered in Leicestershire, UK (Ford, 1958) and Namibia, and Ediacara-type fossils are now known from the White Sea area of Russia, Newfoundland and North-west Territories (Canada), North Carolina (USA), Ukraine, China, and many other places. They all occur in the Ediacaran Period, which was officially recognized by the International Union of Geological Sciences in 2004: the first new geological period declared in 120 years. It ranges from the end of the global Marinoan glaciation (c. 635 Ma) to the first complex trace fossil *Treptichnus* at the start of the Cambrian Period (c. 542 Ma). The main Ediacara biotas can be found in the latter part of the Ediacaran Period, following the regional Gaskiers glaciation of around 580 Ma. Ediacara itself is dated at around 555 Ma.

STRATIGRAPHIC SETTING AND TAPHONOMY OF THE EDIACARA BIOTA

The first fossils found by Sprigg came from the Ediacara Hills area, but rock sequences containing Precambrian fossils also occur in gorges through the Heysen Range to the south (e.g. Parachilna Gorge, Brachina Gorge, Bunyeroo Gorge, Mayo Gorge; **1**), and at the eastern end of the Chace Range. The fossils are confined to a stratigraphic range of no more than 110 m (c. 360 ft) in the Ediacara Member of the Rawnsley Quartzite, which lies 500 m (c. 1,640 ft) below the earliest Cambrian in this area. The Rawnsley Quartzite is part of the Pound Supergroup (**4**), named after Wilpena Pound, a dramatic eroded syncline whose circle of quartzite cliffs faces outwards in a natural fortification.

The Ediacara Member consists of a series of siltstones and sandstones which represent pelagic to intertidal conditions. The implication is that there was a continental edge delivering sediment into deep water, at times by means of turbidite flows and occasionally as a delta which shallowed the water to sub- and inter-tidal levels. Some storm horizons can be seen. It is at these shallow levels, around the storm wave base, that the fossils occur. Aiding preservation are the thin films of clay which

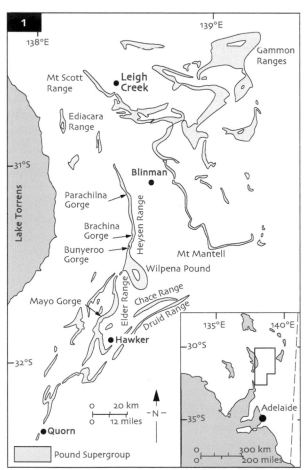

1 Map showing distribution of the Pound Supergroup in South Australia (after Gehling, 1988).

2 Greenwood Cliff in the Ediacara Hills, the site of Sprigg's discovery of fossils in the Rawnsley Quartzite in 1946.

3 Overturned slabs of Rawnsley Quartzite, Ediacara Hills, with fossils preserved on the rippled undersurfaces.

5 Microbial mat preserved on the surface of ripples, Rawnsley Quartzite, Ediacara Hills.

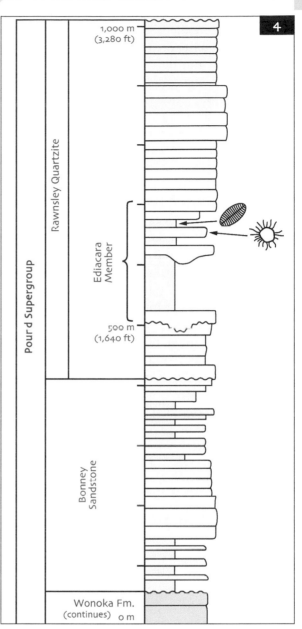

4 Stratigraphy of the late Proterozoic Pound Supergroup of South Australia, showing the position of the Ediacara biota in the Ediacara Member (after Bottjer *et al.*, 2002).

represent gentle deposition from suspension and which occur between the sandstone layers, the latter representing more energetic flow and sediment deposition, possibly during storm events. The clay acted to some extent as a glue, cohering the sands beneath and the fossils lying on the sea bed, and it moulds fine detail of the fossils, enabling interpretation of their morphology. Some of the rippled surfaces also show evidence of a microbial mat on the sea bed (**5**), which could have also aided preservation by binding and stabilizing the sediment.

Being soft-bodied, the fossils are generally preserved squashed. The clay layers compact considerably during diagenesis, so relief is provided by the sandstones. Figure **6** (from Gehling, 1988) shows the effect of different thicknesses of clay on the preservation of the fossils. In some cases, an external mould of the upper surface of the fossil is preserved on the base of the sandstone as an impression (e in **6**). Sometimes, the fossil collapses or decays so that sand fills the space previously occupied by the organism and produces a cast, visible as a positive relief on the base of the sandstone (c in **6**). If the clay layers are thin (**6B**) then the cast can project deeper into the soft sand beneath, forming a counterpart mould (cpm in **6B**); conversely, a counterpart cast (cpc in **6B**) can also form. In this way, dorsal and ventral structures may be superimposed upon one another. Some of the organisms had thin outer walls but more resistant internal organs, which are moulded preferentially, e.g. the gonads of possible medusoids. More recent work has suggested that microbial action may have been important or, indeed, vital for the preservation of the Ediacara biota (Gehling, 1999).

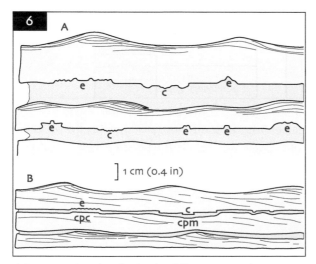

6 Preservational styles in the Ediacara Member;
A: with thick, and **B**: with thin clay interlayers.
c: cast on base of sandstone; **e**: external mould on base of sandstone; **c p c**: counterpart cast on top of sandstone; **c p m**: counterpart mould on top of sandstone (after Gehling, 1988).

Because the fossils are only moulds and casts, no organic matter remains: they are best viewed in low-angle light, either evening light in the field or light from a lamp at low angle in the laboratory. Silicone rubber casts of fossils preserved as moulds, and vice versa, can provide better views of some material.

DESCRIPTION OF THE EDIACARA BIOTA

Ediacaria is a concentric discoidal form first described by Sprigg in the 1940s. It has now been found in other Ediacara localities, e.g. north-west Canada, and also Booley Bay in Ireland, at which locality the strata have been shown on microfossil evidence to be Middle Cambrian in age, thus extending at least this component of the Ediacara biota to younger times. It could be a jellyfish, a benthic form, or possibly the holdfast of another organism such as a sea-pen.

Cyclomedusa (**7, 8**) is a primarily radiate form but with some concentric lines near the centre. It grew to nearly 1 m (c. 3 ft) across. It is probably the commonest and most widespread of Ediacara fossils. Early authors suggested that *Cyclomedusa* was a large, floating jellyfish; Seilacher (1989) envisaged it as benthic; it could be interpreted as a low, cone-like form of sea anemone (Gehling, 1991); and another hypothesis is that it is simply the holdfast of a colonial octocoral. The evidence for the sea anemone form comes from the concentric lines near the centre, which could be artefacts of compression, and the occurrence of a number of specimens closely adpressed, which would be unlikely if they were free-floating creatures. However, its abundance supports the idea that it was a holdfast.

Pseudorhizostomites (**9**) is a radial form with an indefinite edge. It could represent a medusoid, or possibly the impression of the base of a larger organism.

Tribrachidium is a small (c. 20 mm [0.8 in] in diameter) disc-like form with a three-fold radial symmetry consisting of three lobes in the central region, each with a raised leading edge which turns at the outer region of the central area to run along the edge of the area. There is an outer zone composed of three flat areas which appear to emanate from the central lobes; each area bears radiating ridges (**10**). *Tribrachidium* does not fit easily in any extant phylum.

Mawsonites is another radial form, typified by concentric rows of lobate shapes getting larger from the centre outwards. It was interpreted as a medusoid by Glaessner and Wade (1966), but as a complex trace fossil by Seilacher (1989).

Arkarua was described by Jim Gehling from the Chace Range. Its five-fold symmetry immediately suggests it may be an echinoderm – the earliest known – which has yet to develop the calcareous plates typical of most modern members of the phylum.

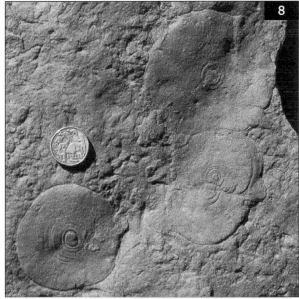

7 Giant *Cyclomedusa* (WAM). About 300 mm (c. 12 in) across.

8 Smaller *Cyclomedusa*, from the locality shown in Figure **3**. Coin 24 mm (c. 0.94 in) in diameter.

9 Cast of *Pseudorhizostomites* (MM). About 50 mm (c. 2 in) across.

10 Cast of *Tribrachidium* (MM). About 20 mm (c. 0.8 in) across.

11 *Inaria*, from the locality shown in Figure 3. Coin 24 mm (c. 0.94 in) in diameter.

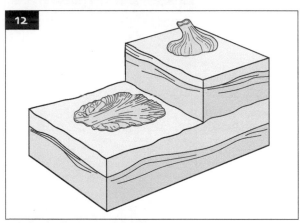

12 Reconstruction of *Inaria* (top) and sketch of fossil as it appears in the rock (below) (after Gehling, 1988).

13 Cast of *Charnia* (MM). Length about 150 mm (c. 6 in).

Inaria would fit with radial forms, except that reconstruction of the organism shows it to have been shaped rather like a garlic bulb in life (11, 12). Its describer, Jim Gehling (1988), suggested that it might be a type of sea anemone.

Charnia (13) is one of two fossils described by Trevor Ford in 1958 from Charnwood Forest, Leicestershire; the other is *Charniodiscus*. *Charnia* is a leaf-shaped structure with a central, apparently stiff shaft and lateral segmented areas. *Charnia* was compared by Ford (1958) to the living sea-pens (Pennatulacea), a group of Cnidaria. However, Antcliffe and Brasier (2007) showed that *Charnia* is unlike modern sea-pens, which grow from the base of the frond, while *Charnia* appears to grow from its tip. When first described, *Charniodiscus* resembled *Ediacaria* in being a simple disc with concentric lines. Later, a specimen was found in which a *Charnia*-like frond was attached to a *Charniodiscus*, and it is now certain that *Charniodiscus* closely resembles *Charnia*. *Rangea* is similar to *Charnia* but shows subsegmentation of the individual segments on each side. *Pteridinium* is another sea-pen-like organism without subsegmentation. The Burgess Shale *Thaumaptilon* (p. 26) appears to be a Cambrian survivor of this group.

Ernietta was sock-shaped in life; its segments were directed upwards and its base appears to have been filled with sediment as an anchor.

Kimberella was originally considered to be a medusoid with four-fold symmetry, but later work suggested that it was benthic and slug-like, possibly an early mollusc.

Small *Dickinsonia* (14) look almost like the radial organisms, but as they grew (up to 1 m [3 ft] in length) they acquired an elongate body and bilateral symmetry. *Dickinsonia* is segmented on each side of a mid-line, but the segments do not match up on either side (15), though this could be a taphonomic artefact. It is difficult to decide which, if any, is head or tail end. The appearance of *Dickinsonia* is reminiscent of a flat worm, and in a recent study by Sperling and Vinther (2010), *Dickinsonia* was likened to the Placozoa. These primitive metazoans were poorly known until recently. They consist of a flat disc of cells which moves across the sea bed feeding through its sole. Discoveries of specimens of *Dickinsonia* apparently at the end of trails show that it did, indeed, move and possibly fed on microbial mats on the sea floor, like modern Placozoa.

Parvancorina (**16**) is a small creature with bilateral symmetry and a definite 'head' end (or holdfast). It could be a juvenile of some larger Ediacara organism. *Praecambridium* is another small, bilateral organism with a 'head' end and a segmented 'body'. Although considered by some to be arthropod-like, Birket-Smith (1981a) suggested it might be a juvenile *Spriggina*. Like *Parvancorina* and *Praecambridium*, *Vendia* is another small organism with bilateral symmetry, a definite head, and body segments, and could also be a juvenile of something else.

Spriggina (**17**) is a small (c. 50 mm [2 in]) organism with a horseshoe-shaped 'head' followed by an elongate, leaf-like body composed of two rows of short segments either side of a medial line. At first, *Spriggina* was thought to resemble a polychaete worm such as *Nereis*, but a close look at the segmentation reveals that the segments do not match across the mid-line, just as in *Dickinsonia*. Seilacher (1989) turned the interpretation upside-down, suggesting that *Spriggina* could be another type of sea-pen, and that the 'head' was actually a holdfast.

One additional fossil type present in the Ediacara biotas is the trace fossil: evidence of crawling and shallow ploughing through the sediment by animals is present in many of the Ediacara localities. For example, a spiralling trail found at Ediacara suggests grazing on sediment rather than simply getting from A to B. Though reflecting fewer body plans than seen at the present day (there are no burrowers), the traces provide evidence for true Metazoa.

First attempts to classify the Ediacara biota, by Sprigg (1947, 1949), Glaessner (1961) and

14 A small *Dickinsonia* (MM). Length about 50 mm (c. 2 in).

15 Larger *Dickinsonia*, from the locality shown in Figure 3. Coin 24 mm (c. 0.94 in) in diameter.

16 Cast of *Parvancorina* (MM). Length about 30 mm (c. 1.2 in).

17 Cast of *Spriggina* (MM). Length about 50 mm (c. 2 in).

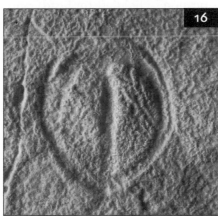

Glaessner and Wade (1966), suggested a variety of extant phyla, including Cnidaria and Annelida. Fifteen species of Ediacara organisms were described as jellyfish (medusoids), e.g. *Mawsonites*, *Cyclomedusa*, *Kimberella*, and *Eoporpita* (the only one which shows tentacles). *Charnia*, *Pteridinium*, and *Rangea* were referred to the Pennatulacea or sea-pens, or soft corals, another class of Cnidaria. *Dickinsonia* and *Spriggina* were allied to the Annelida. However, other organisms such as *Tribrachidium* could not be placed in any modern phylum. Its three-fold symmetry is not seen in the ground plan of any living group of animals.

Seilacher (1989) removed the Ediacara organisms from the Metazoa altogether, preferring instead to regard them as a separate kingdom – Vendobionta – based on different functional design from plants and animals. He looked at the constructional morphology of the organisms and suggested they may have had internal hydrostatic skeletons, their bodies gaining rigidity through internal pressure, as in a car tyre. Compartmental-ization of the body made sense to prevent total loss of pressure (and hence death) when only one or two segments were punctured. They were quilted organisms, similar to inflatable mattresses. The high surface area to mass ratio suggested they respired through the skin; they may have been photosynthetic, as suggested by McMenamin (1998), used photosynthetic symbionts like modern corals, perhaps were chemosymbiotic (with chemosynthetic symbionts so they could survive in deep-water, reducing environments), or may have ingested materials through the body wall. Seilacher suggested that such organisms could survive without bony skeletons or shells because there was little predation.

In the 1980s Fedonkin devised a classification scheme for the Ediacara fossils based on their symmetry and thus independent of any biological interpretation. He divided the organisms into two major groups: Radiata (disc-shaped, no bilateral symmetry) and Bilateria (bilateral symmetry apparent). Radiata can be further subdivided into: Cyclozoa, with a concentric (rather than radial) pattern (e.g. *Ediacaria*); Inordozoa, with a radial pattern of indeterminate radius (e.g. *Cyclomedusa*); and Trilobozoa with a three-rayed symmetry (e.g. *Tribrachidium*). Other radiate forms were considered to belong to the cnidarian classes

Conulata or Scyphozoa. Bilateria may show no particular head or tail (i.e. bidirectional, e.g. *Dickinsonia*), or have a distinct head or rooting structure (i.e. unidirectional, e.g. *Spriggina*). Intermediates exist, for example, between the concentric and indeterminate radial forms: *Cyclomedusa* shows a concentric pattern, especially near the centre of the disc. It is possible to envisage evolutionary trends among the organisms. Inordozoa (e.g. *Cyclomedusa*) could have given rise to more organized, determinate forms (e.g. *Tribrachidium*), or become elongate (e.g. *Dickinsonia*) and then developed a head and tail when directed locomotion evolved (e.g. *Spriggina*). One problem with this scheme, as pointed out by Gehling, is that it only works if the organisms are flattened, yet many reconstructions show the creatures as quite three-dimensional. *Inaria*, for example, was probably bulb-shaped, and *Pteridinium* appears to have had three fronds rather than the two shown by the similar *Charnia*.

In a strange twist to an already bizarre story, Greg Retallack, of the University of Oregon, looked at the indeterminate growth pattern of many of the Ediacara organisms, which suggested that they had no upper limits nor definite bounds to their size and shape (Retallack, 1994). For example, *Dickinsonia* appeared to have no adult size and shape. He also studied taphonomic aspects of the Ediacara biota, particularly their compression. The conclusion? That Ediacara fossils were lichens! Lichens are composite organisms formed by a symbiotic relationship between green algae (which provide photosynthesis) and fungi (which provide bulk). Few palaeontologists have accepted the arguments presented by Retallack. While some Ediacara organisms (e.g. *Spriggina*) show a distinct holdfast (or head), and their growth patterns may be less determinate than worms, for example, they are not random; and the sheer size and bulk of many forms is quite unlike that of modern lichens. Moreover, modern lichens are terrestrial, not marine. The general consensus has always been that the organisms present were metazoan-grade animals; it is possible that some of the frond-like forms actually represent macro-plants.

A study by Dewell *et al.* (2001) suggested that many Ediacara animals were colonial forms (pennatulaceans are colonial), and that they represented a grade of organization more developed than simple sponges, in which few cell types are present, but less evolved than higher Eumetazoa, in which tissues and organs have developed. Thus, the 'ediacarans' are probably not a separate kingdom but part of the Animalia during an early phase of the development of the metazoan body plans that are familiar today.

PALAEOECOLOGY OF THE EDIACARA BIOTA

Early attempts to reconstruct the Ediacara scene (e.g. Glaessner, 1961) showed a preponderance of medusoids and sea-pens in a shallow-water habitat. Later evidence, discussed above, suggests deeper water and different lifestyles. Indeed, Gehling (1991), in a thought-provoking essay on the Ediacara biota, dispelled a number of earlier myths which had grown up about it. First, Ediacara organisms were originally thought to have been large in comparison with Cambrian faunas; in fact, while some large *Dickinsonia* do occur, most specimens are small. The preoccupation with considering the Ediacara fossils as essentially two-dimensional does not take into account preservational factors; after restoration, many organisms are conical, hemispherical, or tubular. Early reconstructions showed medusoids dominating the scene; most organisms were probably benthic sessile or vagile forms. One generalization was that the assemblage was allochthonous (drifted into the site of deposition);

while this may be true for some localities (e.g. Namibia), in most cases the fossils appear to represent autochthonous assemblages with just a few pelagic forms. The trace fossils described were originally thought to have been formed by taxa unrepresented in the body fossil record; Gehling has suggested that a number of traces were formed by animals present in the biota. Studies on the biomechanics of Ediacara organisms (Schopf & Baumiller, 1998) indicated that the flat animals would have been prone to dislodgement from the substrate in the strengths of currents suggested by the sedimentological evidence. They surmised that the animals might have been moved from a quieter environment to their final resting place (in which case why are they so well preserved?), were more dense or more adhesive to the substrate than other researchers had reckoned, and/or lived buried to some degree within the substrate.

COMPARISON OF EDIACARA WITH OTHER LATE PRECAMBRIAN BIOTAS

The map in Figure **18** shows how widespread Ediacara biotas are. The map also shows the distribution of Cambrian Burgess Shale-type faunas (Chapter 2). While many of the organisms present in the Ediacara biotas are similar, there are some important differences between the localities, in taphonomy for example. At Ediacara itself, as discussed, the biota are preserved in shale partings

18 Map showing distribution of Ediacara and Burgess Shale (Chapter 2) biotas worldwide (after Conway Morris, 1990).

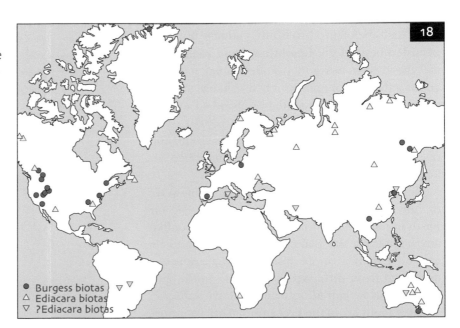

in sandstone beds, in an apparently quiet-water environment. At Mistaken Point, Newfoundland (see Nudds & Selden, 2008), the organisms occur at the bases of volcanic ash layers, while the Namibian fossils show evidence of moving sand and mud during their deposition. Ediacara biotas have also been found in Russia and the Ukraine (White Sea, Podolia, southern Ukraine, Ural Mountains, Siberia); north-west Canada; central Australia; Finnmark, Norway; Charnwood Forest, Leicestershire, UK; south and north China; south-western United States and northern Mexico; and North Carolina. Restricted or doubtful Ediacara-type biotas have also been reported from India, Iran, Ireland, Morocco, Sardinia, and Carmarthenshire, Wales. Discoveries in the Xingmincun Formation, north-eastern China, of flat, flexible, discoidal fossils with concentric or spiral ridges have been called the Jinxian biota (Zhang *et al.*, 2006). These seem unrelated to disc-like forms in the Ediacara biota, and nor do they resemble any living forms.

What happened to the Ediacara biota after the end of the Precambrian? Some apparently continued to the present day. For example, it is possible that some ediacarans are Pennatulacea, and *Thaumaptilon* of the Burgess Shale could be too; Scyphozoa (jellyfish) are an important component of the modern marine plankton. However, there is little doubt that many forms did not survive to the Phanerozoic. At the very beginning of the Cambrian, in the Tommotian, there is a distinctive fauna of 'small shelly fossils' which just pre-date the first Cambrian skeletal body fossils. Some of these small shells belong to hyolithids, and may not have persisted beyond the Cambrian. More interesting are the tiny spines and pastille-shaped sclerites, commonly phosphatic, which are likely to have studded the cuticle of essentially soft-bodied animals like onychophorans or *Wiwaxia* (p. 29). The relationship between the Ediacara biota and the diverse fauna of the Cambrian (described in Chapter 2) is not entirely resolved.

Regardless of what the Ediacara organisms were, or how they lived, the shallow marine scene on Earth in late Precambrian times was utterly unlike any habitat on the planet today. It is only to be expected that, as more fossils of ediacarans are described from different parts of the world, and new ideas as to their affinities and lifestyles arise, so debates about life in the Precambrian will continue.

MUSEUMS AND SITE VISITS

Museums

1. South Australian Museum, Adelaide, Australia.
2. Western Australian Museum, Perth, Australia.
3. Sedgwick Museum, University of Cambridge, Cambridge, England.
4. University of California Museum of Paleontology, Berkeley, California, USA (online exhibit: http://www.ucmp.berkeley.edu/).
5. Humboldt State University Natural History Museum, Arcata, California, USA.
6. Manchester University Museum, Oxford Road, Manchester, England.

Sites

Ediacara biota localities are situated in the Flinders Ranges National Park some 500 km (c. 300 miles) north of Adelaide. One very accessible site is the Brachina Gorge Geological Trail, which starts just north of Oraparinna on the road towards Blinman, and passes through the complete succession between the Trezona Formation (c. 630 Ma) and the Wirrealpa Limestone (c. 520 Ma). Twelve different formations are seen, including the Rawnsley Quartzite with its Ediacara fossils. The trail is 20 km (c. 12 miles) long and is generally passable to conventional vehicles, although the road surface may deteriorate after rain. Camping is permitted in designated areas.

THE BURGESS SHALE

Chapter Two

BACKGROUND: THE CAMBRIAN EXPLOSION

Although multicellular animals had only appeared at the very end of the Precambrian (Chapter 1), their evolution at the beginning of the Cambrian was so rapid that this event is known as the 'Cambrian Explosion'. In an astonishing orgy of evolution, a period of little more than 10 million years at the beginning of the Cambrian saw the appearance of almost every animal phylum and body plan known today, along with some other bizarre forms which soon became extinct, suggesting that this was an experimental phase of evolution. Approximately 35 different animal phyla are known today; in the Cambrian seas there were undoubtedly several more and some authorities have claimed up to 100.

The sudden appearance of this diverse fauna within the geological record has long posed perplexing questions. Darwin supposed that these phyla had been gradually evolving throughout the Precambrian, but had simply not been preserved as they were entirely soft-bodied. Perhaps they simultaneously acquired preservable hard shells or skeletons at the onset of the Cambrian (in response to a critical change in atmospheric oxygen or oceanic chemistry) so that the Cambrian Explosion was no more than an artefact of preservation?

However, the discovery of the Precambrian Ediacara biota (Chapter 1) in the latter half of the twentieth century showed that while soft-bodied animals did exist at the very end of the Precambrian, these were mostly primitive annelids and cnidarians, or were at least on an evolutionary pathway towards cnidarians, and were unlikely ancestors for the characteristic Cambrian animals such as archaeocyathids, brachiopods, trilobites, and molluscs. Moreover, even soft-bodied animals should leave trace fossils, which are also absent from all but the latest Precambrian sediments. It seems that as Conway Morris (2006) concluded in his thought-provoking essay on 'Darwin's dilemma', the Cambrian Explosion was a 'real event'.

The first wave of the evolutionary explosion is evident in the basal Cambrian Tommotian Stage with the sudden appearance of many tiny shelly fossils. It may be, as discussed by Fortey (1997) and Conway Morris (1998), that these did have soft-bodied ancestors with a long Precambrian history, but which were so small that neither their bodies nor their traces would be preserved. Even so it is unlikely that such tiny animals, often only a few millimetres in length, were equipped with the huge range of body plans that were evident later in the Cambrian, and perhaps many of these 'small shellies', as they are known, are simply the hard spines or plates of soft-skinned animals and as such represent the first experiments in the production of hard, mineralized shells and skeletons (see, for example, discussion of *Microdictyon*, p. 36).

Very soon after their appearance the main burst of evolution began and a variety of possible triggers has been put forward. Gradually increasing oxygen levels after 3 billion years of photosynthesis by cyanobacteria and later by plants may have allowed the evolution of more mobile and larger, complex animals which were able to exploit empty niches significantly devoid of competitors. Alternatively, the emergence of the first predators may have initiated the first 'arms race', in which animals had either to escape by moving faster and/or getting bigger or to defend themselves by developing hard shells. Changes in continental configuration (and therefore of ocean currents) have also been suggested. Or maybe genetic mechanisms were simply more flexible at that time, leading to accelerated diversification.

Much of our knowledge of the fauna and flora during the Cambrian Explosion is gleaned from perhaps the best known of all Fossil-Lagerstätten –

the Burgess Shale of British Columbia in Canada. By an accident of geological history, this thin layer of shale has provided a window into the richness of a Middle Cambrian marine ecosystem at a most vital time in the evolution of life on Earth. Here, by a process still not completely understood, decay was arrested so that the complete diversity of the Cambrian seas, including many soft-bodied animals (with their internal organs and muscles) has been exquisitely preserved.

These are the 'weird wonders' popularized by

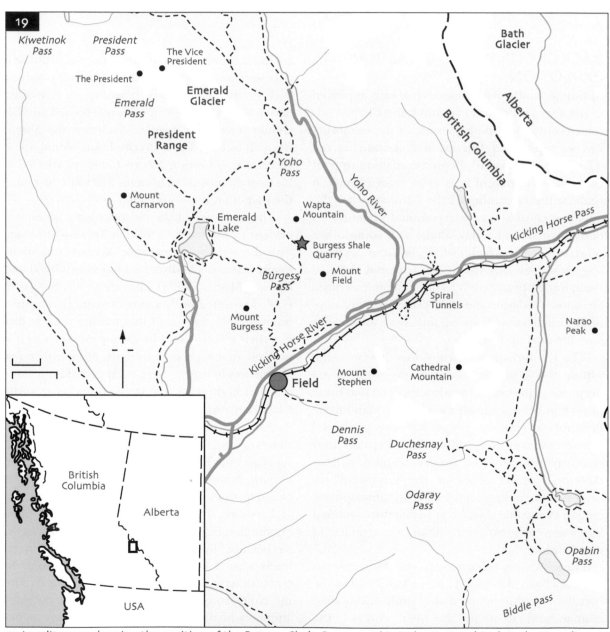

19 Locality map showing the position of the Burgess Shale Quarry within Yoho National Park in the Canadian Rockies.

Stephen Jay Gould (1989) in his book *Wonderful Life*. And if one considers that approximately 85% of Burgess Shale genera are entirely soft-bodied and thus absent from other Cambrian assemblages, it becomes apparent just how misleading the fossil record would be had this particular Lagerstätte not been preserved or discovered.

HISTORY OF DISCOVERY OF THE BURGESS SHALE

It was an American, Charles Doolittle Walcott, then Secretary of the Smithsonian Institution in Washington DC, who first discovered the Burgess Shale, high in the Canadian Rockies (**19**). The romantic story suggests that at the end of the 1909 field season his wife's horse, descending the steep Packhorse Trail that leads off the ridge between Mount Wapta and Mount Field, in what is now the Yoho National Park, stumbled on a boulder. Walcott dismounted to clear the track and on splitting the offending boulder revealed a fine specimen of the soft-bodied 'lace crab', *Marrella*, glistening as a silvery film on the black shale.

Unfortunately his diary does not support this story, but certainly Walcott discovered the first fossils in September 1909 and began excavating in earnest during the summer of 1910. Annual field seasons with his family continued until 1913 and he returned in 1917, 1919, and 1924. By the time of his death in 1927 he had amassed a collection of 65,000 specimens, which were transported back to Washington DC where they remain today in the Smithsonian Institution.

By 1912 Walcott had published many of his finds and, of the 170 species currently recognized from the Burgess Shale, over 100 were described by Walcott himself.

Walcott's Burgess Shale Quarry (**20, 21**) lies at 2,300 m (c. 7,500 ft), just below the top of Fossil Ridge, which connects Mount Wapta and Mount Field. It is a stunning location. Looking west from the quarry, which remains snow-filled even in the summer, a breath-taking panorama of snow-capped mountains, glaciers, lakes, and forests is revealed. Mount Burgess points skywards, the vivid green Emerald Lake is to its right with the Emerald Glacier poised above it. The site of Walcott's camp can be made out on the High Line Trail below.

Walcott also collected higher up the mountain at a site now called the Raymond Quarry after its

detailed excavation in the 1930s by Professor Percy E. Raymond of Harvard University, whose collection now resides at the Museum of Comparative Zoology at Harvard. Apart from the work of Alberto Simonetta, an Italian biologist, who made detailed descriptions of some of the Burgess arthropods in the 1960s, little further work on these remarkable fossils was carried out until the notion of a complete restudy of the Burgess fauna was proposed by Professor Harry Whittington in 1966.

Whittington was an Englishman, but in 1966 was Professor of Paleontology at Harvard University. He managed to persuade the Geological Survey of

20 Walcott's Burgess Shale Quarry at 2,300 m (c. 7,500 ft) in Yoho National Park, Canada; Mount Wapta can be seen in the background.

21 Detail of laminated shale in the Burgess Shale Quarry, showing fining-upwards sequence with coarse, orange layers at the base and finer, grey layers above. (Measurements indicate centimetres below Walcott's 'Phyllopod Bed'.)

Canada that a restudy was timely and collecting trips in 1966 and 1967 produced over 10,000 new specimens, although few new taxa. In 1966 Whittington moved from Cambridge, Massachusetts, to Cambridge, England, and took the project with him. Very soon he recruited two young postgraduates to assist in the huge task of redescribing the Burgess animals, Derek Briggs taking on the arthropods and Simon Conway Morris the worms.

The Cambridge team made painstaking dissections, drawings, and detailed photographs of the Burgess fossils and revealed a wealth of new data unseen by previous workers. Whittington produced a detailed restudy of Walcott's first find, the 'lace crab', *Marrella*, the most common animal from the Burgess fauna, and set new standards for the description of Burgess fossils. Briggs and Conway Morris worked respectively on two of the real enigmas of the Burgess Problematica, *Anomalocaris* and *Canadia sparsa* (later renamed *Hallucigenia*) and by remarkable detective work were able to elucidate the true affinities of these most bizarre Burgess animals.

In 1975 Desmond Collins of the Royal Ontario Museum (ROM) in Toronto obtained permission from Parks Canada to collect loose material from the ridge. A ROM party returned in 1981 and 1982, this time to look for new localities and demonstrated that the fossiliferous beds were actually more extensive than had been thought, with 10 or more new locations, both stratigraphically above and below Walcott's quarry, all along the line of the Cathedral Escarpment.

In 1981 the site was designated a World Heritage Site by UNESCO, but the work of the ROM team continued throughout the 1980s and 1990s and is still ongoing today through the work of Jean-Bernard Caron.

STRATIGRAPHIC SETTING AND TAPHONOMY OF THE BURGESS SHALE

The Burgess Shale Formation is about 270 m (c. 890 ft) thick and includes 10 members (Fletcher & Collins, 1998), one of which, the Walcott Quarry Shale Member, includes the productive layers in Walcott's quarry which have yielded the well-preserved, soft-bodied fossils. This thick formation is the deep-water lateral equivalent of the much thinner platform facies of the Stephen Formation (**22**) and is Middle Cambrian in age, approximately 510 million years old.

Just to the north of Walcott's quarry the dark shales of the Burgess Shale Formation abruptly disappear and abut against much lighter-coloured dolomites belonging to the Cathedral Formation with an almost vertical contact (**22**). The fact that the overlying Eldon Formation passes uninterrupted across this vertical contact shows that it is not a tectonic fault, but instead that it was an original feature of the Cambrian seabed, i.e. a near-vertical submarine cliff. It is significant that the Burgess Shale fossils are always found at the foot of this submarine cliff.

The conventional interpretation is that this cliff represents the margins of an algal reef, the top of which formed a shallow carbonate platform with well-lit waters free of terrigenous sediment. The Burgess animals lived on or in the mud in the deeper, darker water at the foot of the cliff. The seabed sloped away from the cliff into still deeper water which was anoxic and hostile to life.

Various pieces of evidence suggest that the Burgess animals were not preserved in the area in which they were living. The first is the lateral continuity over large distances of the thin beds of shale, with no evidence of bioturbation of the sediment by crawling or burrowing organisms. The second is the fact that the Burgess fossils are found lying at all possible angles within the shale, some even head-first into the sediment. Finally, examination of the thin shale beds in Walcott's quarry reveals that each shows a definite fining-upwards sequence with coarse, orange layers at the base and finer, dark grey layers above, such that each bed represents a separate event, i.e. a separate influx of sediment (**21**).

The preferred explanation of the deposition of the Burgess animals is that from time to time storms, earth movements, or simply instability of the wet sediment pile sent the mud at the foot of the cliff down into the hostile basin in a rapid cloud of sediment, carrying with it the unsuspecting animals which had no time to escape. Conway Morris (1986, text-Figure 1) illustrated a 'pre-slide' and 'post-slide' environment, the former representing the area in which the animals were living, and the latter the area to which they were transported and preserved (now represented by

22 Diagram to show the stratigraphy and structure of the Middle Cambrian succession on Mount Field (after Briggs *et al.*, 1994; Fletcher & Collins, 1998).

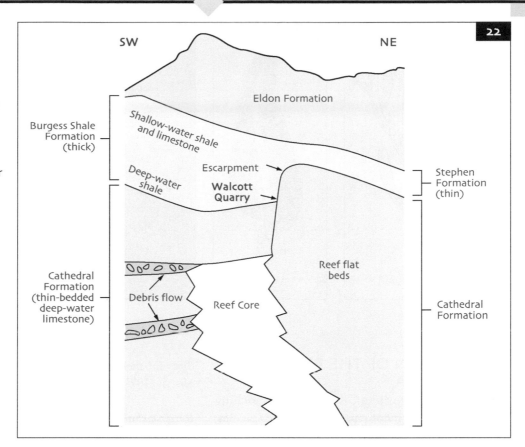

Walcott's quarry). Both appear to have been close to the foot of the cliff, suggesting that the turbidity currents flowed downslope parallel to the cliff. The slide was followed by quiet conditions allowing the fine sediment to settle, giving rise to the fine layering of the shale seen today.

How the Burgess Shale animals have been preserved is still not entirely known. Two of the common prerequisites for soft-tissue preservation were undoubtedly partly responsible, namely rapid, catastrophic burial in a fine sediment, and deposition on a sea floor deficient in oxygen. Such a toxic 'post-slide' environment would have excluded scavengers, and when the cloud of sediment settled the carcasses would have been completely entombed within fine mud. However, anaerobic microbes can break down soft muscle tissue relatively quickly even in the absence of oxygen, and some other factor must have prevented such microbial action.

Butterfield (1995) isolated tissues of various soft-bodied Burgess animals and showed that in many cases they were composed of altered original organic carbon. This was coated, however, by a thin film of calcium aluminosilicate, similar to mica, explaining the silvery appearance of the Burgess fauna under incident light. He suggested that this coating originated from the clay minerals in the mud in which the animals were buried, and that these minerals inhibited bacterial decay – perhaps by preventing reactions of enzymes. This type of preservation is exceptional; normally very soft tissue such as muscle and intestine can only be preserved if it is replaced by another mineral during early diagenesis. This aspect of the story has been extended by a number of other papers such as Gaines *et al.* (2008) who undertook elemental mapping of soft-bodied fossils from 11 other Burgess Shale-type deposits worldwide, and showed that the conservation of organic tissues, rather than early authigenic mineralization, was the primary cause of the preservation of such assemblages, by suppression of the processes that normally lead to organic decay.

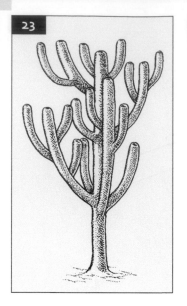

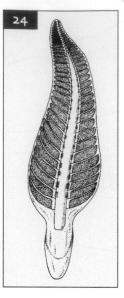

23 Reconstruction of *Vauxia*.

24 Reconstruction of *Thaumaptilon*.

25 The priapulid worm *Ottoia prolifica*, showing anterior proboscis with spines (USNM). Diameter about 10 mm (c. 0.4 in).

DESCRIPTION OF THE BURGESS SHALE BIOTA

Vauxia (Phylum Porifera). A bush-like, branching sponge (**23**), which does not have discrete spicules, but is composed of a tough spongin-like framework, explaining why it is the most common of the Burgess sponges.

Thaumaptilon (Phylum Cnidaria; Order Pennatulacea). A rare constituent of the sessile Burgess fauna, this possible sea-pen (**24**) is nonetheless important as it is a survivor of the Precambrian Ediacara fauna (Chapter 1), resembling the genus *Charnia*. It has a broad central axis with up to 40 branches, each housing hundreds of individual star-like polyps (see Conway Morris, 1993).

Ottoia (Phylum Priapulida). This is the most abundant of the mud-dwelling priapulid worms (**25, 26**), especially in the higher beds of the Burgess Shale. Priapulids are carnivorous animals, rare today, but common in the Cambrian seas. They had a bulbous anterior proboscis surrounded by vicious hooks and spines. At the end of the proboscis is a mouth with sharp teeth; their stomachs often reveal the last meal, which may include hyoliths, brachiopods, and even other specimens of *Ottoia*, for these animals were cannibals. They are commonly fossilized in a U-shape, suggesting that they lived in U-shaped burrows. However, recently discovered straight specimens may suggest that the U-shape was

due to post-mortem contraction (see Conway Morris, 1977a).

Burgessochaeta and *Canadia* (Phylum Annelida; Class Polychaeta). These are polychaete worms, or 'bristle worms' (**27**), segmented animals with paired appendages each bearing numerous bristles or setae (see Conway Morris, 1979).

Hallucigenia (Phylum Onychophora). The most celebrated Burgess animal and the classic 'weird wonder' of Stephen Jay Gould (1989) was placed in this new genus by Conway Morris to reflect its dream-like quality as it seemed unlike any other known animal (**28, 29**). This was partly because it had been reconstructed upside-down, standing on rigid spines and waving its tentacles in the water. Additional specimens suggested reversing this interpretation and suddenly it was apparent that this genus was a marine 'velvet worm', a caterpillar-like group called the lobopodians ('lobed feet'). *Hallucigenia* crawled on its fleshy limbs and used its spines for protection as it scavenged for decaying food. *Aysheaia* is another such Burgess lobopod often found in association with sponges, on which it most probably preyed (see Conway Morris, 1977b).

Marrella (Phylum Arthropoda). A small, feathery arthropod, often known as the 'lace-crab', it is the most common Burgess animal, with over 15,000 specimens discovered, yet it is known from few other

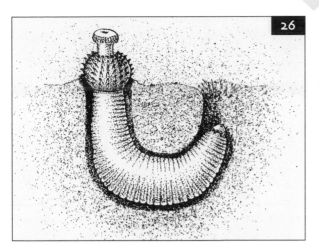

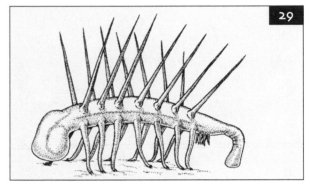

27 The polychaete worm *Canadia spinosa* (USNM). Length about 30 mm (c. 1.2 in).

26 Reconstruction of *Ottoia*.

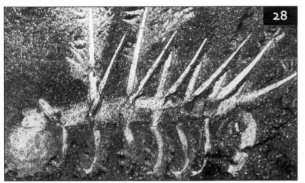

28 The velvet worm *Hallucigenia sparsa* (USNM). Length about 20 mm (c. 0.8 in).

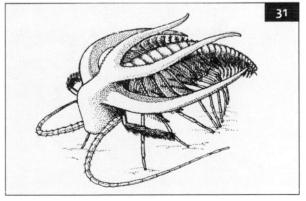

29 Reconstruction of *Hallucigenia*.

30 The arthropod *Marrella splendens* (MM). Length 20 mm (c. 0.8 in).

31 Reconstruction of *Marrella*.

Cambrian deposits (**30, 31**). It was the first to be discovered by Walcott and the first to be redescribed by Whittington. A head shield has two pairs of curving spines, while the head has two pairs of antennae. The 20 body segments each bear a pair of identical legs, suggesting that it is a primitive arthropod and could be ancestral to the three major groups of aquatic arthropods (crustaceans, trilobites, and chelicerates). It is a member of the Marrellomorpha clade, which includes a number of such enigmatic forms (see Whittington, 1971; see also pp. 60 and 71).

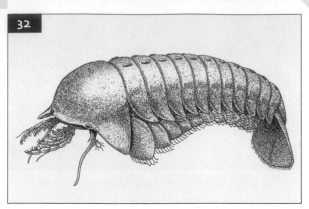

32 Reconstruction of *Sanctacaris*.

33 Reconstruction of *Anomalocaris*.

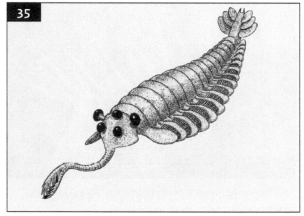

34 The problematical *Opabinia regalis* (USNM). Length about 65 mm (c. 2.5 in).

35 Reconstruction of *Opabinia*.

36 The problematical *Wiwaxia corrugata* (USNM). Length about 35 mm (c. 1.4 in).

37 Reconstruction of *Pikaia*.

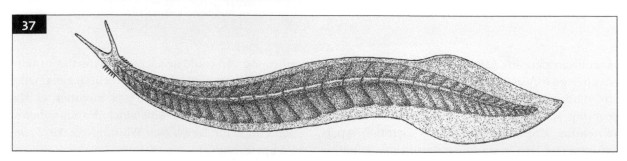

Sanctacaris (Phylum Arthropoda, Class Chelicerata?). This is the most important specimen to be discovered by Collins and the ROM team as it represents a stem arachnomorph and is on the stem line to the chelicerates, the group containing the spiders and scorpions (**32**). The large head-shield protecting six head appendages, five of which were spiny claws to assist in capturing prey, gave it the nickname of 'Santa Claws' (see Briggs & Collins, 1988).

Anomalocaris (Problematica or Phylum Arthropoda?). The monster predator of the Burgess Shale, *Anomalocaris,* does not resemble any known animal and was for a long time considered to be an example of a short-lived experimental arthropod-like phylum (**33**). The prey-grasping anterior appendages were originally thought to represent the segmented abdomen of a crustacean (hence the generic name meaning 'strange crab'), then were interpreted as a set of paired limbs of a giant arthropod, while the circular mouth parts were originally interpreted as the jellyfish *Peytoia.* More complete specimens revealed this to be the largest known of the Burgess animals, reaching lengths of up to 1 m (c. 3 ft). The head has a pair of large eyes, the trunk is covered with flap-like structures, and the tail is a spectacular fan (see Whittington & Briggs, 1985). It is now generally accepted to be a stem arthropod (one of the 'great-appendage arthropods', see p. 38), and did not, apparently, become extinct at the end of the Cambrian as had long been assumed; recently examples have been found in rocks as young as the Devonian (see Chapter 6, The Hunsrück Slate, p. 72). The genus *Hurdia,* thought by Walcott to be a large crustacean, was shown by Daley *et al.* (2009) to be a second genus of anomalocaridid, distinguished by its large, spiked carapace.

Opabinia (Problematica or Phylum Arthropoda?). This truly strange animal had five eyes on the top of its head and a long, flexible proboscis which terminated in a number of spines, apparently a grasping organ (**34, 35**). Each of the body segments possesses lateral lobes with gills and a strange tail was formed by three posterior flaps. Some authorities now believe that it may be related to *Anomalocaris* and that both may be stem arthropods (see Whittington, 1975).

Wiwaxia (Problematica or Phyllum Mollusca?). This strange mud-crawling animal was protected from predators by its dorsal surface being covered by a coat of scale-like sclerites and a double row of pointed spines (**36**). The ventral surface was a soft foot similar to that of a slug or snail and from the open mouth protruded a radula, also reminiscent of molluscs; indeed Caron *et al.* (2006) interpreted this genus as a stem-group mollusc. The micro-structure of the sclerites, however, is more akin to that of polychaete annelids and the true affinities of *Wiwaxia* thus remain in doubt, although it is almost certainly related to the halkieriids (Chapter 3, Chengjiang, p. 41, **58**; Conway Morris, 1985).

Pikaia (Phylum Chordata). An inconspicuous but important element of the Burgess fauna, *Pikaia* possessed a stiff rod along its dorsal margin (**37**), which suggests that it is a primitive chordate, the phylum to which Man and all vertebrates belong, and showing that even our ancestors were present during the Cambrian Explosion. A narrow anterior end has a pair of tentacles, while the posterior is expanded into a fin-like tail. The cone-in-cone arrangement of the muscles is often clearly preserved.

PALAEOECOLOGY OF THE BURGESS SHALE BIOTA

The Burgess Shale represents a marine, benthic community living in, on, or just above the muddy seabed at the foot of a submarine cliff, where the mud was banked sufficiently high to be clear of stagnant bottom waters. The marine basin faced the open sea and was situated within the tropical zone at about 15°N. The presence of photo-synthesizing algae suggests that the depth was not much more than 100 m (c. 300 ft).

Greater in-depth palaeoecological analysis has been undertaken on the Burgess Shale than on any other single geological horizon. Conway Morris (1986) examined over 30,000 slabs of shale with 65,000 fossils, while Caron and Jackson (2008) studied 50,900 specimens belonging to 158 genera representing 17 major taxonomic groups. Approximately 10% of the biota consisted of benthic infauna, i.e. living in the sediment itself, this community being dominated by the burrowing priapulid worms such as *Ottoia, Selkirkia,* and *Louisella,* and polychaete worms such as *Burgessochaeta* and *Canadia.*

The vast majority of Burgess animals (c. 75%) consisted of benthic epifauna, living on the sediment surface, and divided into the fixed or sessile epifauna (c. 30%) and the vagrant epifauna (c. 40%), that walked or crawled across the seabed. The sessile biota consisted mainly of sponges, such as the branching *Vauxia*, and others, such as *Choia* and *Pirania* adorned with sharp, glassy spicules which both supported the skeleton and protected against predators. The sea-pen *Thaumaptilon* was a rare member of the sessile epifauna, as was the enigmatic animal *Dinomischus*, projecting just 10 mm (c. 0.4 in) above the mud on a thin stalk and looking like a small flower (see also pp. 39, 40).

The vagrant epifauna was more varied, but was dominated by arthropods, only a small proportion of which were trilobites, such as *Olenoides*, and closely related arthropods such as the soft-bodied *Naraoia*. It also included the ubiquitous 'lace crab', *Marrella*, and a number of small, scavenging lobopods, most notably *Hallucigenia* and *Aysheaia*. The most bizarre mud crawler was *Wiwaxia*, which moved across the sediment using its slug-like foot.

Animals which lived above the mud surface were fewer in number, comprising only about 10% of the Burgess biota, simply because they were more able to escape the mud flow by swimming away. The nektobenthic animals (near-bottom swimmers) were, however, within the range of turbidity flows and included the tiny chordate *Pikaia*. The medusoid *Eldonia*, which is more probably related to the holothurians ('sea cucumbers') than to true jellyfish, may be regarded as nektobenthos and/or a planktonic floater. Nekton (forms swimming above the substrate) include the giant predator *Anomalocaris* and the enigmatic *Opabinia*, which used its nozzle-like proboscis to search for food.

Trophic analysis by Conway Morris (1986) identified several feeding types including: filter feeders, dominated by the sponges; deposit feeders, dominated by arthropods (deposit collectors) and molluscs (deposit swallowers); scavengers such as the lobopods *Hallucigenia* and *Aysheaia*; and lastly, predators such as the giant *Anomalocaris*, large arthropods (such as *Sidneyia*) and the priapulids. The important role of predation evident in the Burgess Shale ecological framework provides the main contrast with other Cambrian (shelly) assemblages.

COMPARISON OF THE BURGESS SHALE WITH OTHER CAMBRIAN BIOTAS

Middle Cambrian Burgess Shale-type assemblages have since been discovered at about 40 localities worldwide (**18**). Chapter 3 is devoted to the recent surprising discovery of Burgess Shale-type deposits in rocks of Lower Cambrian (Atdabanian) age, and here we discuss two further sites, from the Middle and Lower Cambrian respectively.

House Range, Utah

The slopes of Swasey Peak in the House Range of western Utah expose the Wheeler Shale of Middle Cambrian age. Much of the Wheeler Shale is unfossiliferous, but certain layers contain abundant trilobites and a diverse biota of soft-bodied fossils, including many of the same taxa as in the Burgess Shale. The environment of deposition was very similar; fine sediments accumulated in deeper water offshore, sometimes accumulating rapidly when submarine landslides dropped large amounts of sediment off the shallower carbonate platform.

Fossils include about 15 genera of trilobites, of which the ubiquitous *Elrathia kingi* is so numerous that it is commonly sold throughout the United States as belt-buckles, earrings, bolo ties, fridge magnets, and so on. Agnostid trilobites are also common, usually less than 10 mm (c. 0.4 in) in length. Sponges are known from about 20 species, including the genus *Choia*, along with sponge-like chancellorids and various inarticulate brachiopods. Approximately 20 species of soft-bodied animals have been recorded, including priapulid worms such as *Selkirkia*, the giant predator *Anomalocaris*, a number of the bivalved crustacean arthropods, and even the rare and enigmatic *Wiwaxia*. The fauna is thus very comparable to that of the coeval Burgess Shale.

Kangaroo Island, Australia

Discovered in the 1950s and exposed in Emu Bay on the north-east coast of Kangaroo Island in South Australia, the Emu Bay Shale has been correlated with the Botomian Stage of the Lower Cambrian, approximately 520 million years old, and is thus intermediate in age between the Burgess Shale and the Chengjiang Lagerstätte described in Chapter 3.

Its mode of preservation is similar to that of Burgess Shale, but the larger grain size of the shale results in a poorer quality of resolution. This suggests that the depositional environment of the Emu Bay Shale was probably rather shallower than that of either the Burgess Shale or Chengjiang. The type section on the east side of Emu Bay has yielded a typical Cambrian fauna of redlichioid trilobites, hyolithids, inarticulate brachiopods, and the sponge-like chancellorids, but the well-known Big Gully locality, 3 km (c. 1.9 miles) east of Emu Bay, yields a variety of soft-bodied fossils including the giant predator *Anomalocaris*, the bivalved arthropods *Tuzoia* and *Isoxys*, the palaeoscolecid worms *Palaeoscolex* and ?*Myoscolex*, and the trilobite-like *Xandarella* and *Primicaris*, a fauna which is more comparable to that from Chengjiang (see Chapter 3). Some of the trilobites have antennae and limbs preserved, as in the Burgess Shale and in the Ordovician Beecher's Trilobite Bed (see Nudds & Selden, 2008, Chapter 4). The Emu Bay Shale is the subject of a new project involving a number of specialists and led by John Paterson from the University of New England in New South Wales.

MUSEUMS AND SITE VISITS

Museums

1. National Museum of Natural History, Smithsonian Institution, Washington DC, USA.
2. Royal Ontario Museum, Toronto, Ontario, Canada.
3. Field Visitor Center, Field, British Columbia, Canada.
4. Royal Tyrrell Museum, Drumheller, Alberta, Canada.
5. Sedgwick Museum, University of Cambridge, Cambridge, UK.
6. Manchester University Museum, Oxford Road, Manchester, UK.

Sites

Walcott's Burgess Shale Quarry (**20, 21**) is situated in Yoho National Park and visits to it are strictly controlled. Guided trips may be booked in advance by contacting the Visitor Center at Field, B.C. (+1-250-343-6783; email: yoho.info@pc.gc.ca), or Burgess Shale Guided Hikes (+1-800-343-3006;

www.burgess-shale.bc.ca). The hike is strenuous, involving some steep climbs, and should only be undertaken if you are fit and healthy. Collection of fossils is absolutely forbidden and there are severe penalties for removing fossils from the Park. The Burgess Pass–Yoho Pass hiking trail passes close to the quarries, and in good weather provides a superb day in spectacular scenery. Note that the quarry is visible through good binoculars from Emerald Lake Lodge, situated on the moraine damming Emerald Lake. For further information apply to the Superintendent, Yoho National Park, PO Box 99, Field, British Columbia, Canada. Telephone +1-250-343-6324.

CHENGJIANG

BACKGROUND: BURGESS SHALE-TYPE BIOTAS IN THE EARLY CAMBRIAN

In the previous chapter we discussed *the* classic Fossil Lagerstätte, the Burgess Shale of British Columbia, which showed that the Cambrian Explosion was well under way by middle Cambrian times. In this chapter we abandon the strict chronological order of chapters adhered to elsewhere in this book and turn instead to consider what was happening during the early Cambrian. For in the last quarter of a century two Burgess Shale-type deposits have been discovered in rocks of Lower Cambrian age, one of which forms the subject of this chapter, and which can be argued to surpass even the Burgess Shale itself in its preservation and diversity. Nonetheless it still seemed appropriate to us to consider the Burgess Shale first, if only on grounds of historical priority.

The two new deposits were coincidentally discovered on successive days in 1984, one in the remote region of Sirius Passet in northern Greenland where its inaccessibility has somewhat hindered research, the other in the southern Chinese province of Yunnan, which has led to a blossoming of research papers. It is to this new Chinese Fossil-Lagerstätte of Chengjiang that we now turn our attention to investigate just what effect the Cambrian Explosion was having at the very dawn of the Phanerozoic.

HISTORY OF DISCOVERY OF THE CHENGJIANG BIOTA

Yunnan Province in the extreme south-west of China currently borders Vietnam, Laos, and Burma, but in the early days of the twentieth century these countries were all part of French Indochina, the French colonial empire of south-east Asia (**38**). From the early days of geological

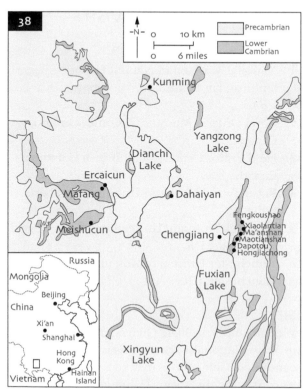

38 Locality map showing the locations of the Chengjiang biota in Yunnan Province (after Hou *et al.*, 2004).

39 The western slopes of Maotianshan (Maotian Hill), Yunnan Province, showing the section through the Chengjiang deposits.

exploration in this region by French scientists, the Lower Cambrian succession of eastern Yunnan Province had been known to be rich in fossils. Jaques Deprat and Henri Mansuy published a major work on the geology and palaeontology of this area in 1912 which featured many new species of trilobites and other arthropods.

It was not, however, until 1984 that the true significance of the palaeontology of this region was realized when Professor Hou Xian-guang, then a member of the Nanjing Institute of Geology and Palaeontology of the Chinese Academy of Sciences, arrived in Kunming City, the capital of the province, to begin a collecting programme of bradoriid arthropods from the Lower Cambrian Heilinpu Formation. After an unsatisfactory stint of field work at Hongjiachong, the field camp was moved to a more complete section on the western slopes of Maotianshan (Maotian Hill) approximately 6 km (c. 3.7 miles) east of the county town of Chengjiang (Hou *et al.*, 2004; **38, 39**).

Professor Hou records in his field notebook that in the afternoon of July 1st 1984 a split slab of mudstone revealed an unknown species, and that in searching for further specimens he soon came across the unmistakable soft-bodied arthropod *Naraoia*, complete with its limbs, which was previously only known from the renowned Burgess Shale Lagerstätte of British Columbia (Chapter 2). Not only had Hou discovered a Burgess Shale-type soft-bodied fauna in China, but these fossils were Lower Cambrian in age – significantly older than the celebrated Canadian deposits. Moreover, the preservation was, if anything, even more exquisite. Hou reported that the specimen, "appeared as if it

were still alive on the wet surface of the mudstone", and, "elated by their discovery, the searchers increased their efforts and other new soft-bodied fossils were revealed one after another" (Hou *et al.*, 2004). It soon became apparent that the Chengjiang biota was also far more diverse, comprising many of the well-known enigmatic Burgess genera such as *Hallucigenia* and *Anomalocaris*, and an equal number of new forms.

Further extensive collecting was undertaken in 1985, 1987, 1989, and 1990, and demonstrated that the soft-bodied fauna was widely distributed in eastern Yunnan Province (especially in Chengjiang County at Hongjiachong, Dapotou, Fengkoushao, Xiaolantian, and Ma'anshan, and further afield on the western and eastern shores of Dianchi Lake, at Meishucun, Ercaicun, Mafang, and Dahaiyan; **38**). This initial collecting programme ceased when Hou moved to Stockholm to begin a scientific collaboration with Swedish palaeontologists to describe this new and unexpected fauna (see Hou & Bergström, 1997 for details), but continued prospecting by Hou's colleague Chen Jun-yuan resulted in thousands of exquisitely preserved fossils, including more new forms which are still being described. Many of the fossils from the Chengjiang Lagerstätte have been beautifully described and illustrated by Hou Xian-guang and colleagues in their 2004 book, *The Cambrian Fossils of Chengjiang, China*, and by Chen Jun-yuan and colleagues in their 1997 publication, *The Cambrian Explosion and the Fossil Record*.

STRATIGRAPHIC SETTING AND TAPHONOMY OF THE CHENGJIANG BIOTA

The Chengjiang Lagerstätte is very restricted stratigraphically occurring (with one exception) in the middle part of the Yu'anshan Member of the Heilinpu Formation (**40**). The Yu'anshan Member is divided into three biozones based on trilobites, and the Chengjiang Lagerstätte occurs only within the middle biozone, characterized by the trilobites *Eoredlichia* and *Wutingaspis* (**40**). (In earlier papers this biozone was sometimes described as the Maotianshan Shale Member of the Yu'anshan Formation.) This horizon can be referred to the Chinese Qiongzhusian Stage, which can be correlated on the basis of small shelly fossils, trilobites, and acritarchs with the Atdabanian Stage of international usage. The Chengjiang biota is therefore Lower Cambrian in age, approximately 525 million years old. (Note, however, that specimens from Malong County, including the iconic Burgess predator *Anomalocaris*, are reported to come from the overlying Canglangpu Formation, which belongs to the slightly younger Canglangpuian Stage; Hou *et al.*, 2004.)

The Yu'anshan Member in total comprises 100–150 m (c. 330–500 ft) of greenish-grey, thinly bedded, graded mudstones that assume a distinctive orange hue on weathering and which illustrate millimetre-scale rhythmic couplets of lighter and darker layers (**41**). They are underlain by black siltstones of the Shiyantou Member, which in turn are underlain by thick phosphorites of the Yuhucun Formation (**40**). The latter are associated with a late Neoproterozoic to early Cambrian transgression that flooded the South-west China Platform (Hou *et al.*, 2004) and the Yu'anshan Member is thus generally considered to have accumulated in shallow marine conditions.

Early ideas suggested that the coarsening upwards Heilinpu and Canglangpu formations were deposited by a prograding delta and that the Chengjiang animals lived adjacent to the delta front in a distal marine environment and periodically succumbed to catastrophic, live burial in episodic turbidity flows associated with the delta, each sediment couplet representing a separate flow. More recent work based on detailed sedimentary logging suggests that the sediment couplets (**41**) reflect instead a seasonal periodicity and that the Chengjiang animals lived in a lower shoreline to proximal offshore environment (Zhu *et al.*, 2001). In this scenario the upper, lighter layers of the couplets, which include the exceptionally preserved fossils, represent storm-generated muds (tempestites) that settled from suspension in low salinities resulting from the frequent input of fresh water during wetter seasons ('event mudstones' of Zhao *et al.*, 2009). The darker layers of the couplets, which are almost devoid of fossils, were deposited in normal salinities during dry seasons ('background mud-

Chinese Stage	Chinese Biozone	Formation	Member	Lithology	Events
Canglangpuian	*Sichuanolenus –Paokannia*	Canglangpu	Guanshan	siltstone	
	Metaredlichioides –Chengkouia				
	Drepanuroides				
	Yunnanaspis –Yiliangella				
Qiongzhusian	*Yunnanocephalus –Malungia*	Heilinpu	Yu'anshan	siltstone	
	Eoredlichia –Wutingaspis				Chengjiang Biota
	Parabadiella			mudstone	← first trilobites and bradoriids
Meishucunian	*Sinosachites –Tannuolina*		Shiyantou	siltstone	
	barren interval				
	Heraultipegma		Dahai	phosphorite	
	Paragloborilus –Siphogonuchites		Zhongyicun		← major radiation of small shelly fauna
	Anabarites –Protohertzina	Yuhucun			
					← first small shelly fauna
Dengyingxian	no formal zones recognized		Baiyansho	dolomite	

40 Diagram to show the stratigraphy of the Lower Cambrian succession in the Chengjiang region, Yunnan Province (after Hou *et al.*, 2004).

stones' of Zhao *et al.*, 2009). Whichever scenario is correct, rapid deposition is critical in the exceptional preservation of the Chengjiang biota (Hou *et al.*, 2004).

The complex taphonomy of the Chengjiang fossils is being elucidated by the detailed work of Sarah Gabbott and colleagues at Leicester University, and is very different to that of the Burgess Shale (see p. 25). Fossils are generally preserved flattened on bedding planes, only rarely showing any three-dimensionality. Their reddish-purple colour, which contrasts with the orange matrix, is due to a concentration of iron oxide pseudomorphs after iron pyrite. Gabbott *et al.* (2004) have shown that the clay-rich host sediment was deficient in organic carbon but replete in available iron, so that a decaying carcass acted as a local substrate for iron-reducing bacteria. Rapidly decaying tissues are preserved by framboidal pyrite, while more recalcitrant tissues are preserved by larger, euhedral pyrite crystals. Soft tissues, such as gills, muscles, and intestines, as well as arthropod cuticles, are often preserved simply as impressions moulded in the very fine mud, and usually show no sign of decomposition, suggesting that fossilization took place rapidly after death.

Unlike the Burgess Shale the Chengjiang biota was most probably buried in the area in which it was living and thus is autochthonous in nature. Inarticulate brachiopods are often preserved with their pedicles in life position, lobopods are preserved associated with their hosts, and almost all fossils lie parallel to the bedding in contrast to the jumbled nature of the Burgess animals preserved at all angles within the muds. Minimal transportation seems to have occurred and the Chengjiang biota thus represents the earliest known record of a complete Phanerozoic sea-floor community.

DESCRIPTION OF THE CHENGJIANG BIOTA
This section should be read in conjunction with the section on the Burgess Shale biota (pp. 26–29) by way of a comparison.

Poriferans. Sponges are a major component of the Chengjiang fauna, second only to arthropods in their diversity. Most are demosponges, a group with siliceous spicules. Some genera are unique to Chengjiang, while others, such as the disc-like

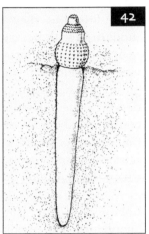

41 Millimetre-scale rhythmic couplets of lighter and darker layers in the Yu'anshan Member, Heilinpu Formation.

42 Reconstruction of *Paraselkirkia*.

Choia, are also known from the Burgess Shale (p. 30). However, the common Burgess genus *Vauxia* (p. 26) does not occur here.

Cnidarians and ctenophores. Both of these groups (which were formerly included together in the phylum Coelenterata) are represented at Chengjiang by just one or two species. The cnidarian *Xianguangia* is a soft-bodied anemone-like animal bearing about 16 tentacles, while the ctenophore *Maotianoascus*, with eight petaloid lobes, appears remarkably similar to the modern-day gelatinous comb jellies or sea gooseberries. The Burgess cnidarian *Thaumaptilon* (p. 26), a relic from the Ediacara seas, is not known at Chengjiang.

Priapulid worms. The mud-dwelling 'penis worms' are represented by at least five genera at Chengjiang, of which *Paraselkirkia* (**42**) is the most common and can be compared to the Burgess genus *Selkirkia*, which has been interpreted as living vertically in the sediment with its bulbous anterior proboscis projecting from the sediment. The much rarer *Acosmia*, however, is more similar to the common Burgess genus *Ottoia* (p. 26), and is usually preserved in a tightly curved manner, suggesting it lived in a U-shaped burrow.

Palaeoscolecid worms. Three genera, known from thousands of specimens, have been allied to the palaeoscolecidans, a group of worms with distinct papillate annuli. Their exact affinities are unclear; Chen and Zhou (1997) included them with the priapulids, but they also have morphological similarities with nematodes ('roundworms') and nematomorphs ('horsehair worms'). The former view now seems to be the consensus, but whichever group they belong to, they were probably infaunal deposit feeders. The abundant *Maotianshania* (**43**) constitutes about 5% of the fauna and often shows fine details of its soft parts including an anterior retractable, spiny proboscis and a posterior region coiled through 300°.

Lobopods. The Chengjiang fauna has an abundance of lobopodian animals ('velvet worms') related to the iconic Burgess genus *Hallucigenia*. It was the discovery of one of these that first suggested that the original interpretation of *H. sparsa* from the Burgess Shale was upside-down (see p. 26). *Hallucigenia* itself is represented at Chengjiang by a second species, *H. fortis*, which has an eighth leg-pair at the posterior. Closely related is *Microdictyon* (**44, 45**), which had long been known from Cambrian deposits around the globe from isolated sieve-like sclerites, the zoological relationships of which were a mystery. Complete specimens from Chengjiang have shown that these enigmatic phosphatic microfossils are actually the plates of a lobopod, and some authorities have even suggested that these are homologous to the compound eyes of arthropods.

Arthropods. These form the major component of the Chengjiang fauna, with at least 60 valid species. They include the three main groups of Palaeozoic aquatic arthropods, the chelicerates, crustaceans, and trilobites, but the true affinities of many of Chengjiang's arthropods remain in doubt and many are basal forms.

The chelicerates are possibly represented by one of the most common genera, *Fuxianhuia* (**46, 47**), which has a broad head shield and anterior, succeeded by a narrow elongate abdomen. The head carries a pair of uniramous grasping appendages suggesting its affinity with the chelicerates, although some authorities consider it to be a basal euarthropod. A mismatch of the number of leg pairs with the number of body segments suggests a primitive condition. The much rarer genus *Chengjiangocaris* may be related to *Fuxianhuia* and may also be related to *Sanctacaris* from the Burgess Shale, which had previously been considered to be the earliest known chelicerate (p. 29).

Many of the Chengjiang arthropods are characterized by the possession of a large, bivalved carapace covering the head and part of the abdomen. These are mainly crustaceans and some are congeneric with those from the Burgess Shale. For example, *Canadaspis* (**48**) has a shrimp-like appearance with a pair of stalked eyes, while *Waptia* has a pair of very long multisegmented antennae. Again the true affinity of both genera is still in some doubt, as their limb morphology differs from most crustaceans. One of the most common of the bivalved arthropods is *Kunmingella*, one of the bradoriid arthropods researched by Hou, and which led him to discover the Chengjiang fauna. Bradoriids have traditionally been regarded as ostracod crustaceans, but the soft parts of *Kunmingella* preserved here at Chengjiang suggest that this is not so, and that instead they represent early derivatives of the crustaceans, ancestral to ostracods.

43 The palaeoscolecid worm *Maotianshania cylindrica* (MM). Length about 19 mm (c. 0.75 in)

The Chengjiang biota includes several trilobites, many of which belong to the Redlichioidea, a relatively primitive group. Perhaps most common is *Eoredlichia*, which gives its name to the trilobite biozone to which the Chengjiang biota belongs (see section on Stratigraphy). Soft parts, including limbs and the alimentary canal are often preserved. Other arthropods, closely related to trilobites, are entirely soft-bodied, such as *Naraoia* (**49**), also known from the Burgess Shale (p. 30).

44 The velvet worm *Microdictyon sinicum*, showing sieve-like sclerites (NIGP). Length 20 mm (c. 0.8 in).

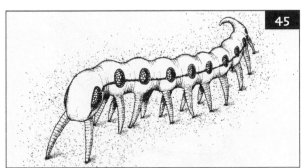

45 Reconstruction of *Microdictyon*.

46 The possible chelicerate *Fuxianhuia protensa* (MM). Minimum length 30 mm (c. 1.2 in).

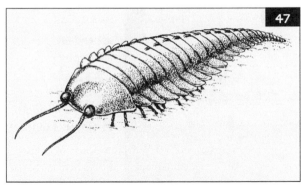

47 Reconstruction of *Fuxianhuia*.

48 The crustacean *Canadaspis laevigata* (NIGP). Length 23 mm (c. 0.9 in).

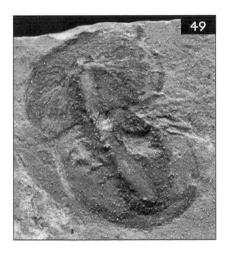

49 The soft-bodied trilobite-like *Naraoia* showing the branching alimentary system (MM). Length 11 mm (c. 0.4 in).

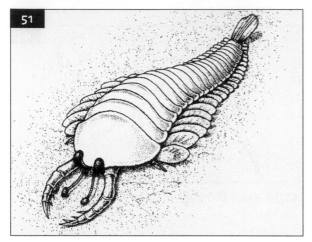

50 The 'great-appendage arthropod' *Leanchoilia illecebrosa* (RCCBYU). Length 27 mm (c. 1.1 in).

51 Reconstruction of *Fortiforceps*.

52 The grasping appendage of the 'great-appendage arthropod' *Anomalocaris saron* (RCCBYU). Length 90 mm (c. 3.5 in).

pions. The common genus *Leanchoilia* (50) and the much rarer *Fortiforceps* (51) belong to this group, the former also occurring in the Burgess Shale.

Anomalocaridids. For many years these iconic Burgess Shale predators (p. 29; 33) were considered to belong to a separate phylum, but information gleaned in particular from complete specimens from Chengjiang now suggests that they are probably arthropods and should be included with the 'great-appendage arthropods' (see previous text). Four distinct genera of anomalocaridids are known from Chengjiang, which differ mainly in the morphology of the grasping appendages (52). One of these has been estimated to be at least 2 m (c. 6 ft) in length based on the size of the circlet of plates surrounding the mouth (Chen & Zhou, 1997).

Vetulicolians. The genus *Vetulicola* (53, 54), known only from Chengjiang, was originally described as an arthropod, but Shu *et al.* (2001) allied it with the Burgess genus *Banffia*, and placed them both in the new phylum Vetulicolia. The vetulicolians have their anterior covered by a solid carapace consisting of four large plates and bearing four to five openings along each side, which have been interpreted as gill slits. This would define them as primitive deuterostomes, thus having affinities to chordates (Shu *et al.*, 2009). On the other hand perhaps they are a sister group of the arthropods that has lost limbs but gained gill structures analogous to those of deuterostomes (see Aldridge *et al.*, 2007)? Posteriorly is a flattened tail divided into seven or eight telescoping segments; eyes and appendages are unknown.

Chordates. A number of possible chordates have been described from the Chengjiang biota including both a urochordate and a cephalochordate, but these records remain highly doubtful (Hou *et al.*, 2004). Far more important however, is the record by Shu *et al.* (1999) of *Haikouichthys* and *Myllokunmingia* (55), which are not only confirmed chordates, they are also the earliest known vertebrates in the fossil record. The presence of V-shaped muscle blocks, a dorsal fin, filamentous gills, and sensory structures in the head define these as agnathan (jawless) fish comparable to present-day hagfish. Prior to this record the agnathans were unknown prior to the Lower

The so-called 'great-appendage arthropods', characterized by a prominent limb at the front of the head, include the anomalocaridids (see later text) and some other Cambrian euarthropods, all thought to be basal to chelicerates, with the great appendage homologous with the chelicerae of living chelicerates such as the spiders and scor-

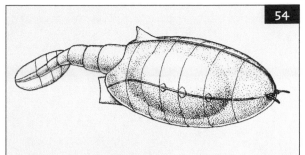

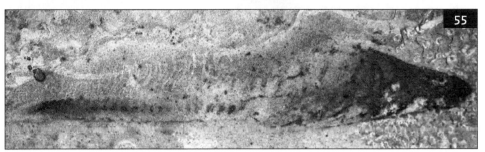

Ordovician, while the earliest vetebrates were represented by late Cambrian conodonts.

Others. The remaining invertebrate fauna mainly comprises two groups with hard shells that are a common constituent of 'normal' Cambrian faunas throughout the globe, namely the brachiopods and the hyoliths. Hyoliths have long, cone-shaped, calcareous shells, but their soft parts are poorly known and their affinities are thus uncertain; they may represent a separate phylum. The Chengjiang brachiopods are exclusively inarticulate, but both major subgroups are represented. The phosphatic linguliform brachiopods were attached to the substrate by a soft pedicle which is sometimes preserved (e.g. *Lingulella*). The common genus *Heliomedusa*, originally interpreted as a medusoid, is now thought to be a craniiform brachiopod, but unlike most members of this group, which have calcareous shells, its shell was unmineralized and often shows soft deformation. It had a reduced or absent pedicle and simply lay on the sea floor.

There are also several enigmatic members of the Chengjiang biota including the stalked *Dinomischus* (**56**) and the discoidal *Eldonia* (**57**). The former has some similarities with the crinoids, including the presence of a distal excretory tube, while the latter, originally described as a medusoid, may be related to the holothurians ('sea cucumbers'). Both are known from the Burgess Shale (p. 30).

53 The basal deuterostome *Vetulicola cuneata* showing the carapace with gill slits and the segmented tail (MM). Length 85 mm (c. 3.3 in).

54 Reconstruction of *Vetulicola*.

55 The agnathan fish *Myllokunmingia fengjiaoa* (RCCBYU). Length 25 mm (c. 1 in).

56 The stalked *Dinomischus venustus* showing the distal excretionary tube (NIGP). Length 73 mm (c. 2.9 in).

57 The discoidal *Eldonia eumorpha* (MM). Diameter about 42 mm (c. 1.6 in).

58 The halkieriid *Halkieria evangelista*, from Greenland (GMC). Length 60 mm (c. 2.4 in).

PALAEOECOLOGY OF THE CHENGJIANG BIOTA

The Chengjiang biota, as with the Burgess Shale, represents a marine benthic community living in, on or just above a muddy seabed, possibly in a lower shoreline to proximal offshore environment. The marine basin was within the South-west China Platform and formed part of the South China Plate, which in early Cambrian times straddled the equator in the tropical zone, just to the north of the Gondwana supercontinent. Being thousands of kilometres to the east of the Laurentian continent, which included North America, it is remarkable that the two biotas are so similar.

As with the Burgess Shale the fauna can be divided into benthic infauna, which lived within the sediment, benthic epifauna, living on the sediment, either sessile or vagrant, and the nekton and plankton (swimmers and floaters), which lived above the sediment surface. The infauna was again dominated by the carnivorous priapulid worms such as *Paraselkirkia*, and by the numerous palaeoscolecidans, but unlike the Burgess Shale, polychaete worms are not represented. Some of the inarticulate lingulid brachiopods, however, would have also lived in burrows and, just like their modern counterparts, were filter feeders.

The sessile epifauna were also mainly filter feeders and, as in the Burgess Shale, were dominated by sponges, rare anemones, and the enigmatic, stemmed *Dinomischus*, which strangely were significantly larger than at Burgess. There were also several epifaunal brachiopods such as the common *Heliomedusa*, which lacked a pedicle and simply lay on the surface.

The arthropods, which dominate the biota, were mainly members of the vagrant epifauna, crawling or walking across the seabed, scavenging or preying on smaller animals. These include a number of trilobites and the possible early chelicerate *Fuxianhuia*, which was a common member of this group, as were the lobopods such as *Hallucigenia* and *Microdictyon*. These genera probably fed on dead carcasses, while the much rarer *Luolishania*, often associated with sponges, probably crawled up and preyed on them, just as did *Aysheaia* in the Burgess Shale (p. 26).

Floating and swimming animals were perhaps more common than in the Burgess Shale. Floaters include the ctenophores (comb jellies) and the medusoid-like *Eldonia* and *Rotadiscus*, while swimmers include a variety of arthropods, the anomalocaridids (which were more diverse than at Burgess), the vetulicolians, and the first known vertebrates, *Myllokunmingia* and *Haikouichthys*, both possible agnathan fish.

A detailed quantitative taphonomical analysis of almost 12,000 specimens by Zhao *et al.* (2009) revealed that the tempestite horizons were dominated by arthropods (44%), priapulids (26%), and brachiopods (24%), with poriferans and lobopods making up the majority of the remainder. As Hou *et al.* (2004) concluded, the Chengjiang biota illustrates that even at this early stage in the Cambrian Explosion complex, structured ecosystems with different trophic groups had already evolved, with a diversity far greater than that of the Ediacara seas (Chapter 1).

COMPARISON OF CHENGJIANG WITH OTHER LOWER CAMBRIAN BIOTAS

Sirius Passet, northern Greenland

Discovered in the same year, 1984, by the Geological Survey of Greenland near J. P. Koch Fjord in Peary Land, north Greenland, this locality, now known as Sirius Passet, was first seriously collected by a 1989 expedition led by John Peel, a member of the Greenland Survey, and Simon Conway Morris of Burgess Shale fame (see p. 24). It soon became apparent that this was another predominantly soft-bodied Cambrian fauna, and like the Burgess Shale was seemingly deposited in deep-water muds adjacent to a shallow carbonate bank. More than 3,000 specimens were collected from a 5 m (c. 16.5 ft) thick unit of fissile shale within the Buen Formation, with a further 4,000 specimens retrieved in a subsequent 1991 expedition. A Danish expedition, with David Harper, revisited Sirius Passet in 2009 and made a further substantial collection.

The fauna is again dominated by arthropods (although only one species of trilobite), sponges, polychaete and priapulid worms, and palaeo-scolecidans. In addition are rare hyoliths and a few small brachiopods, but significantly the diversity of taxa with hard skeletons is lower than that of the Burgess Shale, perhaps suggesting reduced oxygen levels (Conway Morris, 1998). Aside from this, the two faunas are remarkably comparable, which is not surprising given that in Cambrian times, North America and Greenland were both part of the Laurentian continent. Like Chengjiang, however, the Sirius Passet biota also dates from the Lower Cambrian Atdabanian Stage, and provides further evidence of the onset of the Cambrian Explosion.

For example, one of the first and most intriguing specimens to be discovered at Sirius Passet was *Halkieria*, a slug-like animal (**58**) with a dorsal coat of scale-like sclerites just as in the Burgess genus *Wiwaxia* (p. 29). But, at either end of the long body is a shell looking remarkably like an inarticulate brachiopod (see Conway Morris & Peel, 1990; Conway Morris, 1998, Figure 86). *Wiwaxia* has affinities with the molluscs and the annelids; perhaps these enigmatic halkieriids are telling us that at the dawn of the Phanerozoic, groups as diverse as the molluscs, annelids, *and* brachiopods were at that time phylogenetically very close? Molecular biology supports this conclusion.

MUSEUMS AND SITE VISITS

Museums

The Chinese National Fossil Law, passed in 2002, now strictly forbids export of any fossils from Chengjiang, thus preventing most European museums from acquiring specimens even though they are frequently seen on sale quite cheaply at fossil shows around the world. Manchester University Museum in the UK, and the Tokyo National Museum of Nature and Science, have a few specimens on display which were fortunately acquired prior to this new law being passed. We are unaware of any other displays of this material apart from those in China. For example, there is a museum dedicated to this fauna in Chengjiang town, while the museum at the Nanjing Institute of Geology and Paleontology also includes a good display from Chengjiang, as does the National Geological Museum of China in Beijing.

Sites

Access to the sites can only be obtained after the appropriate permission is obtained from Yunnan Province and Chengjiang County governing bodies. The various sites that yield the Chengjiang biota are protected by law. It is illegal to access the sites without permission, and it is also illegal to take even a single fossil from the outcrop, for whatever reason, unless permission is obtained. It is really only bone fide research workers who have collaboration with appropriate Chinese colleagues who would get such permission.

THE SOOM SHALE

BACKGROUND: EARLY PALAEOZOIC LAGERSTÄTTEN

Between the extraordinary biotas of the Lower and Middle Cambrian of Chengjiang (Chapter 3) and Burgess Shale (Chapter 2), and the Silurian Herefordshire Nodules (Chapter 5), the fossil record of Lagerstätten is sparse, and the Ordovician Period is fairly barren of exceptional biotas. However, one Ordovician horizon, the Soom Shale Member of the Table Mountain Group, Cape Supergroup, of Lower Palaeozoic sedimentary rocks of the Western Cape, South Africa, sprang to fame in the 1990s with the discovery of giant conodont animals with pre-served musculature and feeding apparatuses. The Soom Shale is the most important Ordovician Lagerstätte and is unique in that it comes from a high-latitude (60°S), cool, glacially influenced, marine habitat. In evolutionary terms, some of the animals of the Soom Shale hark back to the Middle Cambrian biotas – for example, there is a trilobite-like naraoiid *Soomaspis* – yet other animals of the Soom Shale, such as the eurypterid *Onychopterella*, presage a greater diversity to come.

HISTORY OF DISCOVERY OF THE SOOM SHALE

Because it is relatively soft, the Soom Shale forms a distinct bench in the landscape of rugged sand-stone escarpments and plateaus which dominates the scenery of the south-western Cape region, so its existence has been well known to geologists for many years. Few fossils occur in the predominantly sandstone rocks of the Table Mountain Group, to which the Soom Shale belongs and, at first sight,

the Soom Shale appears to be equally barren. Indeed, it was not until 1958 that fossil tracks and trails were discovered in the Table Mountain Group. The presence of some brachiopods and fragments of trilobites was provisionally announced in 1967, and later published by Cocks *et al.* in 1970, which gave the first indication of the Ordovician age of these shales. This was the first Lower Palaeozoic fauna to be described from South Africa.

The fossils of the Soom Shale first rose to fame in 1993 with the report of eyes in a conodont animal. It was only in 1983 that conodonts – usually small, phosphatic tooth-rows, sometimes found arranged in basket-like structures – were found in association with soft tissues, which later (1993) were shown to belong to chordates. It was the evidence of cartilage supporting large eyes in the Soom Shale conodont *Promissum pulchrum* that helped to determine the chordate nature of conodonts. Following this discovery, a concerted effort was made to find more specimens in the Soom Shale. This is a daunting task because fossils are rare in the shale, but dedicated hunting by Dick Aldridge and Sarah Gabbott, of Leicester University, UK, and Johannes Theron, of the Geological Survey of South Africa, has since unearthed a wide variety of animals and plants. Further specimens of *Promissum* turned up, which added more information about the morphology of conodonts and, in 1995, a specimen with muscle tissue was reported. Other exciting animals from the Soom Shale include the eurypterid ('sea scorpion') *Onychopterella*, which also shows details of musculature as well as the gut, enigmatic naraoiids, and orthocone cephalopods (relatives of squids and ammonites with straight, conical shells).

STRATIGRAPHIC SETTING AND TAPHONOMY OF THE SOOM SHALE

The Soom Shale Member is maximally 10 m (c. 30 ft) thick, characteristically finely laminated and yellow-brown to light or dark grey (see below). Above it lies the coarser, buff Disa Siltstone Member, about 130 m (c. 425 ft) thick. Together, these two members comprise the Cedarberg Formation (**59**), named after the beautiful Cedarberg Mountains (**60**), which form a north–south ridge between Citrusdal and Clanwilliam in

59 Stratigraphic section showing the position of the Soom Shale Member of the Cedarberg Formation, Table Mountain Group (after Theron *et al.*, 1990).

60 San rock art on the wall of a cave in Cedarberg Formation sandstone, with a view to the Cedarberg Mountains across the Brandewyn River valley.

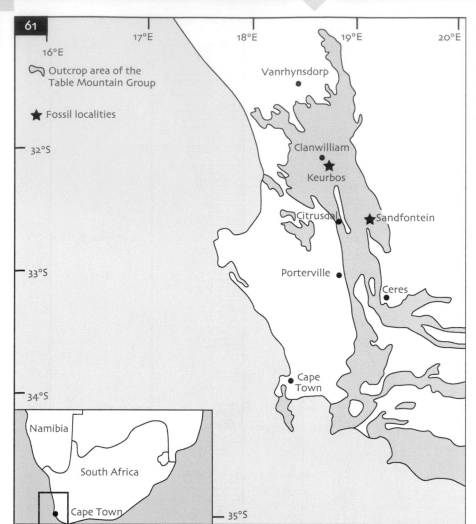

61 Map showing the outctrop of the Table Mountain Group in the Cape region of South Africa (after Theron *et al.*, 1990).

62 Quarry in Soom Shale at Keurbos Farm, 13 km (c. 8 miles) south of Clanwilliam, Western Cape.

the Western Cape, some 150 km (c. 90 miles) north of Cape Town (**61**). The Cedarbergs are named after the endemic Clanwilliam Cedar tree, *Widdringtonia cedarbergensis*, which, after extensive logging in the eighteenth and nineteenth centuries, is now sadly represented only by gnarled and twisted specimens on mountain tops above 1,000 m (c. 3,300 ft). The age of the Soom Shale has been determined as Hirnantian (444–446 Ma) on the basis of the occurrence of the trilobite *Mucronaspis olini* by Cocks and Fortey (1986) in an outcrop in the Hexrivier Mountains, some 100 km (c. 60 miles) from the main locality at Clanwilliam. Most of the biota has come from the locality (**62**) at Keurbos Farm, about 13 km (c. 8 miles) south of Clanwilliam; some animals come from the farm of Sandfontein, 25 km (c. 15 miles) east of Citrusdal.

The Table Mountain Group is a 4,000 m (c. 13,000 ft) thick pile of predominantly sandy

63 Glacial striae and pock-marks beneath the glacial tillite of the Pakhuis Formation, Cedarberg Mountains; the striae and pock-marks show lateral push-up ridges which imply their formation in soft, sea-bed sediment.

64 Dropstone deforming silt laminations, Pakhuis Formation, Cedarberg Mountains. The lens cap is 77 mm (c. 3 in) in diameter.

sediments. Two of the lower formations, the Graafwater and Peninsula, form the distinctive Table Mountain at Cape Town, but the higher beds do not appear until further north. The Pakhuis Formation lies between the Peninsula and Cedarberg Formations and is the clue to understanding the environment of the time. The Pakhuis Formation consists predominantly of tillite – fossilized till or boulder clay, the product of glaciers. In many places beneath the Pakhuis Formation, the tillites can be seen to rest on grooved Peninsula Formation – glacial striae which were carved as ice moved across the Peninsula Formation sediments. The grooves and pock-marks in 63 have distinct lateral mounds, indicating that the grooving at this locality was created when the Peninsula Formation was soft, probably when ice was moving across the sea floor. Because it is an erosion surface, the base of the Pakhuis Formation is an unconformity, but its top grades gradually into the finely laminated Soom Shale. Therefore, the Soom Shale appears to represent quiet water not far from the ice front. The mud and silt laminae, which are generally in the order of 1 mm (c. 0.04 in) thick, rarely up to 10 mm (c. 0.4 in), and are laterally persistent, represent the gentle settling of fine particles derived from the ice meltwater. These silt layers may represent distal turbidites – the final settling-out of a flurry of sediment which spread out across the floor of the sea (or possibly brackish bay), or possibly seasonal freezing and thawing (varves). There is some evidence for floating ice in the area

too, in the form of dropstones – pebbles dropped from passing icebergs onto the muddy seabed as the icebergs melted (64). The overlying Disa Siltstone Member is composed of coarser siltstones with evidence of occasional wave and current activity; thus it represents shallow marine water with a greater input of sediment as the ice sheets melted.

The Soom Shale shows no signs of bioturbation (animal activity in the sediment), which suggests that it was not possible for animals to live within the sediment or directly on the sea floor. This may have been because the temperature was too low although, as we shall see later, there were animals living in the waters above. Perhaps the mud and water on the sea floor were anoxic (lacking oxygen) or toxic in some way. Many of the laminate surfaces have ribbon-like algae ('seaweeds') on them, though these algae were not necessarily benthic (bottom-living); they could have descended to the sea floor after dying at the surface.

The Soom Shale animals are preserved as thin films of clay minerals replacing the original organic and/or mineral tissues of the carcasses. Shale is highly compressible, so the fossils are quite flattened. The sequence of events which led to the preservation of the fossils is interesting, and some aspects of it are unique. From the evidence given above, we can conclude that the water was cool, shallow (or not too deep), and there were no currents. Had there been currents, we would expect to see not only some rippling of the shale surfaces perhaps, but also alignment of elongate fossils such as the enigmatic spines called *Siphonacis*. The bottom waters were probably anoxic; there are virtually no benthic animals, no bioturbation, and the animals appear to be nearly complete and not scavenged or much decayed, as would normally occur on an oxygenated sea floor. It is possible that the sea-floor waters were not always anoxic, but that biota were only preserved during times of anoxia. Cold water appears to inhibit decay, and to increase the dissolution of calcium carbonate, which helps to explain the lack of calcitic shells but the preservation of organic tissues. At an early stage, organic material was replaced by the clay mineral illite. Later, apatite, as found in conodont teeth and parts of some brachiopod shells, was replaced by silica. There has been some debate (see Gabbott, 1998; Gabbott *et al.*, 2001) as to how organic tissues can be replaced by clay minerals, and whether the original replacement was by kaolinite which later changed to illite, or whether the illite replaced the organic molecules directly. Gabbott prefers the latter process.

What makes the preservation of the Soom Shale unique is that there are no other Lagerstätten in which direct replacement of organic materials by clay minerals has been demonstrated. The Burgess Shale (Chapter 2) animals are preserved as films of shiny micaceous minerals, similar to clays, as well as some clays adhering to the cuticles of trilobites. However, these minerals appear to have been adsorbed onto the organic surfaces, and formed rather later in the diagenetic process than the clay replacement in the Soom Shale. It is possible that clay mineral adsorption preceded replacement in the Soom Shale. Further work is necessary before we can be sure whether the Soom Shale is unique in this respect, or whether it represents an end-member of a geochemical spectrum.

DESCRIPTION OF THE SOOM SHALE BIOTA

Compared to other Ordovician shallow-water communities, as found for example in localities in north Wales and Scotland, the Soom Shale biota is considerably restricted in diversity. There are microfossils such as plant spores, acritarchs, and chitinozoans (the last two are organic-walled cysts of unknown affinity).

Microfossils. Chitinozoa are particularly well preserved in the Soom Shale. Like acritarchs, chitinozoans are organic-walled, generally flask-shaped microfossils of uncertain affinity, but they are known only from rocks of Ordovician to early Carboniferous age, and whereas acritarchs appear (from their biochemical composition, similar to sporopollenin) to be plant-related, chitinozoans are composed of a chitin-like substance, which suggests an animal affinity. Moreover, while chitinozoans are usually found as isolated individuals after maceration of the rock matrix, they can also be found on rock surfaces in strings, radial aggregates, and clusters inside a membrane suggestive of a cocoon. Search of the rock surface on and around the Soom Shale conodont animals and orthocone cephalopods by Gabbott *et al.* (1998) revealed all four types of occurrence – individuals, strings, aggregates, and cocoons. The most likely hypothesis of the zoological affinity of chitinozoans is that they are eggs and egg-masses or cocoons of an animal group which had the same stratigraphic range as the Chitinozoa: Ordovician to early Carboniferous. This narrows the possibilities down somewhat, and then if it is assumed that the producer of the chitinozoans is to be found among the restricted fauna of the Soom Shale, the possibilities are reduced to conodont animals or orthocone cephalopods. Other possible contenders, such as graptolites and gastropods, can be ruled out because they do not occur, or are found only rarely, in the Soom Shale. There is more association between chitinozoans and cephalopods than conodonts in the Soom Shale occurrences, so it is slightly more likely that Chitinozoa represent egg-masses of orthocone cephalopods.

Brachiopods. Few macrofossils are abundant in the Soom Shale, but some brachiopods are found more commonly than other fossils. These include orbiculoid and lingulate inarticulates (with phosphatic shells), and a few ribbed articulates (with calcified shells), belonging to the Rhynchonelliformea (Bassett *et al.*, 2009). Lingulates occur in shallow water and are infaunal, so need an oxygenated substrate; their presence suggests that the sea floor during Soom Shale times was at least occasionally oxygenated because there is no evidence that they drifted in from elsewhere (i.e. it is autochthonous). Two lingulate brachiopods preserve traces of organic tissue: the orbiculoid *Trematis* shows bands of periostracum preserved around the pedicle notch, and clay mineral casts of the pedicle are preserved in many specimens of *Kosoidea*.

Over 100 orbiculoid brachiopod specimens have been collected scattered in the shale at the Keurbos locality, but vastly more have been found in association with orthocone cephalopods. The orbiculoids occur attached as epizoans to the orthocone shells and, because the brachiopods occur all over the orthocones with no preferred side or orientation, it is presumed that they colonized the orthocones when they were alive and swimming in the nekton. Furthermore, if the brachiopods colonized the orthocone in life, then one would expect them to be larger nearer the apex of the shell (the end which formed earliest) than the aperture (which is the part of the shell where growth continues). One particularly large and well-preserved orthocone specimen and its associated epizoa was studied by Gabbott (1999) (**65**). She found that while there was no great difference in the size of brachiopods from the apex to the aperture of the orthocone, there were fewer brachiopods near the apex, and the loose brachiopods found in the nearby rock matrix were generally larger than those on the orthocone. To explain this distribution it is necessary to understand what might happen to an orthocone when it died. The carcass would fall to the sea floor and begin to decay. Buoyancy gases in the shell, perhaps combined with gases produced by decay, would buoy up the shell apex while the aperture would become buried in the sediment, so the brachiopods near the aperture would be covered up and die, while those (larger ones) near the apex might well also die and drop off the exposed shell since they were now in a poorly oxygenated environment.

Lobopods. These animals are generally regarded as onychophorans, related to these animals or tardigrades, or in the stem group of the Arthropoda. Lobopods were discussed in regard to *Hallucigenia* and *Microdictyon* in Chengjiang (Chapter 3) and *Aysheaia* and *Hallucigenia* in the Burgess Shale (Chapter 2). Lobopods are rare in rocks younger than the Cambrian, having been described only from Mazon Creek (Chapter 8) and amber

(including Baltic amber, Chapter 19), but one is known from the Soom Shale (**66**) (Whittle *et al.*, 2009). It shows features similar to those from the Burgess Shale, Chengjiang, and also some from the Lower Cambrian Sirius Passet biota (see p. 41). The fossil shows an annulated trunk, lobopod appendages with annulations, and curved claws, which suggest a benthic mode of life. Thus it is a rare record of a benthic organism from the Soom Shale, and demonstrates occasional oxygenation of the bottom waters during the deposition of the shale.

65 Orthocone nautiloid with associated orbiculoid brachiopod epizoans. (GSSA). Length 243 mm (c. 9.5 in).

66 Lobopod specimen, part (left) and counterpart (right) (GSSA). Counterpart slab is 30 mm (c. 1.2 in) wide.

Trilobites and relatives. The widespread Upper Ordovician trilobite genus *Mucronaspis* has been identified in the Soom Shale, some 100 km (c. 60 miles) away from the main Keurbos locality, and it is this trilobite, together with the brachiopods, which provided the dating of the Soom Shale. Of greater interest is the strange animal *Soomaspis splendida* (**67**), which was described by Fortey and Theron (1995). *Soomaspis* resembles an agnostid trilobite in having no eyes, just three thoracic segments, and cephalon and pygidium about the same size (isopygous). However, its uncalcified cuticle places it with the Nektaspida (see *Naraoia* from the Burgess Shale, Chapter 2, p. 30). The discovery of *Soomaspis* prompted Fortey and Theron to re-examine the relationship between the nektaspids and the rest of the trilobites. Because of the reduced number of thoracic segments, agnostids and nektaspids superficially resemble young stages of typical trilobites, which begin life without segments and add them one by one through ontogeny. In addition, the small adult size and reduced number of appendages of agnostids suggests that they attained their state by early onset of maturation in a juvenile form of their ancestors – an evolutionary process termed progenesis. The evolutionary benefit this gave the agnostids was to enable them to bypass a long existence as part of the benthos and instead enjoy a rapid turnover of generations entirely within the plankton. Nektaspids, on the other hand, had few or no thoracic segments, were larger trilobites and, as evidenced from *Naraoia* from the Burgess Shale, had a considerable number of appendages beneath their uncalcified shields. Thus, to produce a nektaspid only the production of thoracic segments is inhibited, while growth rate is normal or enhanced and maturation occurs at usual adult size. This process is termed hypermorphosis. Because of these different developmental routes, the small size of agnostids and their calcified exoskeleton, Fortey and Theron (1995) argued that nektaspids and agnostids were unrelated to each other, though possibly both could still be accommodated in the Trilobita. While agnostids can be derived from normal Cambrian trilobites, and their small size, loss of eyes, and so on are secondary, nektaspids appear to be more primitive than trilobites, and never developed dorsal eyes, a calcified cuticle, or the numerous segments of the typical trilobites. More recent cladistic analysis of the Nektaspida by Paterson *et al.* (2010) has suggested that this group is not part of the trilobites proper, but the relationships between them remain unclear.

Eurypterid. A few myodocopid ostracods have been found in association with the orthocone cephalopods, but the most interesting other arthropod in the Soom Shale is the eurypterid *Onychopterella* described by Braddy *et al.* (1995) (**68**). Eurypterids ranged from the Ordovician to the Permian periods, with their acme in the Silurian. Early eurypterids were marine animals, but by the late Silurian they had apparently invaded brackish and fresh waters, even becoming amphibious. They ranged up to 2 m (c. 6.6 ft) in length (there is evidence of a giant eurypterid trackway made by an animal of about this size in Permian sediments near Laingsburg, just 150 km (c. 90 miles) south-east of the Cedarberg range), and were thus the largest arthropods that ever lived. They were the top predators on Earth for some 200 million years. However, of the two Soom Shale eurypterid specimens, the larger would have measured less than 150 mm (c. 6 in) in life. Several features make the Soom eurypterid important. First, that few southern hemisphere eurypterids are known, and *Onychopterella* had previously only been recorded from the Silurian of North America. Therefore the Soom occurrence extends the genus into the Ordovician, and also the family Erieopteridae, to which it belongs, into the former Gondwanaland. Second, the eurypterid has a peculiar spiral structure preserved between the giant coxae of the last pair of appendages (the swimming legs). This structure was interpreted by Braddy *et al.* (1995) as a spiral valve in the gut, a feature normally associated with animals which scavenge on detritus in mud, but also known from another South African eurypterid *Cyrtoctenus*. Clearly, since nothing lived on the sea floor in Soom times, it is presumed that *Onychopterella* was not a mud-grubber; moreover, the spiral valve found in fish and *Cyrtoctenus* is further back in the alimentary system, so its presence in *Onychopterella* is problematical. Third, *Onychopterella* also has gill lamellae preserved. It has been known for a long time that eurypterids used gills for respiration in water, but all that had ever been preserved was a rather characteristic spongy material whose surface area for gas exchange is too small to support the respiratory needs of such large animals. Selden (1985) discussed this palaeophysiological enigma, concluding that the spongy material was actually an accessory organ for breathing air when on land, such as can be found in amphibious crabs and woodlice, and predicted that true gill lamellae would be thin and only preserved in exceptional circumstances. The Soom Shale has provided the exceptional preservation required to see these essential organs (see Braddy *et al.*, 1999).

Orthocone cephalopods. Fourteen orthocone cephalopods from the Keurbos farm locality were described by Gabbott (1999) (**65**). All specimens preserve the body chamber and some also retain the phragmocone (the rest of the shell, which contains gas chambers). The largest body chamber is 103 mm (c. 4 in) long, and the longest phragmocone is 243 mm (c. 9.5 in) long and 44 mm (c. 1.7 in) at its widest point, so the largest specimen would have been just less than 350 mm (c. 14 in) long. All are greatly flattened and preserved only as moulds because the original

67 The nektaspid arthropod *Soomaspis splendida* (GSSA). Length 30 mm (c. 1.2 in).

68 The eurypterid *Onychopterella* (GSSA). Length 150 mm (c. 6 in).

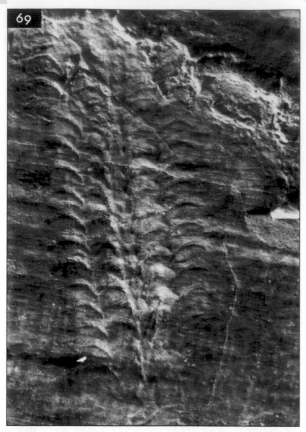

69 Radula of orthocone nautiloid (GSSA). Length 9 mm (c. 0.35 in).

70 The conodont anterior apparatus *Promissum pulchrum* (GSSA). Length 22 mm (c. 0.9 in).

aragonite has dissolved. The adhering epizoans have already been described, but there are other features of the orthocones which are exceptionally preserved. Radulae, rows of teeth which the orthocone used for feeding, are preserved as external moulds of teeth (**69**). Radulae are rarely preserved in fossil cephalopods, and are more commonly found in Mesozoic ammonites than Palaeozoic cephalopods. Indeed, radulae have only so far been reported in Palaeozoic non-ammonoids from the Upper Carboniferous Mazon Creek of Illinois (Chapter 8) and the Silurian of Bolivia. So the Soom Shale radulae are the oldest so far recorded for this group. There are two types of radulae known from cephalopods: a row of 13 elements in nautiloids and a row of seven in coleoids (squids and cuttlefish) and ammonoids. The Mazon Creek nautiloid shows 13 elements, so is presumed to be a nautiloid, and the Silurian orthoceratid from Bolivia appears to show seven, suggesting a closer relationship with coleoids and ammonoids than nautiloids. Only four radular elements can be seen in the Soom Shale cephalopod, and it is possible that there were more elements which were not preserved or are hidden by the four visible elements, so the Soom Shale material does not help in understanding the evolution of cephalopod radulae.

Conodonts. The Soom Shale is perhaps best known for its exceptional preservation of conodont animals (**70**). From their discovery well over a century ago until just 20 years ago, the animal to which the simple to complex, spiny, phosphatic structures known as conodonts belonged remained tantalizingly obscure. A number of, mostly worm-like, contenders had been put forward but in most cases it was difficult to prove that the enclosing soft-bodied creature had not simply eaten a conodont animal. In 1982 a fossil was discovered in which it could be shown that the conodont elements were truly part of the anatomy of the animal (Briggs *et* *al.*, 1983), and that animal was a chordate, perhaps related to the primitive jawless amphioxus and hagfish.

The Soom Shale conodont animal *Promissum pulchrum* was first described in 1986 as a very early land plant which could hold an important position in the evolution of land plants (the name translates as 'beautiful promise') (Kovács-Endrödy, 1987). The reason that *Promissum* was mistaken for a plant is that, while conodonts are most commonly found as isolated elements after acid-digestion of rocks, when they are large enough to be seen on bedding planes, conodont elements are sometimes found arranged into apparatuses which are thought to represent feeding structures. While it was obvious to some who studied the specimens (e.g. Rayner, 1986) that the fossils were not plants, it was not easy to say what they really were: giant graptolites perhaps? The graptolite expert Barrie Rickards studied the specimens and eventually concluded that they could be large conodonts, and so brought in Dick Aldridge, conodont expert, to confirm their identity. Thus, *Promissum pulchrum* became known as the largest conodont assemblage known, and clearly belonged to a giant animal (Theron *et al.*, 1990). An intensive search for more specimens followed and, in 1993, the hunt was rewarded when preserved soft tissues of *Promissum pulchrum* were found (Aldridge & Theron, 1993). Later finds from the locality at Sandfontein gave more information about the type of animal which bore the conodont elements. It was up to 400 mm (c. 16 in) long, with a pair of eyes at the front end, a notochord (stiff rod) along its trunk, and myomeres (muscle blocks), which suggests that it propelled itself with an eel-like wave passing down its trunk and tail. The conodont elements are positioned behind and below the eyes, and were clearly involved in feeding. However, among many aspects of conodont biology which have yet to be resolved, one remains as it has ever since the realization that conodont apparatuses were a feeding mechanism: how did it function? Broadly, one school of thought contends that the elements were used as teeth or raking devices, while the other (pointing to the lack of wear and the growth pattern) maintains that they were internal supports, perhaps for tentacles of some kind. So far, neither the Soom Shale occurrence nor any other reported conodont animal has provided the answer to this puzzle.

Miscellanea. Like most other Fossil-Lagerstätten in this book, the Soom Shale contains some enigmatic creatures. There are scattered, organic-walled spines given the name *Siphonacis*, which could belong to a spiny animal of some sort. Also, among the epizoa attached to the orthocone shells are some cornulitids – Palaeozoic cone-shaped fossils of uncertain affinity.

PALAEOECOLOGY OF THE SOOM SHALE BIOTA

Gathering together the evidence provided by the Soom Shale biota, and the data given by the sedimentology, a picture can be painted of a shallow sea with a muddy, cold, mainly lifeless bottom, although with occasional oxygenation events, as evidenced by the benthic lobopod and the infaunal lingulate brachiopods. In the water body there were, however, many swimming organisms (nekton). They occupied a number of feeding niches: the eurypterids, conodonts, and orthocones were presumably predators and/or scavengers, while the brachiopods and cornulitids were filter feeders. Occasionally the sea floor allowed some benthos, such as the lingulate brachiopods (filter feeders) and scavenging myodocopid ostracods. There is also evidence of undiscovered large predators or scavengers in the form of coprolites, crushed brachiopod shells, and broken conodont elements.

COMPARISON OF THE SOOM SHALE WITH OTHER LOWER PALAEOZOIC BIOTAS

As mentioned in the introduction to this chapter, there are very few other exceptionally preserved biotas in the Ordovician with which to compare the Soom Shale. One, the Winneshiek Lagerstätte, in a shale from the middle Ordovician St Peter Formation in north-eastern Iowa (Liu *et al.*, 2006), contains a variety of animals, preserved with soft tissues, impressions, and three-dimensional preservation. The community apparently inhabited the margins of the Laurentian seaway during the mid-Ordovician transgression. Included among the fossils are several conodont animals, phyllocarid crustaceans, and early jawless fish. The Fezouata Formation, described recently from the Ordovician of Morocco (Van Roy *et al.*, 2010), preserves a Burgess Shale-type biota. The preliminary report illustrates some exciting animals, which are yet to be described in detail.

The Soom Shale lies within the Hirnantian Age, the last in the Ordovician. The end of the Ordovician Period is marked by a mass extinction which ranks second only to that at the end of the Permian (Chapter 9, Karoo) in its effect on marine life; approximately 85% of marine species living at the time were wiped out. Two stages of extinction have been recognized. The first was linked to the onset of a major glaciation centred on African Gondwana, but its effects were more wide-ranging. The lowering of sea levels and changes in oceanic circulation aerated the deep oceans and brought nutrients, and possibly toxic material, up from oceanic depths. In the British Hirnantian, it is noticeable that there was a severe decline in the diversity of graptolites, plankton which are normally associated with low, subtropical latitudes. Three-quarters of all trilobites and a quarter of all brachiopods became extinct, to be replaced by the so-called Hirnantia Fauna of animals more suited to life in cool waters.

The second extinction phase has been related to the melting of the Gondwanan ice sheet and its associated transgression. This event is marked by the development of anoxic water conditions, which led to the deposition of black shales close to the Ordovician–Silurian boundary in the offshore environments. The deposition of this black shale coincides worldwide with the extinction of faunas adapted to the oxic glacial conditions (Delabroye & Vecoli, 2010). What caused the ice age in the first place is not known, but an intriguing idea of Melott *et al.* (2004) is that the glaciation could have been initiated by a gamma ray burst from an exploding star (a supernova) within 6,000 light years of Earth (i.e. within a nearby arm of our Milky Way galaxy). A 10 s burst of gamma rays would have stripped the Earth's atmosphere of half of its ozone almost immediately, causing surface-dwelling organisms, including those responsible for photosynthesis, to be exposed to high levels of UV radiation. This would have killed many species and caused a drop in temperatures. While plausible, there is no unambiguous evidence that such a nearby gamma ray burst has ever actually occurred, although scientists predict that one close enough should occur two or more times every billion years.

MUSEUMS AND SITE VISITS

Museums
The sparse Soom Shale fauna is deposited in the Geological Survey of South Africa and is not on exhibition anywhere at present.

Sites
The Soom Shale quarry at Keurbos Farm is situated some 13 km (c. 8 miles) south of Clanwilliam on the dirt road heading towards Algeria. Care should be taken on this road, especially in wet weather when it is rather slippery, because 4WD farm traffic uses it at greater speeds than ordinary cars. The quarry (**62**) is unmistakable, being on the east side of the road where a bend crosses over a small valley about 2 km (c. 1.25 miles) north of Keurbos. There is a detailed locality map in Theron *et al.* (1990). There is little to see apart from the grey shale itself. However, the surrounding area has spectacular sandstone scenery (**60**), and Clanwilliam is a wonderful centre for exploring the northern Cedarberg.

THE HEREFORDSHIRE NODULES

BACKGROUND: THE SILURIAN PERIOD

The Silurian Period is relatively poor in Fossil-Lagerstätten, but this situation has improved in recent years with the discovery of the Hereford-shire Nodules described here (Briggs *et al.*, 1996). Herefordshire is a pleasant, rural English county in the Welsh Borders (**71**), which was the favourite haunt of a number of geologists in the late 18th and early 19th centuries. Sir Roderick Impey Murchison, latterly Director-General of the Geological Survey, studied the rocks and fossils of the region and published his monumental work *The Silurian System* in 1839. He named his new period after the Silures: a British tribe who inhabited the Welsh Borderland during Roman times.

Following the end-Ordovician mass extinction (Chapter 4), faunas recovered and, by mid-Silurian times, high sea levels had produced warm, shallow epicontinental seas which provided a hospitable environment for marine life of all kinds. Reefs predominated in the Wenlock Edge area, with deeper water to the west and shallower to the east and south, as evidenced by their shelly faunas. Organisms were beginning to colonize the land, but that is the subject of Chapter 7, The Rhynie Chert.

HISTORY OF DISCOVERY OF THE HEREFORDSHIRE NODULES

Robert J. King, mineralogist and retired curator at the Department of Geology, University of Leicester, visited Herefordshire in the summer of 1990. Intrigued by the fascinating geology, he returned later that year to collect rock samples. On splitting open a hard, near-spherical concretion on a quarry floor, he caught sight of some sparkling mineralization that seemed to preserve a fossil. King collected nine concretions, four of which revealed fossils when cracked open and, in December 1990, he donated these specimens to

71 Herefordshire: the rolling hills of the Welsh Borderland.

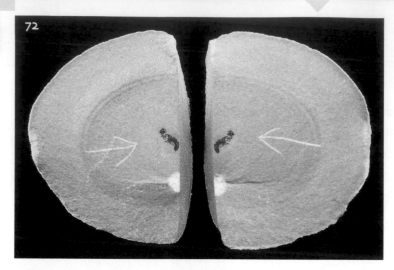

72 A split Herefordshire Nodule containing the arthropod *Offacolus* (OUMNH). Each half-nodule is about 30 mm (c. 1.2 in) wide.

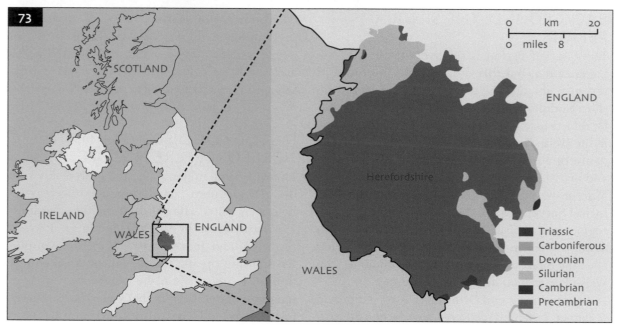

Triassic
Carboniferous
Devonian
Silurian
Cambrian
Precambrian

73 Map of the location of Herefordshire (after Briggs *et al.,* 2008).

the collections at the University of Leicester. In 1994, King's successor as curator, Roy G. Clements, asked David Siveter, a professor at Leicester University, to look at one of these finds. It showed an arthropod (later to be recognized as *Offacolus*) with its limbs preserved! That discovery prompted the involvement of David's twin brother, Derek Siveter, a professor at the University of Oxford. Derek photographed the material and both Siveters went with King to Herefordshire in December 1994 to find the source of these concretions. The following year, Derek Briggs of Yale University, a specialist in exceptionally preserved fossils, joined the team.

The Herefordshire fossils are not visually spectacular in hand specimen, and the calcite that gives them their sparkle would normally not preserve fine details, but the preservation of arthropod legs is so unusual, it was worth the team persevering and finding a way to extract more information from the specimens. At first, they studied these concretions by splitting them into pieces with a hydraulic vice until a fossil appeared. About half of the concretions examined in this way contained fossils. They attempted to discern the morphology of the commoner species, such as the tiny arthropod *Offacolus kingi*, by studying them in several hundred randomly split concretions (72).

74 Stratigraphic section of Silurian strata of Herefordshire.

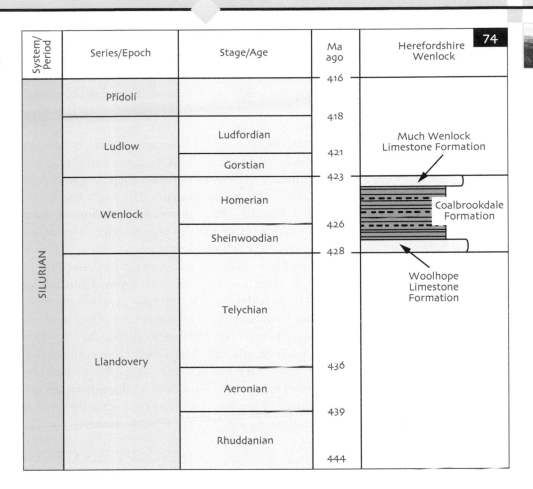

System/Period	Series/Epoch	Stage/Age	Ma ago	Herefordshire Wenlock
SILURIAN	Přídolí		416 / 418	
	Ludlow	Ludfordian	421	Much Wenlock Limestone Formation
		Gorstian	423	
	Wenlock	Homerian	426	Coalbrookdale Formation
		Sheinwoodian	428	Woolhope Limestone Formation
	Llandovery	Telychian	436	
		Aeronian	439	
		Rhuddanian	444	

But, despite their best efforts, the information obtained was woefully incomplete, and it became clear that they needed to find a way to extract the fossils from these rocks.

The calcite casts were too small and delicate to be dug out physically, and they could not be dissolved out by acids because they are so similar in composition to the rest of the concretion. They were not visible in x-ray photographs nor in the other scanning methods available because they have the same density as the rock. So the team resorted to grinding away the fossil in very fine increments, up to 50 per millimetre, and recording each exposed surface as a digital image. This serial-sectioning method has been used by palaeontologists since the beginning of the last century, but in those days reconstructing the three-dimensional fossil from serial drawings was laborious. By the late 1990s, however, it was possible to put together hundreds of digital images using powerful computers and reconstruct a virtual fossil from the many slices. Mark Sutton, of Imperial College

London, developed computer software to do this easily (see Sutton *et al.*, 2001a). Using this software, the virtual fossils can be manipulated on the computer screen, using stereo glasses to add depth; they can be rendered as rotating animations; even a three-dimensional representation can be made, which makes a replica of the fossil by fusing powdered plastic resin with a laser beam.

STRATIGRAPHIC SETTING AND TAPHONOMY OF THE HEREFORDSHIRE NODULES

The Herefordshire fossils were deposited during middle Silurian (Wenlock) times, 425 million years ago (**73, 74**), in a marine basin that extended across what is now central England and Wales. This basin first formed at the beginning of the Cambrian Period, some 120 million years earlier. The fossils are preserved in a soft, cream-coloured volcanic ash (bentonite) that mixed with some of the normal marine sediment. This ash deposit is known from a single locality, where it is exposed over a distance of about 30 m (c. 100 ft). Measuring more than a metre (3 ft) thick in places, it is unusually unconsolidated and can be dug by hand without difficulty. The concretions that carry the fossils, on the other hand, are hard; they vary in size from about 20 mm (c. 0.8 in) to 200 mm (c. 7.9 in) in diameter, and occur randomly within the ash.

The volcanic ash was deposited on top of a thin layer of mud covering a thicker sequence of lime-stones: the remnants of a reef that was effectively dead and had probably sunk well beneath the surface of the sea. Indeed, the animals that became fossilized here most likely lived 150–200 m (c. 490–650 ft) down, below the photic zone. The evidence for this is the lack of photosynthetic algae, which are common in contemporaneous rocks laid down at shallower depths on the sea floor to the east. It is not clear whether a volcanic eruption entombed these Silurian animals in a single fall of ash, or whether they were buried many years after the eruption, covered in ash that had been transported along the seabed by a turbidity current. Either way, it is clear that some special circumstances caused their remarkable preservation.

First, rapid accumulation/precipitation of clay minerals occurred around the dead organisms, which decayed over time, leaving empty spaces behind. Then calcite filled these natural moulds, faithfully replicating the shape, and occasionally the internal features (e.g. gut), of the animals. Even spines and other structures just a few microns across were preserved in this way. About the same time, the concretions began to form, being cemented by calcite. The early hardening of the concretions prevented squashing of the fossils as the ash layer slowly compacted. The evidence for this is that the fossils are undeformed and the concretions are spherical rather than ellipsoidal (see Orr *et al.*, 2000 for more detailed information).

Freshly exposed concretions are hard with a blue-grey core. The Herefordshire concretions are unusual in that they do not correspond in size to the fossils they contain (see Chapter 8, Mazon Creek). Also, the fossils are commonly not in the middle, so the nucleus of the concretions appears to have been something other than the fossil itself, but no trace remains.

DESCRIPTION OF THE HEREFORDSHIRE NODULES BIOTA

Protists. Radiolaria are planktonic, single-celled amoeboid organisms which eat other plankton, such as bacteria, diatoms, copepods, crustacean larvae, and other protists, and many have symbiotic algae. They construct beautiful skeletons of silica and can be found in rocks from the Cambrian to the Recent, and are used extensively for biostratigraphy. The Herefordshire concretions have yielded one of the few recorded Silurian radiolarian faunas worldwide and the only one known from the Silurian of Britain. The fauna consists of just a few species, and has affinities with Silurian radiolarian assemblages of the Urals, the Canadian Arctic, and Alaska (Siveter *et al.*, 2007a).

Worms. Whole body fossils of polychaete annelids (bristle worms) are rare and nearly always compressed and poorly preserved. On the other hand, *Kenostrychus clementsi* (**75**), the second commonest animal in the Herefordshire Nodules, while neither the oldest nor the most unusual fossil polychaete, is beautifully preserved in three dimensions. It is a rather generalized polychaete in many respects, with large, biramous parapodia, unspecialized anterior segments, and a small

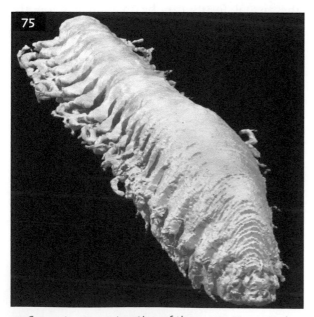

75 Computer reconstruction of the worm *Kenostrychus* (OUMNH). Length about 6 mm (c. 0.24 in).

prostomium with median and lateral antennae and ventral prostomial palps, but its gills are coiled tentacles attached to an unusual part of its notopodia (trunk appendages). This configuration has implications for the early evolution of polychaete respiratory structures, and *Kenostrychus* is interpreted as a member of a stem group within the polychaetes (Sutton *et al.*, 2001b). Polychaete fossils also occur in the Burgess Shale (Chapter 2), the Hunsrück Slate (Chapter 6) and Mazon Creek (Chapter 8).

Brachiopods. These are among the commonest fossils in rocks of Palaeozoic age, and are studied by many palaeontologists, but they rarely preserve any internal soft tissues. A single specimen from the Herefordshire Nodules is unique in the enormous amount of information on brachiopod soft-part morphology it preserves (**76**). Three tiny, post-larval brachiopods, representing at least two different species, are attached to a third, larger species of brachiopod, named *Bethia serraticulma* (Sutton *et al.*, 2005). All three species preserve the fleshy pedicle (attachment stalk), and one also shows sensory hairs. *Bethia* also preserves internal structures including the mantle that secretes the shell, and the lophophore used for filter feeding. In spite of its spectacular state of preservation, *Bethia* is difficult to place taxonomically, ironically because the soft tissues obscure parts of the shell that are critical for classification. The original authors assigned it to the order Orthida, but other brachiopod workers have interpreted the preserved pedicle as a pedicle sheath, which occurs in young specimens of the Strophomenida.

More recently, another epibiont (attached organism) has been found on *Bethia*: an apparently unmineralized, bilaterally symmetrical animal about 2 mm (c. 0.08 in) in length, with a pair of coiled tentacles which have been interpreted as a lophophore (Sutton *et al.*, 2010). This animal most likely belongs to the brachiopod stem group, though the absence of a shell is probably secondary. The find also suggests that unmineralized, sessile lophophorates might have been significant members of Palaeozoic ecosystems.

Molluscs. Gastropods (snails) are common in the fossil record, but fossilization of their soft tissues is almost unknown, and had not been reported from the Palaeozoic until they were discovered in the Herefordshire Nodules. In an unnamed platyceratid gastropod (Wenlock platyceratids are sorely in need of taxonomic revision), the digestive system is preserved in detail, and morphological data on the gonads, digestive gland, pedal muscle, radula, mouth, and foot are also preserved (Sutton *et al.*, 2006). However, the specimen lacks an operculum (the disk which seals the opening and protects the soft parts when the animal retracts into its shell), and this suggests the gastropod was sessile, perhaps a commensal coprophage (dung-eater) on larger organisms such as starfish (see p. 61).

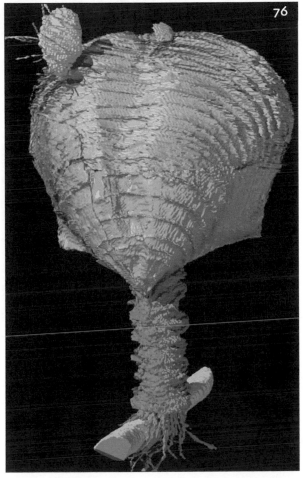

76 Computer reconstruction of the brachiopod *Bethia* (OUMNH). The shell is about 6 mm (c. 0.24 in) in diameter.

77 Computer reconstruction of the unusual mollusc *Acaenoplax*; left: anterior end; right: rest of the body (OUMNH). Total length of animal about 35 mm (c. 1.4 in).

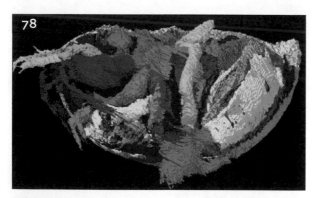

78 Computer reconstruction of the male ostracod crustacean *Colymbosathon* (OUMNH). Total length about 5 mm (c. 0.2 in).

Acaenoplax hayae (**77**) resembles a spiny worm. Its ventral side is smooth, with flexible lobes arranged in a chevron pattern which probably helped it to grip the sediment during locomotion. The wrinkled body bears an array of sharp spines

behind the head that were probably defensive. The largest spines project from fleshy ridges on its back, which also has seven plate-like shells along its length, most of which are also spiny. At the rear, two shells enclose a respiratory cavity from which fleshy gills protrude. The presence of seven shells, together with an obvious space where an eighth seems to have been lost in evolution, suggests that *Acaenoplax* is related to the chitons: polyplacophoran molluscs (see p. 96), whose shells are known from the Silurian. However, *Acaenoplax* is not a chiton because it has no foot. The body plan of *Acaenoplax* is closer to aplacophorans, which are simple, worm-like molluscs that lack a foot and have a rear respiratory cavity, although no living aplacophoran is known to have shells. *Acaenoplax* appears to be a missing link between polyplacophorans and aplacophorans. This conclusion implies that recent aplacophorans are secondarily simplified, having lost the shells present in their *Acaenoplax*-like ancestors. Not everyone agrees with this conclusion, however; Steiner and Salvini-Plawen (2001) interpreted *Acaenoplax* as closer to polychaete annelids.

Crustaceans. Ostracods are tiny, aquatic crustaceans with hinged, bivalved shells enclosing a shrimp-like body. They are by far the commonest arthropods in the fossil record, known from thousands of species and innumerable empty shells from at least the Ordovician onwards, and are valuable palaeoenvironmental and biostratigraphic indicators, but their fossilized soft parts are extremely rare. They are abundant today, with an estimated more than 20,000 living species, and have been very successful in adapting to a wide range of marine, brackish, and freshwater environments, from the deep ocean to forest soils. They have been studied by palaeontologists for more than 150 years, but because ancient forms have been known only from their shells, there has been some dispute regarding whether they truly represent the ancient counterparts of extant ostracods. Inside the shell of the first ostracod investigated (**78**), Siveter *et al.* (2003) found the head and trunk, with specialized appendages for feeding and locomotion, and a pair of lateral eyes. Also preserved are the gut, the gills, and evidence of the circulatory system. Moreover, the male copulatory organ is preserved, which is the earliest unequivocal evidence for gender in animals – the oldest known penis. The animal was given the name *Colymbosathon ecplecticos*, which translates to 'swimmer with a large penis'.

A second, exceptionally well-preserved ostracod, again with preserved gut, lateral eyes, and appendages, turned out to be a female containing eggs and possible juveniles. *Nymphatelina gravida* (**79**) provides a unique view of maternal care in a fossil invertebrate and demonstrates that an egg-brooding reproductive strategy has existed from the Silurian to the present day (Siveter *et al.*, 2007c). Another beautifully preserved ostracod was described by Siveter *et al.* (2010). In this species, *Nasunaris flata*, the medial eye is preserved with a Bellonci organ, which has sensory and secretory functions, and is the first known from the fossil record. Also preserved are the gills, the gut, appendages, including a worm-like (vermiform) seventh limb, and lateral eyes. In ostracods, the Bellonci organ, a vermiform seventh limb, and lateral eyes are known only in the Myodocopa, so all of the specimens found so far belong here. Moreover, the presence of gills places *Colymbosathon* and *Nasunaris* in the modern family Cylindroleberididae, thus demonstrating remarkably conserved morphology over a period of 425 million years.

The soft-part anatomy indicates that the Herefordshire ostracods belong to the myodocopes, but the shell of *Nymphatelina* resembles that of a different group of fossil ostracods, the palaeocopes, whose soft-part anatomy is not known. Similarly, the shell morphology of *Nasunaris* more closely resembles several families of myodocopes other than the Cylindroleberididae, especially the Cypridinidae and Sarsiellidae, which are also known only from their preserved hard parts. This demonstrates that the shell alone may be a poor clue to the true nature of the animal within.

The Herefordshire deposit also contains a number of other crustaceans, including barnacles. These animals begin life in the plankton as a small, free-swimming larva (the cyprid) which eventually settles down to become either a stalked goose barnacle or a more familiar balanomorph form. Both of these are attached to rocks in the intertidal zone. The adult barnacle grows a mineralized shell of several parts, which generally separate when the animal dies and so barnacle fossils are normally found as isolated plates. Without mineralization,

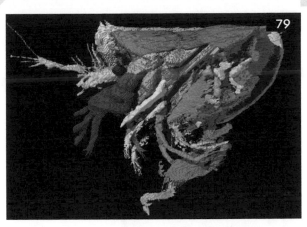

79 Computer reconstruction of the female ostracod crustacean *Nymphatelina* (OUMNH). Total length about 6 mm (c. 0.24 in).

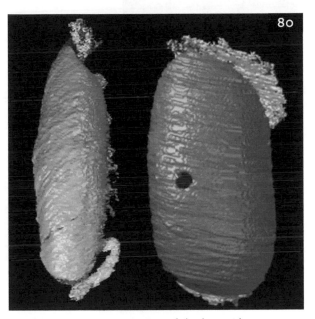

80 Computer reconstruction of the barnacle *Rhamphoverritor*; left: free-swimming stage in lateral view; right: juvenile in lateral view (OUMNH). Length about 5 mm (c. 0.2 in).

the larva is much less likely to become fossilized at all. The Herefordshire Nodules preserve the first fossil cyprid known, *Rhamphoverritor reduncus* (**80**).

More familiar living crustaceans are the Malacostraca, which includes shrimps, lobsters, and crabs. Crabs and lobsters have a good fossil record because of their thick, calcified cuticles. More primitive malacostracans, such as the

81 Computer reconstruction of the stem crustacean *Tanazios*; left: ventral view; right: dorsal view (OUMNH). Length about 30 mm (c. 1.2 in).

phyllocarids, are unmineralized and so their fossil record is poor. This is unfortunate, since the primitive, early forms hold clues to the origin and early diversification of the group. Step forward the Herefordshire Nodules again! The earliest completely preserved phyllocarid (and the earliest completely known malacostracan), *Cinerocaris magnifica*, was described by Briggs *et al.* (2003). It has two pairs of antennae and prominent eyes projecting forward from the head, which also bears feeding appendages. The body is divided into a thorax and abdomen, the latter being muscular and ending in the forked tail typical of this group.

Another crustacean, *Tanazios dokeron* (81), has two pairs of antennae and a mandible, which are characteristic of crustaceans, but its rear head appendages were not specialized for feeding – they appear very similar to those of the trunk. The head shield of *Tanazios* is a strange horned structure, and the long trunk is made up of more than 60 short segments, each with a similar pair of unusual, digitate appendages. A pair of antenniform appendages projects from the tail. Siveter *et al.* (2007d) concluded that *Tanazios* was a stem-group crustacean.

Other arthropods. Pycnogonids (sea-spiders) are marine arthropods known from nearly 1,200 living species. They are widespread in the oceans today, down to 6,000 m (c. 19,700 ft) depths, but have an extremely sparse fossil record of just a few tens of specimens belonging to nine species. Minute specimens from the Cambrian Orsten deposits of Sweden have been referred to planktonic larvae of pycnogonids (though not everyone agrees with this assignment), but the pycnogonid described from the Herefordshire Nodules, *Haliestes dasos* (82), is the oldest adult sea-spider, some 35 Ma older than those from the Hunsrück Slate (Chapter 6), and the most completely known fossil species (Siveter *et al.*, 2004). The relationship of sea-spiders to other arthropods is debated. There are two main viewpoints: that they belong with the chelicerates, or they are separate from all other arthropods. The large, chelate, first appendage of *Haliestes* provides evidence for a chelicerate affinity for the pycnogonids, and cladistic analyses place the new species near the base of the pycnogonid crown group, implying that modern pycnogonids had arisen by Silurian times.

One of the first animals to be described from the Herefordshire Nodules was *Offacolus kingi* (Sutton *et al.*, 2002). *Offacolus* (83, 84) possesses prosomal appendages similar to those of the modern horseshoe crab *Limulus*, but also bears robust setae on the second to fifth appendages which are unlike those found in any other arthropod. Its opisthosomal appendages are similar in number and morphology to the book-gills of *Limulus*. *Offacolus* was placed near the base of the Chelicerata, as a sister taxon to the eurypterids (see Chapter 4, The Soom Shale) and living chelicerates, but more advanced than other middle Palaeozoic horseshoe crabs.

Marrellomorphs are a poorly known group of fossil arthropods that include *Marrella* from the Burgess Shale (Chapter 2), and *Mimetaster* and *Vachonisia* from the Hunsrück Slate (Chapter 6). Marrellomorphs branched off the main arthropod lineage earlier than did the trilobites, chelicerates, crustaceans, and their relatives. *Xylokorys chledophilia* (85) from the Herefordshire Nodules (Siveter *et al.*, 2007b) provides the most complete three-dimensionally preserved marrellomorph, and the first from the Silurian. It is closest in form to the Hunsrück *Vachonisia* (p. 71). In both

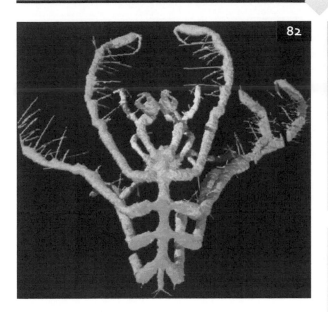

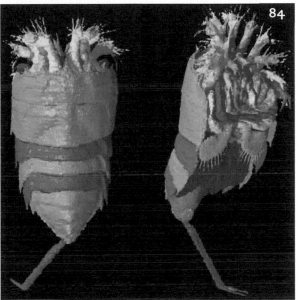

82 Computer reconstruction of the sea-spider (pycnogonid) *Haliestes* (OUMNH). Body length 35 mm (c. 1.4 in).

83 The chelicerate *Offacolus* as seen in the nodule (OUMNH). Width about 2 mm (c. 0.08 in).

84 Computer reconstruction of the chelicerate *Offacolus*; left: dorsal view; right: ventral view (OUMNH). Width about 2 mm (c. 0.08 in).

85 Computer reconstruction of the marrellomorph arthropod *Xylokorys* (OUMNH). Carapace width about 1.2 mm (c. 0.05 in).

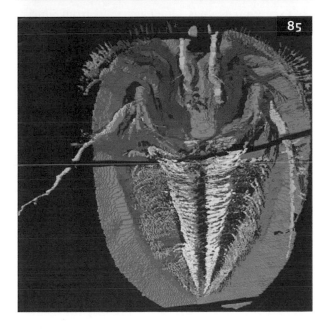

animals, the head and trunk are almost completely covered by a dorsal shield. The head bears five pairs of appendages and the trunk approximately 35 pairs of biramous appendages.

Echinoderms. Starfish (asteroids) are well known in the fossil record from their calcified hard parts, but in the Herefordshire Nodules, soft tissues are preserved in a specimen of *Bdellacoma* (Sutton *et al.*, 2005). This genus had been placed in the ophiuroids (brittle stars), but the soft-part anatomy, including amazingly preserved tube-feet, radial canals, and parts of the gut, place *Bdellacoma* at the base of the asteroid crown group.

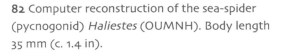

PALAEOECOLOGY OF THE HEREFORDSHIRE NODULES BIOTA

The Herefordshire biota contains a variety of marine invertebrates such as worms, molluscs, starfish, and brachiopods, together with a range of arthropods, and some forms of unknown affinity. The fossils with hard parts in the Herefordshire Nodules, which would be preserved under normal fossilization processes, include brachiopods, the trilobite *Tapinocalymene*, small gastropods, and monograptid graptolites. This fauna also occurs in the overlying shales, the overall fauna from which is best allocated to the *Visbyella* community of Hancock *et al.* (1974), which was characteristic of deep water (about 200 m [650 ft] depth) at the edge of the continental shelf in the Welsh Basin during Silurian times. The community includes epifaunal suspension-feeding brachiopods, scavenging and predatory echinoderms, and arthropods of various kinds, shallow-burrowing worms and bivalves, and planktonic graptolites.

COMPARISON OF THE HEREFORDSHIRE NODULES WITH OTHER MID-PALAEOZOIC LAGERSTÄTTEN

Silurian Lagerstätten are particularly rare, although because the Silurian Period is the shortest in the Palaeozoic, it is likely to contain the fewest examples of exceptional biotas. One Lagerstätte from Silurian rocks can be compared with Herefordshire: the Waukesha biota from Wisconsin, USA (Mikulic *et al.*, 1985). This biota occurs in a local development of finely laminated mudstone and dolomite within the dolomitic Brandon Bridge Formation of late Llandovery age. Preservation seems to have occurred by phosphatization within an anoxic environment. Like Herefordshire, this biota contains polychaete worms, trilobites, graptolites, ostracods, and other crustaceans, and some bizarre arthropods. It additionally preserves one of the first conodont animals (see p. 50) to be described. However, the Waukesha biota lacks certain typically fully marine faunal elements such as brachiopods and molluscs, though these are present in the Brandon Bridge Formation of which the Waukesha is a member. Similarities between the Herefordshire and Waukesha biotas stem from the range of organisms which lived at the time, while the contrasts are due to both preservational and ecological differences.

The Eramosa Lagerstätte from Ontario, Canada (von Bitter *et al.*, 2007) preserves a diverse biota including articulated conodonts, heterostracan fish, annelids, arthropods with soft body parts, and a diverse marine flora. Significantly, an *in-situ* fauna of marine shelly fossils and trace fossils is also preserved. This association of exceptionally preserved remains together with more typical fossils distinguishes the Eramosa from other Silurian Lagerstätten, and suggests that the Eramosa is not the product of exceptional preservation in an atypical environment, but high-quality preservation of a normal Silurian shallow-marine ecosystem.

MUSEUMS AND SITE VISITS

Museum

Since each nodule is ground away, only virtual fossils exist. Nevertheless, the Oxford University Museum of Natural History, Parks Road, Oxford OX1 (tel. +44-1865-272950) has a small display with some publications, computer images, and nodules.

Sites

At present, the location of the site is not publicized, and no visits are possible. Being very soft, outcrops of the Coalbrookdale Formation are few and far between, but it can be seen in a number of quarries and road cuttings. The Herefordshire & Worcestershire Earth Heritage Trust manages some sites, and they should be contacted for information (Herefordshire & Worcestershire Earth Heritage Trust, Geological Records Centre, University of Worcester, Henwick Grove, Worcester WR2 6AJ. Tel: +44-1905-855184. E-mail: abigail.brown@worc.ac.uk).

THE HUNSRÜCK SLATE

BACKGROUND: THE RISE OF THE VERTEBRATES AND THE AGE OF FISHES

Vertebrates (animals with backbones) are members of the phylum Chordata, which all share a notochord (a tough rod of collagen running down the length of the body), and V-shaped blocks of muscle called myotomes. It had long been considered that the oldest chordate in the geological record was the tiny *Pikaia* from the Middle Cambrian Burgess Shale of British Columbia (Chapter 2), and thus that the chordates originated during the Cambrian Explosion.

However, the discovery in 1999 of small, jawless, agnathan fish from the Lower Cambrian of Chengjiang in southern China (Chapter 3), has pushed the origin of the vertebrates themselves back into the melting pot of the Cambrian Explosion. The jawless agnathans, which include the present-day lampreys and hagfish with their sucker-like mouths, are certainly a primitive group of vertebrates, but the ancestors of higher vertebrates are not to be found among them.

During the early Silurian a second group of fish appeared, the enigmatic acanthodians (inappropriately named the 'spiny sharks' because of their prominent caudal fin). Generally small and with numerous spines on the dorsal and ventral margins, they became abundant in Devonian freshwaters and persisted until the end of the early Permian. They are significant as they are the first jawed vertebrates in the fossil record.

However, from the Cambrian through the Silurian the fossil record of fish is very sparse. The major episodes of early fish evolution most probably took place at this time, but did so in freshwater habitats, and it is not until the Devonian Period that freshwater deposits (mainly on the Old Red Sandstone continent) become common in the geological record. The Devonian Period is known as 'the age of fishes' because by its close all five major groups of fish were locally abundant (see Benton, 2000 for details).

The third group to appear were the placoderms – heavily armoured fish with poorly developed jaws. They were almost confined to the Devonian and are now extinct. For a time, however, they were dominant and a huge variety of weird forms with massive head armour included giants such as *Dunkleosteus*, over 20 m (c. 65 ft) long. Its jaws had no teeth, but the upper edges of the lower jaw were sharpened into blade-like plates.

Finally, the two major extant groups of fish, the osteichthyans and the chondrichthyans, evolved and flourished at this time. Osteichthyans (the bony fish) appeared early in the Devonian and by the end of the Palaeozoic had almost sole possession of lakes and streams and had invaded the seas. They include two lineages: the sarcopterygians (lobe-finned fish), such as lungfish and coelacanths, which in Devonian times were more significant; and the actinopterygians (ray-finned fish), which today include almost all freshwater fish and the vast majority of marine forms.

Chondrichthyans (the cartilaginous fish) – the sharks, skates, rays, and the chimaeras (ratfish) – were the last of the five classes of fish to evolve. They are not known before the late Middle Devonian, suggesting that their lack of bone is not a primitive condition, but rather that evolution is towards bone reduction.

63

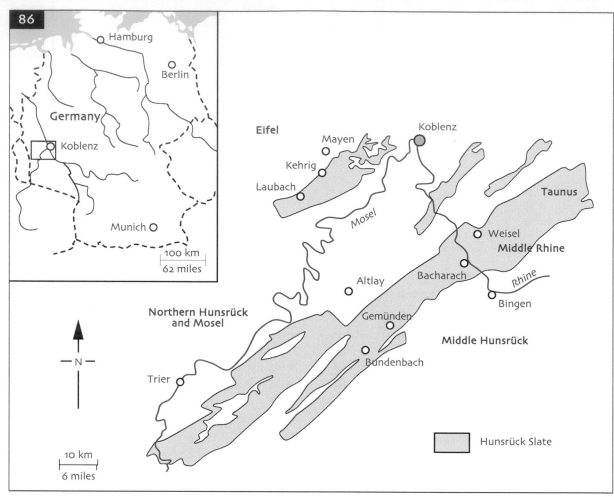

86 Locality map to show the Hunsrück Slate region in the Rheinisches Schiefergebirge of western Germany (after Bartels *et al.*, 1998).

By the end of the Devonian the early dominance of seas and freshwater by the armoured fish had given way to the modern sharks and bony fish. Moreover, by the Middle Devonian Givetian Stage the sarcopterygians had given rise to a second major group of vertebrates, and the first to conquer land, when their lobe-fins, supported by a single basal bone and strong muscles, were modified into the tetrapod limbs of the first amphibians.

Several rich fossil deposits within Devonian rocks record these important chapters in vertebrate evolution, such as the late Devonian marine deposits of Gogo in Western Australia and the Old Red Sandstone lake deposit of Achanarras in Caithness, Scotland (pp. 73–74). The Lower Devonian Hunsrück Slate of the Rhenish Massif in western Germany, however, not only includes a rich invertebrate biota in addition to vertebrates, but is

also a true Conservation Lagerstätte with soft tissues exquisitely preserved in many groups, most notably the echinoderms (especially the starfish) and arthropods. Moreover, their peculiar preservation by pyritization enables their examination by x-ray techniques, which often reveal minute detail.

HISTORY OF DISCOVERY AND EXPLOITATION OF THE HUNSRÜCK SLATE

The Hunsrück Slate has been an important source of roof slates for several centuries in the Rheinisches Schiefergebirge (Slate Mountains) of the Rhine and Mosel valleys (**86**). Certainly the slate was used in Roman times as evidenced from numerous Roman sites in west Germany, but the earliest documented evidence of mining in this area comes from the fourteenth century (Bartels *et* *al.*, 1998). Extensive mining continued throughout the succeeding centuries, but the Industrial Revolution of the late eighteenth century saw a huge expansion in the production of roofing slates. Slate was exported along the Rhine and Mosel rivers, but a collapse in the industry during the German depression (1846–49) led to poverty and hardship in the mining settlements.

Economic revival and a new sense of nationalism after the Franco-Prussian war of 1870–71 led to renewed growth in the slate industry and extensive mines were developed by the bigger companies. In the early twentieth century deeper shafts were sunk, railways were employed, and the use of technology generally increased. Production continued through the years of the Second World War into the 1960s, but competition from synthetic and cheaper imported slate has since led to a decline. In recent years only one mine (Eschenbach-Bocksberg Quarry) has been operating in the Bundenbach region and, since 1999, this has merely been preparing imported slate from Spain, Portugal, Argentina, and China and has ceased mining local rock.

The mining of the Hunsrück Slate was vital to the discovery of its fossils. Although they are not uncommon, fossils are only readily recognized by handling large quantities of rock and many of the fine specimens on display in museums were originally found by the slate splitters. The first scientific paper on these fossils was by Roemer (1862), who described and illustrated asteroids and crinoids from the Bundenbach region. Distinguished German palaeontologists such as R. Opitz (1890–1940), F. Broili (1874–1946), R. Richter (1881–1957), and W. M. Lehmann (1880–1959) made extensive studies of Hunsrück fossils from the 1920s to the 1950s, but Lehmann's death, coupled with the decline in the slate industry, led to a corresponding decline in research, especially as few new specimens were being discovered.

At the end of the 1960s Wilhelm Stürmer, a chemical physicist and radiologist at Siemens Corporation, combined these skills with his interest in palaeontology and developed new methods of examining the Hunsrück fossils using x-rays (Stürmer, 1970). His beautiful radiographs of unprepared slates, using soft x-rays (25–40 KV) and stereoscopic exposures combined with high-resolution films and image processing, show

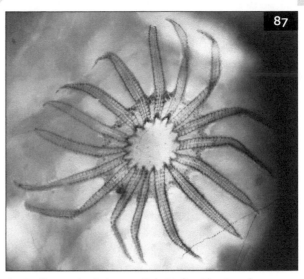

87 Radiograph of the starfish *Helianthaster rhenanus* (DBMB). Width approx. 150 mm (c. 6 in).

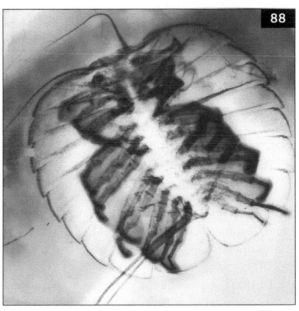

88 Radiograph of the arthropod *Cheloniellon* (BKM). Width 120 mm (c. 4.8 in).

intricate detail of soft tissue not revealed by conventional techniques (**87, 88**). More recently Christoph Bartels, Michael Wuttke, and Derek Briggs initiated 'Project *Nahecaris*' which reinvigorated research on the Hunsrück fauna (Bartels *et al.*, 2002) and which has recently been extended into a new project led by Jes Rust in Bonn. Bartels *et al.* (1998) give an extensive bibliography for all of this research.

STRATIGRAPHIC SETTING AND TAPHONOMY OF THE HUNSRÜCK SLATE

The Hunsrück Slate (Hunsrückschiefer) is a thick sequence of muddy marine sediments of Lower Devonian age which by low-grade metamorphism have been altered to slates. The sequence should strictly be regarded as a facies, rather than a stratigraphic unit, as the deposition of mud was diachronous, beginning earlier in the north-west and migrating south-eastwards. The precise age of the slate is therefore variable, but ranges from late Pragian to early Emsian and is therefore approximately 407 million years. Its outcrop in the Hunsrück hills (**89, 90**) occurs in a belt approximately 150 km (c. 90 miles) in length and covers an area of 400 sq. km (c. 150 sq. miles).

Deposition of the mud occurred in a narrow north-east–south-west offshore marine basin lying between the recently uplifted Old Red Sandstone Continent to the north and the Mitteldeutsche Schwelle (see Chapter 13, The Solnhofen Limestone, p. 159) to the south. Immediately after its uplift, at the end of the late Silurian/early Devonian Caledonian Orogeny, large volumes of sand and muddy detritus were shed into rivers and transported south. Finer sediment was carried in suspension and deposited offshore in the Central Hunsrück Basin. The total thickness of the Hunsrück Slate has been estimated at 3,750 m (c. 12,300 ft) (Dittmar, 1996), although the roofing-slate sequence in the area around Bundenbach and Gemünden is somewhat less than 1,000 m (c. 3,300 ft) (**91**).

During the subsequent Variscan Orogeny of the Carboniferous Period the muddy sediments were subjected to low-grade metamorphism, producing the characteristic slaty cleavage. On the steep limbs of the folds, such as in the Bundenbach–Gemünden area, the cleavage lies parallel to the bedding, enabling the fossils to be revealed (**90**).

The ecological setting of the Hunsrück Slate biota has only recently been investigated in detail. The presence of photosynthesizing red algae and the well-developed eyes of certain fish and arthropods (Stürmer & Bergström, 1973; Briggs & Bartels, 2001), suggest that the community was living within the photic zone, i.e. less than 200 m (c. 650 ft) below sea level (Bartels *et al.*, 1998). Average sedimentation rates have been estimated

89 The Hunsrück hills near Bundenbach in the Rheinisches Schiefergebirge of western Germany.

90 Steep limbs of folds near Gemünden with vertical beds and cleavage parallel to bedding.

at only 2 mm (c. 0.08 in) per year, but Brett & Seilacher (1991) suggested that rapid sedimentation occurred episodically, caused by tropical storms disturbing sediment in shallower water and sending sediment-laden turbidity currents down into deeper water. Communities living on the

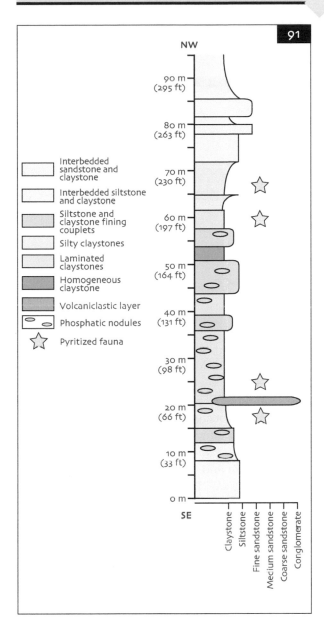

91 Generalized log of part of the Hunsrück Slate sequence in the Bundenbach area (after Sutcliffe *et al.,* 1999).

exactly as postulated for the Burgess Shale (Chapter 2). However, the burial of crinoids with rooting structures *in situ* and the preservation of syndepositional arthropod trails confirm that the Hunsrück community was living in the area in which it was fossilized, and that it was buried in life position (Sutcliffe *et al.,* 1999).

The bottom waters were clearly well-oxygenated, allowing the benthic community to become established, including a thriving infauna, as shown by diverse trace fossils. Preservation of soft tissue requires that it is not disturbed by burrowers after burial, so the sediment must have rapidly become anoxic and inhospitable, eliminating both benthos and scavengers. Burial events were, however, only of limited lateral extent, maybe confined to a few hundred sq. metres (several hundred sq. feet) (Bartels *et al.,* 1998, p. 50), and living communities survived adjacent to them.

The preservation of the Hunsrück Slate fossils is remarkable in that mineralized skeletons and unmineralized soft tissue have both been preserved by the process of pyritization. Moreover, fragile skeletons (of echinoderms, for example) are often preserved as complete, articulated individuals. Soft tissue preservation, which includes arthropod limbs, eyes, and intestines and the tentacles of cephalopods, is confined, however, to four restricted horizons within the thick Hunsrück sequence, and only occurs in a narrow band between the villages of Bundenbach and Gemünden in the Hunsrück region (**86, 91**). Outside this band the sequence represents more normal conditions of quiet, low-energy sedimentation. It yields a depauperate and partly different fauna (dominated by crinoids and corals, but lacking starfish and chelicerates), and is usually preserved as fragmented or disarticulated hard parts only (Südkamp, 2007). Conditions for rapid pyritization must only have existed for limited periods of time within a restricted area.

Pyritization of soft tissue is unusual in the fossil record; the only other Fossil-Lagerstätten to show such preservation are Beecher's Trilobite Bed (Ordovician) of New York State (see Nudds & Selden, 2008, Chapter 4), and the Jurassic beds of La Voulte-sur-Rhône in France. Briggs *et al.* (1996) showed that the pyritization of soft tissue can only occur in exceptional circumstances of sediment chemistry when there is a low organic content, but

muddy sea floor were quickly overcome and buried, explaining the dominance of benthic organisms.

Earlier authors (e.g. Koenigswald, 1930) suggested that the currents transported communities from the area in which they were living into a hostile environment favourable for preservation,

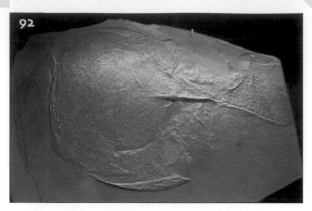

92 The agnathan fish *Drepanaspis* (SM). Length 220 mm (c. 8.7 in).

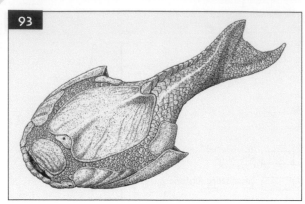

93 Reconstruction of *Drepanaspis*.

a high concentration of dissolved iron. When a carcass is buried in such sediment, sulphate-reducing anaerobic bacteria break down its organic matter producing sulphide. The high concentration of iron in the sediment converts this to iron monosulphide. Finally, aerobic bacteria convert this by oxidation to pyrite. If the organic content of the sediment is too high the dissolved iron precipitates in the sediment and not in the carcass.

The pyrite that preserves the soft tissue of the Hunsrück Slate fossils thus grew as the tissue decayed, but unfortunately microstructure is not preserved as it is when the tissue is replaced by calcium phosphate (see Chapter 16, The Santana and Crato Formations, p. 208). Fossils are usually compressed, but pyritization does preserve some relief, which is not usually the case in argillaceous deposits such as the Burgess Shale (Chapter 2). It has been reported by Allison (1990) that, although the formation of pyrite initially involves reduction, it also requires oxidation at a later stage, and therefore occurs in the upper layers of the sediment near the aerobic–anaerobic interface.

DESCRIPTION OF THE HUNSRÜCK SLATE BIOTA

Fish. As Bartels *et al.* (1998, p. 229) pointed out, the Hunsrück fossils provide an unrivalled picture of the diversity of fish in early Devonian seas. Four of the five main groups of fish are present with only the chondrichthyans, which had not yet evolved,

unrepresented. Most of the fish present are agnathans and placoderms. Agnathans are represented by *Drepanaspis* (**92, 93**), which is locally abundant and suggests temporary brackish conditions, and the much rarer *Pteraspis*. The flattened form of *Drepanaspis* suggests that it was nektobenthic and that it was easily caught in the turbidity flows. Placoderms comprise several genera, such as *Gemuendina* (**94, 95**) and *Lunaspis*, but are less common. The former was shaped like a modern ray and was probably also a bottom dweller – some individuals are up to 1 m (c. 3 ft) long. Acanthodians are known from their fossilized spines up to 400 mm (c. 16 in) long and in one case in association with the pectoral girdle (Südkamp & Burrow, 2007). Finally, a single specimen of a sarcopterygian (lobe-finned fish) represents the earliest record of a lungfish.

Echinoderms. 'Starfish' (including asteroids and ophiuroids) are perhaps the most familiar and most abundant Hunsrück fossils and are often found intact with the soft skin preserved between the arms. Asteroids (true starfish) comprise some 14 genera, most with five arms, but some (e.g. *Palaeosolaster*, **96**) had more than 20. *Helianthaster* (**87, 97**) was one of the largest asteroids known, with 16 arms up to 200 mm (c. 8 in) long. Ophiuroids ('brittle-stars'), also with 14 genera, are sometimes very common, such as the graceful *Furcaster* (**98**) and *Encrinaster* (**99**), with their long slender arms.

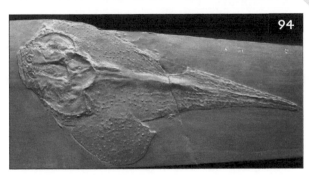

94 The placoderm fish *Gemuendina stuertzi* (SM). Length 220 mm (c. 8.7 in).

95 Reconstruction of *Gemuendina*.

96 The starfish *Palaeosolaster gregoryi* (BKM). Width approx. 250 mm (c. 10 in).

97 The starfish *Helianthaster rhenanus* (BM). Width approx. 150 mm (c. 6 in).

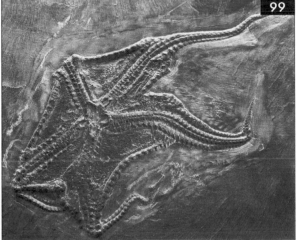

99 The brittle star *Encrinaster roemeri* (MM). Maximum width 120 mm (c. 4.8 in).

98 The brittle star *Furcaster palaeozoicus* (BM). Length of arms approx. 75 mm (c. 3 in).

Crinoids are also common (**100**), but echinoids, blastoids, cystoids, and holothuroids ('sea-cucumbers') are rare. Almost all of the 65 species of crinoid were sessile forms attached to the substrate and many of the Hunsrück specimens are preserved intact and articulated.

Arthropods. The Hunsrück arthropods are spectacular with soft-part preservation of appendages and internal organs; all three major aquatic groups (trilobites, crustaceans, and chelicerates) are represented. Trilobites are numerous, dominated by the phacopids (e.g. *Chotecops*, **101**), but

crustaceans are much rarer, the most common being the malacostracan *Nahecaris* (**102, 103**). This bivalved phyllocarid has a head bearing a pair of large eyes and two pairs of biramous antennae. The eight thoracic segments each had a pair of biramous limbs, while the cylindrical abdominal segments supported five flap-like limbs probably used in swimming.

The chelicerates include rare xiphosurans, eurypterids, and scorpions, as well as some remarkable fossil sea-spiders (pycnogonids; see p. 60), represented by the genus *Palaeoisopus* (**104, 105**). The cephalosoma (head region) of these

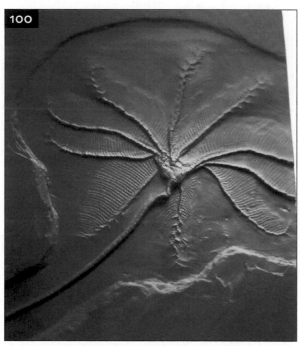

100 The crinoid *Imitatocrinus gracilior* (BKM). Length of arms approx. 60 mm (c. 2.4 in).

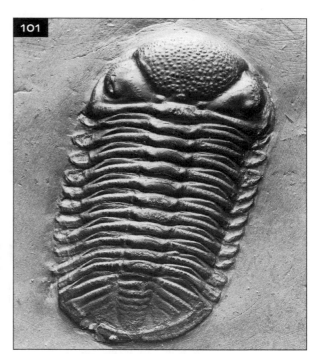

101 The phacopid trilobite *Chotecops* (DBMB). Length 85 mm (c. 3.4 in).

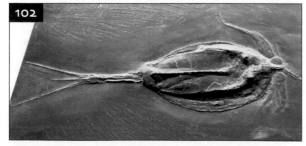

102 The crustacean *Nahecaris stuertzi* (BM). Total length 157 mm (c. 6.2 in).

103 Reconstruction of *Nahecaris*.

impressive animals supported four pairs of appendages, the cheliciphores (to cut up food), the palps (which are sensory), the ovigers (for carrying eggs), and the first locomotory limb with a span of up to 400 mm (c. 16 in). Until recently this was the only known fossil record of a group which is quite common in today's seas, but in the last few years they have turned up in the Jurassic Lagerstätte of La Voulte-sur-Rhône, near Lyon in southern France, and also in the Silurian beds of Herefordshire, UK (Chapter 5). These new finds will help fill in the enormous gaps in the evolutionary history of these animals.

The Hunsrück fauna also includes some rather enigmatic arthropods, such as *Mimetaster* with a star-shaped dorsal shield (**106, 107**), and *Vachonisia* with a carapace looking like a large brachiopod (Stürmer & Bergström, 1976). *Mimetaster* had two pairs of robust walking legs on its head and a short trunk of up to 30 segments, each with a pair of biramous limbs. *Vachonisia* was similar, but with four pairs of large head limbs and a tapering trunk with up to 80 pairs of smaller limbs. These peculiar arthropods recall the more ancient body plans of some of the Burgess Shale forms (Chapter 2), and can be referred to the Burgess clade Marrellomorpha (see p. 27).

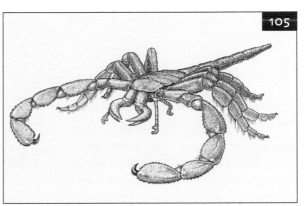

104 The sea-spider *Palaeoisopus problematicus* (BM). Length of longest leg approx. 180 mm (c. 7 in).

105 Reconstruction of *Palaeoisopus*.

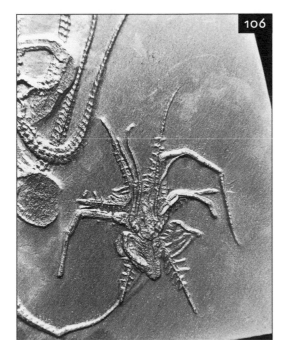

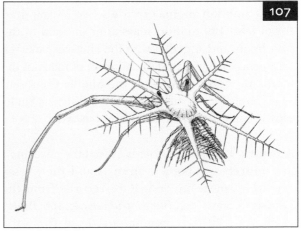

107 Reconstruction of *Mimetaster*.

106 The enigmatic arthropod *Mimetaster hexagonalis* (PC). Width of specimen (left to right) 46 mm (c. 1.8 in).

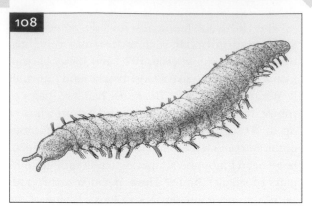

108 Reconstruction of *Bundenbachochaeta*.

109 The rugose coral *Zaphrentis* (DBMB). Width of calyx 48 mm (c. 2 in).

They are important for they prove that successors of the archaic Burgess arthropods did survive at least until Devonian times (Briggs & Bartels, 2001) and have recently been redescribed by Gabriele Kühl and Jes Rust of the University of Bonn (Kühl *et al.*, 2008; Kühl & Rust, 2010).

Even more surprising was the discovery in 2009 of an anomalocaridid in the Hunsrück Slate (Kühl *et al.*, 2009). *Schinderhannes bartelsi* illustrates that the giant Cambrian anomalocaridids (see **33**) survived for at least 100 million years after the end of the Cambrian, and, together with the Hunsrück marrellomorphs, suggests that the absence of many of the iconic Burgess genera after the Middle Cambrian may simply be a result of the lack of Burgess Shale-type preservation from this point onwards.

Annelids. Polychaete worms (the 'bristle worms') are entirely soft-bodied animals and their fossil record is sparse indeed. Rare records from the Hunsrück Slate (e.g. *Bundenbachochaeta*, **108**; Briggs & Bartels, 2010) complement those from the Cambrian Burgess Shale (Chapter 2), the Silurian Lagerstätte of Herefordshire (Chapter 5) and the Carboniferous of Mazon Creek (Chapter 8).

Other invertebrates. None of the other invertebrate groups form dominant members of the Hunsrück fauna, but several groups are represented. Siliceous sponges (cf. *Protospongia*) are restricted to two genera, but cnidarians are more variable. They include chondrophorans ('by-the-wind-sailors'), solitary rugose corals (common Devonian types such as *Zaphrentis*, **109**), colonial tabulate corals (e.g. *Pleurodictyum* and *Aulopora*), scyphozoan conulariids (**110**), and ctenophores ('comb jellies' or 'sea gooseberries'). Molluscs include gastropods, bivalves, and cephalopods, the latter being an important element, mainly comprising goniatites and orthoconic nautiloids, some even with their tentacles preserved extending beyond the chambered shell. Brachiopods are also relatively common, and again may show soft-part preservation, including the pedicle (see Südkamp, 1997).

Plants. The calcareous alga *Receptaculites* is the only autochthonous marine plant. Fragments of terrestrial vascular plants also occur, washed out to sea from the coast, including members of the rhyniophytes (see Chapter 7, The Rhynie Chert).

Trace fossils. These include coprolites (from fish), epifaunal tracks (of arthropods, ophiuroids, and fish), mobile infaunal traces (of bivalves, echinoderms, and polychaetes), and constructed infaunal burrows (Sutcliffe *et al.*, 1999).

110 A group of conulariids (with discinid brachiopods and an edrioasteroid) attached to an unidentified linear object (SM). Length of slab 160 mm (c. 6.3 in).

PALAEOECOLOGY OF THE HUNSRÜCK SLATE BIOTA

The Hunsrück Slate, like the Burgess Shale (Chapter 2), represents a marine, benthic community living in, on, or just above the muddy seabed of an offshore basin situated at about 20°S. Bottom waters were oxygenated and subjected to currents and, as at Burgess, the presence of photosynthesizing algae suggests that the depth was certainly less than 200 m (c. 650 ft).

No statistical analysis of the Hunsrück Slate biota has been attempted, but of the 400 species of macrofossils described, the majority were certainly benthic. A small percentage consisted of benthic infauna, living in the sediment itself, as shown by pyritized burrows such as *Chondrites*, and the infaunal traces of deposit feeders such as the polychaete worm *Bundenbachochaeta* together with some bivalves and echinoderms.

The majority of Hunsrück animals consists of benthic epifauna, living on the sediment surface, the sessile epifauna being dominated by meadows of crinoids, with sponges, corals, conulariids, brachiopods, and bryozoans, plus most of the bivalves. These animals shared the seabed with calcareous receptaculitid algae. The vagrant epifauna, walking or crawling across the seabed, was dominated by starfish (asteroids and

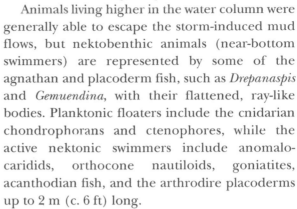

ophiuroids) and arthropods (trilobites, crustaceans, chelicerates, and the archaic forms), but also included some gastropods.

Animals living higher in the water column were generally able to escape the storm-induced mud flows, but nektobenthic animals (near-bottom swimmers) are represented by some of the agnathan and placoderm fish, such as *Drepanaspis* and *Gemuendina*, with their flattened, ray-like bodies. Planktonic floaters include the cnidarian chondrophorans and ctenophores, while the active nektonic swimmers include anomalocaridids, orthocone nautiloids, goniatites, acanthodian fish, and the arthrodire placoderms up to 2 m (c. 6 ft) long.

Trophic analysis identifies the full range of feeding types including: filter feeders, dominated by the crinoids and sponges; deposit feeders, including the gastropods, polychaetes, some arthropods (such as the enigmatic *Mimetaster* and *Vachonisia*), and possibly the starfish (their large mouths suggest that they were deposit feeders unlike modern predators: Bartels *et al.*, 1998, p. 43). Other arthropods were either scavengers, such as the phyllocarid crustacean *Nahecaris*, with its robust mandibles, or predators, such as the anomalocaridid *Schinderhannes*, with its spiny 'great appendages', and the giant sea-spider, *Palaeoisopus*, which was armed with large chelicerae, or pincers, and probably preyed on crinoid meadows (Bergström *et al.*, 1980). The largest predators were undoubtedly the orthoconic nautiloids, plus the shark-like acanthodians and the arthrodire placoderms, which possibly preyed on orthocones. All of the cnidarians would have captured small organisms with their tentacles.

COMPARISON OF THE HUNSRÜCK SLATE WITH OTHER DEVONIAN FISH BEDS

Gogo Formation, Western Australia
This locality in the Kimberley district of north-west Australia has yielded an exceptional fauna of fossil fish from a late Devonian marine (reef) setting. First discovered in the 1940s, the fossils are preserved in limestone concretions which formed during early diagenesis, thus preventing compaction and preserving the fish uncrushed in three dimensions (compare with Santana

Formation, Chapter 16). Careful preparation using acetic acid can free the fish entirely from their matrix. Preservation is spectacular and the fauna includes abundant armoured placoderms, including a new group, the camuropiscids, shark-like predators with a torpedo-shaped skull and tooth plates evolved for crushing. Several new ray-finned fish and lungfish were also discovered during the Gogo Expedition of 1986 (see Long, 1988; Long & Trinajstic, 2010). Crustaceans are also common in the concretions and were probably the prey of the placoderms. As yet no acanthodians or cartilaginous fish have been discovered.

Achanarras Fish Bed, Caithness, Scotland

Fossil fish from the Devonian Old Red Sandstone of Scotland have been well known since Agassiz's classic work, *Recherches sur les Poissons Fossiles*, of the 1830s. They were deposited in the Orcadian Lake, a huge subtropical lake on the southern margins of the Old Red Sandstone Continent, which covered much of Caithness, the Moray Firth, Orkney, and Shetland during the Middle Devonian. The fish lived in the shallow water of the lake margins where waters were warm and well oxygenated. On death their carcasses drifted to the centre of the lake and sank into deeper, colder, anoxic waters beneath a thermocline, where they were preserved in laminated muds on the lake floor, a situation not unlike that of the Crato Formation (Chapter 16).

Mass mortality events, possibly caused by deoxygenation of the water due to algal blooms, led to high concentrations of fish carcasses on the lake floor, which were preserved in good condition due to the lack of scavengers on the lake bed. The fauna includes agnathans, placoderms, acanthodians, and bony fish including ray-finned and lungfish. Small arthropods may have been a food source for the smaller fish, which themselves fell prey to the large carnivorous placoderms and lungfish (Trewin, 1985, 1986).

MUSEUMS AND SITE VISITS

Museums
1. Lehr-und Forschungsgebiet für Geologie und Paläontologie der Rheinisch Westfälischen Technischen Hochschule, Aachen, Germany.
2. Schloßparkmuseum und Römerhalle, Bad Kreuznach, Germany.
3. Hunsrückmuseum, Simmern, Germany.
4. Museum für Naturkunde der Humboldt-Universität zu Berlin, Germany.
5. Institut für Paläontologie der Universität Bonn, Bonn, Germany.
6. Hunsrück-Fossilienmuseum, Bundenbach, Germany.
7. Museum für Naturkunde der Stadt, Dortmund, Germany.
8. Naturhistorisches Museum Mainz, Mainz, Germany.
9. Naturmuseum und Forschungsinstitut Senckenberg, Frankfurt am Main, Germany.
10. Bergbaumuseum, Bochum, Germany (Bartels Collection).

Sites
The most spectacular pyritized fossils come from the area around the small villages of Bundenbach and Gemünden in the Hahnenbach and Simmerbach valleys to the southwest of Koblenz. The Hunsrück-Fossilienmuseum in Bundenbach has an adjacent disused slate mine open for visitors between April and September. Outside the mine the extensive spoil tips give excellent access for fossil hunting. Fossils are not uncommon, but are difficult to recognize in the field and require expert preparation. Another good outcrop is the Eschenbach-Bocksberg Quarry to the south-west of Bundenbach. It is necessary to contact the owner, Johann Backes (info@Johann-Backes.de) to make an appointment. Excursions for individuals and groups, including a fossil preparation class, are also arranged by local geologist Wouter Südkamp (Gartenstraße 11, D-55626 Bundenbach; fax +49-6544-9093; website www.hunsrueck.com/suedkamp). From Gemünden a 4 km (c. 2.5 miles) nature trail, Geologischer Hunsrück-Lehrpfad, can be followed from the village, and includes a viewing platform above the extensive spoil heaps of the opencast Kaiser Mine. The mine is open to visitors from April through September (telephone +49-6765-1220), but is closed on Mondays, as are most of the museums in the area.

THE RHYNIE CHERT

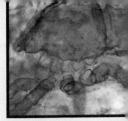

BACKGROUND: COLONIZATION OF THE LAND

The consequences of plants and animals leaving the marine realm and colonizing the surface of the land were far-reaching, not least for the evolution of the human race, which belongs to the Earth's terrestrial biota. Only very few of the metazoan phyla which emerged from the great Cambrian Explosion produced terrestrial forms, but these have been very successful in terms of diversity. The Arthropoda includes some terrestrial crustaceans (e.g. woodlice), but of much greater importance are the terrestrial chelicerates (spiders, scorpions, mites, and their allies) and the insects (which make up 70% of all animals alive today). From fish arose the tetrapods – amphibians, reptiles, birds, and mammals. Molluscs, too, in the form of slugs and snails, have been remarkably successful on land, as any gardener will testify. In order to live successfully on land, plants developed features such as stiff trunks and reproductive devices which gave rise to the familiar trees and flowers we see on land today. Terrestrialization was thus a major episode in the evolution of life on Earth. It was neither an instantaneous event nor restricted to a particular geological period; indeed, some animals (such as crabs) may be considered to be terrestrializing today. Nevertheless, when the physical conditions on the land surface became sufficiently favourable to life in the Silurian Period, the invasion began in earnest.

Among the physical barriers to be overcome when a plant or animal adjusts to life on land, having previously lived in the sea, is that of water supply. Water is necessary for all biological processes but its supply is variable on land, compared to the sea. Plants and animals adopt four main strategies to cope with under- (or over-) supply of water. Some, like microbes, ostracods, and algae, live in permanent water on land (between soil particles and in ponds); they are effectively aquatic. Others, like amphibians, millipedes, slugs, and woodlice, live in moist habitats and only venture out into dry air for short periods. Some plants and animals are poikilohydric, that is they can tolerate desiccation and rehydrate when necessary; examples are bryophytes (mosses and liverworts), and 'resting' stages such as plant spores, fairy shrimp eggs, and tardigrade ('water-bear') tuns. This group was probably the first onto land. Finally, the most successful of all are the homoiohydric forms, which maintain permanent internal hydration mainly by having a waterproof cuticle or skin. These are the familiar land plants (tracheophytes), tetrapods, and arthropods mentioned previously.

Another physiological barrier to be overcome when crossing the threshold from water to land is breathing or, more specifically, exchange of the gases oxygen (O_2) and carbon dioxide (CO_2). Both plants and animals exchange O_2 and CO_2 with the outside across a semi-permeable membrane but, because the O_2 and CO_2 molecules are both larger than the water molecule (H_2O), this membrane will leak, resulting in a loss of water. To overcome this problem, land plants have small holes (stomata) in their waterproof cuticle, which can be closed to prevent excessive water loss. Animals, which might have used external gills when living in water, have lungs or tracheal systems (in insects, for example) for breathing air enclosed within the body, and connected to the air by small holes (called spiracles in insects).

Other adaptations which evolved during terrestrialization include: the development of strong legs and better balance to compensate for the loss of buoyancy; sense organs which would operate in a medium which had different optical and acoustic properties (sound is used more for communication on land); more careful ionic balance, which is linked with the reduced availability of water on land; and the development of direct copulation in mating (in water gametes can simply be emptied close to the opposite sex without contact). In spite of all the potential problems, organisms swarmed onto the land. It was, after all, an unexploited ecological niche and there was, at least to begin with, some respite from predation in the sea.

HISTORY OF DISCOVERY OF THE RHYNIE CHERT

The Rhynie Chert was discovered in 1912 by Dr William Mackie, who studied at the University of Aberdeen in Scotland and later became a physician and keen amateur geologist based in Elgin. In a field outside the Aberdeenshire village of Rhynie (**111**) he found lumps of chert, a siliceous rock produced by hot springs, containing obvious plant axes (stems) and rhizomes (underground stems) (**112, 113**). Thin sections of the rock reveal exquisite preservation of cells, including the water-conducting vessels typical of vascular land plants (**114**). In October 1912 a trench was dug by Mr D. Tait, a fossil collector to the Geological Survey and, as a result of the material discovered, between 1917 and 1921 a series of five papers was published by Kidston and Lang in which they described the plants *Rhynia*, *Aglaophyton*, *Horneophyton*, and *Asteroxylon*. A period of some 30 years went by without any major contributions, but in the late 1950s interest was rekindled through the work of Dr A. G. Lyon of Cardiff University in Wales, who described spores fossilized in the process of germination. Further trenches were dug in the 1960s and 1970s. Dr Lyon later bought the Rhynie Chert site and donated it to Scottish Natural Heritage in 1982. There is little evidence of these trenches to be seen at Rhynie today; just a grassy field and grazing livestock. Now, in addition to the vascular plants, algae, fungi, and

lichens are known from the site, the new material having been discovered by a team of palaeobotanists at the University of Münster in Germany, led by the late Professor Remy.

Shortly after the discovery of the plants, in the 1920s animal remains were described from the chert. Mites and other arachnids (trigonotarbids) were reported by Hirst (1923), fairy shrimps by Scourfield (1926), and springtails by Hirst and Maulik (1926). With few exceptions, the accuracy of the descriptions and the detailed drawings of arthropods in these papers has been corroborated by later workers. In 1961 Claridge and Lyon described book-lungs in the trigonotarbid arachnids, thus proving that they were, indeed, land animals. More recently, centipedes have been found in the chert, as well as an arthropod with gut contents which tell us that it was a detritus feeder (Anderson & Trewin, 2003).

While early studies concentrated on the remarkably well-preserved early land plants and animals – the site held the record for the earliest land animals for some 70 years – little work was done on the geological aspects until recently. Interest in the area was renewed in 1988 when Rice and Trewin from the University of Aberdeen demonstrated that the silicified rocks in the area are enriched with gold and arsenic, thus confirming their hot-spring origin. The ensuing mineral exploration revealed much about the subsurface geology but little exploitable gold. The discovery by the Aberdeen group in the 1990s of a new fossiliferous chert, the Windyfield Chert, some 700 m (c. 0.5 mile) from the original locality, resulted in recognition of part of a geyser vent rim. Work continues, in part by comparison with modern siliceous hot-spring systems such as those at Yellowstone National Park, Wyoming (**115, 116**).

111 The Rhynie Chert, Aberdeenshire, Scotland, lies beneath this field. The hills behind are formed of older rocks beyond the inlier.

112 A piece of Rhynie Chert, about 120 mm (c. 5 in) long.

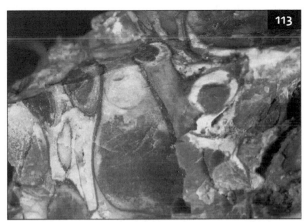

113 Closer view of the piece of Rhynie Chert shown in Figure 112. Note the tubular structures, about the size of macaroni, which are stems of vascular plants.

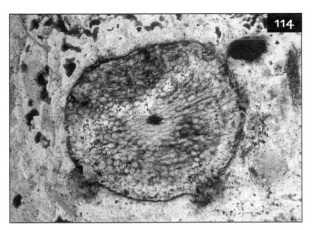

114 Thin section of Rhynie Chert showing detailed preservation of cellular structure in vascular plant stems. Stem is about 2 mm (c. 0.08 in) across.

115 Castle Geyser, Yellowstone National Park, Wyoming. The white deposit around the geyser (hot spring) is sinter; the different colours result from different species of algae and cyanobacteria which each live in different water temperatures.

116 *Triglochin maritimum* tolerates life on the edge of hot springs and, though only distantly related botanically, resembles the kind of vegetation which grew around the Rhynie hot spring. Fountain Paint Pots, Yellowstone National Park, Wyoming. The plant grows up to 300 mm (c. 12 in) tall.

STRATIGRAPHIC SETTING AND TAPHONOMY OF THE RHYNIE CHERT

The host rocks of the Rhynie Chert are Early Devonian (Pragian, 407–411 Ma) in age, and are included in a sequence of sedimentary rocks deposited by streams, rivers, and lakes during the Devonian Period when Scotland, much of northern Europe, Greenland, and North America were joined together in a large continent called Laurussia, located between 0° and 30° south of the equator. At Rhynie, the Devonian rocks are surrounded by older Dalradian metamorphic and Ordovician plutonic igneous rocks (**117, 118**). The Rhynie sediments were deposited in a relatively narrow, north-east–south-west trending basin within these older rocks. The basin is a half-graben; the western edge is marked by a fault which was active at the time of deposition, and at the eastern edge the sedimentary rocks lie unconformably on basement rocks. The palaeoenvironment envisaged for deposition of the chert is one of rivers and lakes depositing a complex of cross-bedded sands in the high-flow regions and muds on flood plains and in shallow ephemeral lakes. Hydrothermal activity centred on the fault system altered the sub-surface rocks in the vicinity of the fault and deposited sinter around hot springs and geysers at the surface in the Rhynie area. Later Earth movements have caused strata within the basin to dip towards the north-west, while the chert-bearing rocks near Rhynie village are folded into a syncline which plunges to the north-east.

The preservation of the biota is variable, and depends mainly on two things: the condition of the organisms at the time of fossilization (the amount of decay) and the degree and timing of silica replacement. There is a whole array of preservation types, ranging from the exquisitely preserved three-dimensional internal anatomy of those

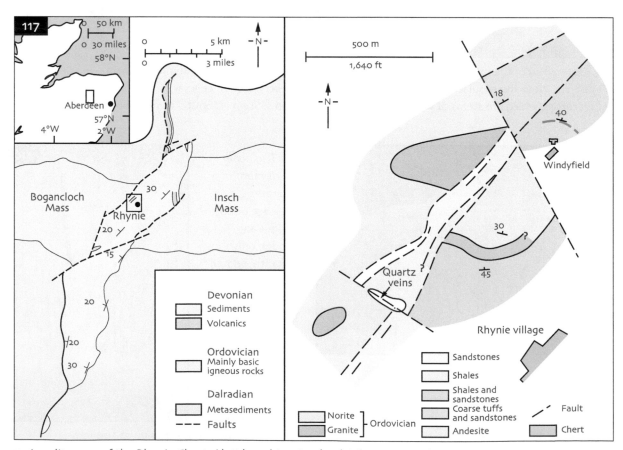

117 Locality map of the Rhynie Chert, Aberdeenshire, Scotland (after Rice *et al.*, 2002).

organisms that were completely silicified at or soon after death, to organisms preserved as compressed, unidentifiable, coalified strips silicified after decay and burial. In a few cases where the whole bodies (cf. moulted skins) of arthropods have been found, gut contents have been preserved, and even delicate features such as book-lungs are perfectly silicified. Some arthropod fossils clearly represent moults: **119** shows a cross-section of a trigonotarbid opisthosoma (abdomen) containing a leg! The leg

moult must have become lodged in the shed skin of the opisthosoma during moulting. Orientation of the plants varies from upright stems with terminal sporangia (spore-cases) and horizontal rhizomes (therefore presumably in growth position) to collapsed, decayed, prostrate straws which form a layer of litter. Some plant sporangia have been found to contain arthropod remains (**120**), but because these sporangia are among the horizontal plant debris, it is thought most likely

118 Stratigraphy of the Rhynie Chert, Aberdeenshire, Scotland (after Rice *et al.*, 2002).

119 Cross-section of abdomen of a trigonotarbid arachnid containing a trigonotarbid leg, thus proving that this specimen is a moulted skin. Abdomen about 1.5 mm (c. 0.06 in) across.

120 Trigonotarbid remains (including pedipalp, left, and chelicera) inside the sporangium of a vascular plant. Note the thickened sporangium wall, about 0.5 mm (c. 0.02 in) thick.

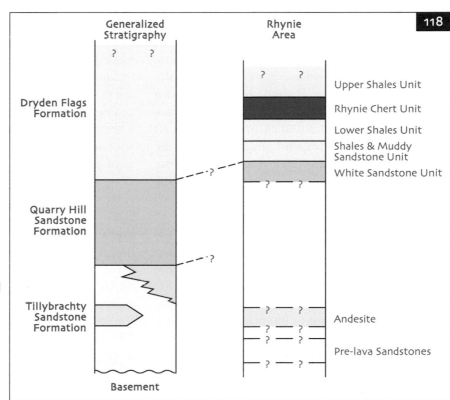

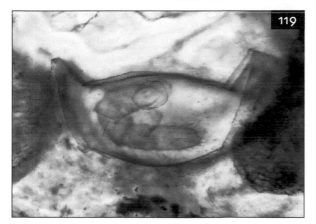

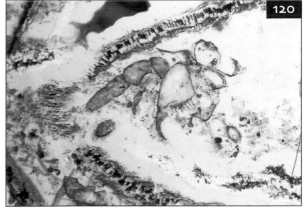

that the animals entered empty (dehisced) sporangia when they were lying on the ground, as a sheltered place to moult, rather than climbing the plants to feed on the spores.

In hot-spring environments silica is precipitated as sinter: an amorphous, hydrated opaline form called opal-A. When the hydrothermal solution, supersaturated with respect to opal-A, erupts at the surface as a geyser or hot spring it cools and opal-A is precipitated. As well as the drop in temperature and evaporation of the hydrothermal fluids, other factors may affect precipitation, such as pH, and the presence of dissolved minerals, organic matter, and living organisms such as cyanobacteria. Silicification of plant material is a permeation and void-filling process called permineralization, in contrast to the direct replacement of cell walls (petrifaction) in which the organic structure acts as a template for silica deposition. Silicic acid (the common soluble form of silica) polymerizes with the loss of water, and opaline silica forms. Nucleation of amorphous silica on wood and plant material involves hydrogen bonding between the hydroxyl groups in the silicic acid, and cellulose and lignin within the organic tissue. Silica precipitation within the cells and openings between cell walls preserves the plant's histology. Around geysers in Yellowstone National Park, opal-A is deposited in plant tissues within a month of immersion. After about a year, the opal-A deposition has created a strong external and internal matrix that stabilizes the plant tissues against collapse and replicates the structure of the plant (Channing & Edwards, 2004). In the Rhynie hydrothermal area, regular outflow of solutions from hot springs and geysers will have produced similar rapid and pervasive silicification before significant decay, and thus led to the exquisite preservation of the plants in the Rhynie Chert. Sinter is a light, porous substance which bears little resemblance to chert rock. However, during burial and over time the amorphous silica phase opal-A becomes unstable and gradually changes to the more stable crystalline form of silica: quartz. As the sinter is being transformed from opal-A to quartz, percolating silica-bearing fluids may precipitate yet more crystals in voids and fractures in the rock so that the resulting chert retains almost none of its original porosity.

DESCRIPTION OF THE RHYNIE CHERT BIOTA

Cyanobacteria. Cyanobacteria are simple, photosynthetic microbes. They may be unicellular or form filamentous chains of cells. They are prokaryotes, so the cells do not contain nuclei. A number of cyanobacteria are found in the Rhynie Chert, some forming distinct stromatolitic laminae or structured colonial growth (Krings *et al.*, 2007a), having grown up from mats on sinter surfaces. Other types are found in more aquatic parts of the chert, and still others within decaying plants. The Rhynie cyanobacteria may have played a significant role in fixing atmospheric nitrogen in the soil.

Chlorophytes. Chlorophytes, or green algae, are photosynthetic eukaryotes (their cells contain nuclei). They may be unicellular, form filamentous chains of cells, or more complex structures as in the stoneworts. A number of filamentous and unicellular green algae are known from the Rhynie Chert, particularly in chert beds deposited in aquatic environments, although commonly their poor preservation makes identification difficult. Charophytes (stoneworts) are large, structurally complex, green algae comprised of a series of multicellular nodes and long single cells or internodes. Branching occurs at the nodes and may be repeated. Charophytes occur in fresh to brackish water. One charophyte has been described from the Rhynie Chert, *Palaeonitella cranii*, and its reproductive parts suggest it was in the stem lineage to the modern charophyte group Chareae (Kelman *et al.*, 2004). *Palaeonitella* inhabited the freshwater alkaline pools and streams which formed on sinter aprons created by hot-spring activity, and is found in the chert along with the crustacean *Lepidocaris*.

Vascular plants. In many layers of the chert, plants are preserved so well that details of their cellular structure can be studied (**114**). It can be demonstrated that these are true land plants by the following features: cuticles, stomata, intercellular air spaces (for gas diffusion), vascular strands with lignin (for water conduction and support), sporangia with a well-developed dehiscence, and spores. Land plants include the bryophytes (mosses and liverworts), and the tracheophytes (vascular plants), which have true water-

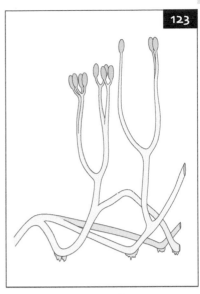

121 The living, primitive, leafless vascular plant *Psilotum nudum*, with lateral, subspherical sporangia, which shows the grade of organization of the Rhynie Chert vascular plants. Waimangu Valley, Rotorua, New Zealand. The plant is about 100 mm (c. 4 in) tall.

122 Top view of *Psilotum nudum* showing dichotomous branching pattern.

123 Reconstruction of *Aglaophyton*. Maximum height about 200 mm (c. 8 in).

conducting vessels. The early tracheophytes were relatively simple in their level of organization, and they showed alternation of generations: a haploid gametophyte which reproduces by means of gametes to produce the diploid sporophyte generation which, in turn, produces haploid spores that grow into a gametophyte. The sporophytes of the Rhynie tracheophytes have a simple construction, few exceeded 200 mm (c. 8 in) in height, and some resemble the living primitive plant *Psilotum* (**121, 122**). Six genera of Rhynie tracheophytes are known from both the sporophyte and gametophyte generations, and another from the Windyfield Chert. These fossils have provided botanists with a unique glimpse into the evolution of the reproductive cycle of early land plants (Taylor *et al.*, 2005). All essential stages of the reproductive cycle, sporophytes and male and female gametophytes, are now known for three of the six Rhynie Chert tracheophytes: *Rhynia gwynne-vaughanii*,

Aglaophyton major, and *Horneophyton lignieri*.

Aglaophyton major (**123**) has creeping rhizomes and smooth, naked, upright axes up to 6 mm (c. 0.24 in) in diameter. The rhizomes show bulges bearing tufts of rhizoids for taking up water and nutrients. Branching is dichotomous and fertile axes terminate in pairs of cigar-shaped sporangia. The gametophyte is known as *Lyonophyton rhyniensis*; it was much smaller and the upright axis ended in a cup-like structure that bore either antheridia (male organs) or archegonia (female organs). The systematic position of *Aglaophyton* is uncertain because it shows a mixture of features of different groups of plants. It has many characteristics of the rhyniophytes, a group of primitive plants known only from fossils, which have simply branched, naked stems (*Rhynia* is an example). The vascular cells (xylem) do not show thickenings and are more reminiscent of the hydroids of bryophytes. The xylem of *Aglaophyton* does not possess true tracheids, which suggests it is not a true tracheophyte. *Aglaophyton* apparently grew mainly on dry, litter-covered substrates, on its own or with other plants, though it probably required wet conditions for germination.

Asteroxylon mackiei (**124**) is one of the more advanced and complex Rhynie plants. It had an extensive system of branching rhizomes; the upright axes grew to about 400 mm (c. 16 in) in height with a maximum diameter of 12 mm (c. 0.5 in) and with dichotomous branching. The aerial axes possessed scale-like enations. On the fertile axes, each kidney-shaped sporangium was attached by a stalk between an enation and the axis. The water-conducting strand of *Asteroxylon* is characteristically star-shaped in cross-section (hence its Latin generic name); smaller strands radiate from each star point to meet the bases of the enations. *Asteroxylon* belongs to the lycophytes, a group which includes modern clubmosses. It grew mainly in organic-rich soils as part of a diverse community together with other plants, and could probably tolerate quite dry habitats.

Horneophyton lignieri (**125**) has smooth, naked, upright axes and a bulbous rhizome bearing tufts of rhizoids. The aerial axes grew up to 200 mm (c. 8 in) in height, had a maximum diameter of 2 mm (c. 0.08 in), and showed repeated dichotomous branching. Fertile axes terminated in branched, tubular sporangia with a central columella. The gametophyte stage of *Horneophyton*, *Langiophyton mackiei*, was much smaller, the upright axis ending in a cup-like structure that bore the archegonia or antheridia. The vascular cells in *Horneophyton* possessed thickenings, which suggest it was a tracheophyte. However, the presence of a columella in the sporangium shows similarities with bryophytes. *Horneophyton* seemed to prefer sandy, organic-rich substrates, often on its own, and flourished in damp to wet conditions.

Nothia aphylla (**126**) had an extensive, branching network of rhizomes with ventral ridges bearing

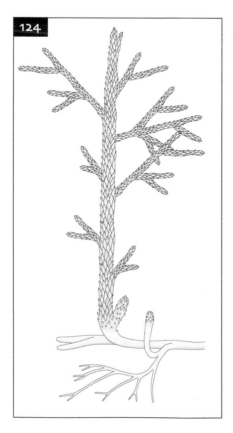

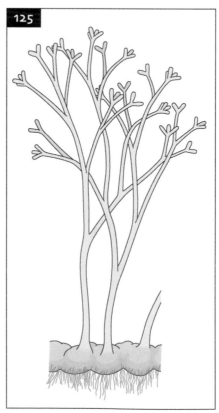

124 Reconstruction of *Asteroxylon*. Maximum height about 400 mm (c. 16 in).

125 Reconstruction of *Horneophyton*. Maximum height about 200 mm (c. 8 in).

rhizoid tufts. Branches turned upright to form the aerial axes, which were naked but had an irregular surface. Repeated dichotomous branching gave the plant a dense growth form. Fertile axes bore lateral, kidney-shaped, stalked sporangia. The male gametophyte, *Kidstonophyton discoides*, was much smaller, with upright axes bearing cup-like structures with tubular outgrowths bearing the antheridia; the female gametophyte is unknown. The vascular cells did not have thickenings and were similar to those seen in some modern bryophytes, the sporangia are comparable to those of a group of primitive fossil plants called zosterophylls, and the simply branched naked axes resemble those of rhyniophytes, so its systematic position is uncertain. *Nothia* grew in sandy soils and plant litter, on its own or with other plants.

Rhynia gwynne-vaughanii (**127**) was one of the commonest plants in the Rhynie ecosystem. Like *Aglaophyton*, it had creeping, branched rhizomes and smooth, naked upright axes. It grew to a height of about 200 mm (c. 8 in), with dichotomously branching axes up to 3 mm (c. 0.1 in) in diameter. *Rhynia* possessed curious hemispherical projections on its axes, and those on the rhizomes bore tufts of rhizoids. Fertile axes had terminal cigar-shaped sporangia. The gametophyte of *Rhynia* is *Remyophyton delicatum* (Kerp *et al.*, 2004); both male and female are known and it is a small structure with projections bearing antheridia or archegonia in a terminal cup. *Rhynia* is the typical member of an extinct group of primitive plants called rhyniophytes, which are characterized by their simple branching and naked stems. *Rhynia* commonly grew in thickets, typically on its own, and was also an early colonizer of well-drained sinter and sandy substrates. It is also found with other plants in a wide range of habitats.

Trichopherophyton teuchansii is a rare plant in the Rhynie ecosystem. Its height is unknown but the dichotomously branched, aerial axes had a maximum diameter of 2.5 mm (c. 0.1 in). The

126 Reconstruction of *Nothia*. Maximum height about 200 mm (c. 8 in).

127 Reconstruction of *Rhynia*. Maximum height about 200 mm (c. 8 in).

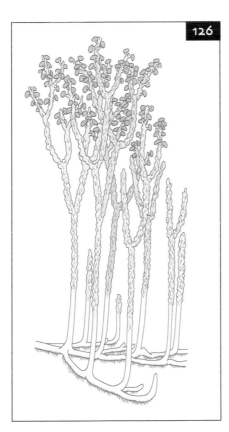

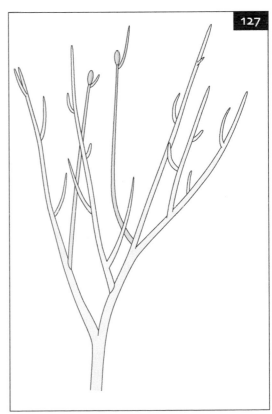

subterranean rhizomes were smooth with small, blunt structures which probably acted as rhizoids. The aerial axes bore spiny projections. Fertile axes bore lateral, stalked, kidney-shaped sporangia, also with spiny projections. The vascular cells in *Trichopherophyton* possessed thickenings which, together with the shape and position of the sporangia, suggests it belongs to the zosterophylls. *Trichopherophyton* was a late colonizer of organic-rich substrates, always growing with other plants in a diverse flora.

Ventarura lyonii is a higher land plant from the Windyfield Chert. Its height is uncertain but was at least 120 mm (c. 5 in). The repeatedly dichotomously branched aerial axes had a maximum diameter of 7.2 mm (c. 0.28 in). The subterranean rhizomes were smooth with small blunt-tipped structures acting as rhizoids. The aerial axes bore peg-like projections. The axes internally show a lignified middle layer to the cortex called the sclerenchyma. Fertile axes bore lateral, stalked, kidney-shaped sporangia. The vascular cells possessed thickenings which, together with the type of sporangia, suggest *Ventarura* belongs to the zosterophylls. The palaeoecology of *Ventarura* is not generally known, but it probably grew in patches near freshwater ponds and probably in sandy and organic-rich substrates.

Nematophytes. Nematophytes are an extinct group of plants consisting of an inner lattice of spirally coiled tubes which may be smooth or spirally thickened. The tubes are closely packed near, and may be orientated perpendicular to, the edge of the plexus where there is an outer cuticular envelope. Their internal structure shows similarities to those of certain algae, and they contain spirally thickened tubes which are commonly found in palynological preparations (Taylor & Wellman, 2009). *Prototaxites* (formerly *Nematophyton*) and *Nematoplexus* are known from Rhynie but both are poorly preserved. There is an ongoing debate about the nature of nematophytes. Originally described as a conifer, the huge trunks of *Prototaxites* with a distinctive anatomy are quite unlike any living or fossil land plant. Subsequent interpretations have had it as a lichen, a red alga, green alga, brown alga, or a fungus (Hueber, 2001). An intriguing interpretation, based on carbon isotopes as well as anatomy and physiology, suggests that *Prototaxites* represents rolled-up liverworts (Graham *et al.*, 2010). The idea is that large mats of these bryophytes covered the Silurian–early Devonian land surface, and that once the edge of the mat became loosened, it could become rolled up like a rug by the action of wind, water or gravity, and thus resemble a large log.

Fungi. Fungi are multicellular, non-photosynthetic eukaryotes. They feed saprophytically (on dead organic matter), parasitically (on live organisms), symbiotically with green plants (endotrophic mycorrhizae), or with an alga or cyanobacterium in a lichen. Numerous fungi are known from the Rhynie Chert. Tom Taylor (University of Kansas) and collaborators described parasitism by a fungus on *Palaeonitella*, arbuscular mycorrhizal fungi in symbiotic relationship with *Aglaophyton*, and the oldest known ascomycete fungus within *Asteroxylon* (Taylor *et al.*, 2004); host response to fungal infection has been described in *Nothia* (Krings *et al.*, 2007b). Lichens are non-vascular organisms formed by the symbiotic relationship between a fungus and an alga or a cyanobacterium. The lichen thallus comprises distinct layers of fungal hyphae (the mycobiont) and the alga/cyanobacterium (the photobiont). The oldest known lichen is recorded from the Rhynie Chert: *Winfrenatia reticulata* (Taylor *et al.*, 1997). *Winfrenatia* most likely colonized hard substrates such as degrading sinter surfaces. It may have weathered rock surfaces and thus contributed to soil formation.

The Rhynie faunal list consists of crustaceans, trigonotarbid arachnids, mites, collembolans, euthycarcinoids, and myriapods.

Crustaceans. The commonest arthropod in the Rhynie Chert is *Lepidocaris rhyniensis* (**128**). It was first described by Scourfield (1926), who erected a new crustacean order, Lipostraca, for the new animal. He later described young stages of the animal (Scourfield, 1940). *Lepidocaris* is a tiny, multi-segmented form with 11 pairs of phyllopods (leaf-like limbs), long, branched antennae and a pair of caudal appendages. It was aquatic, living on detritus in ephemeral pools in the hot-spring environment like fairy shrimps do today.

Another crustacean, known only from its distinctive univalved carapace, was described by Anderson *et al.* (2004). The unnamed crustaceans are commonly found clustered together around

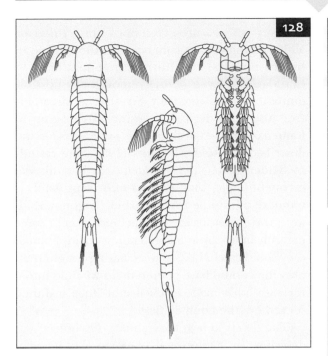

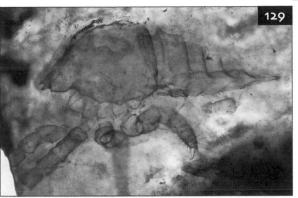

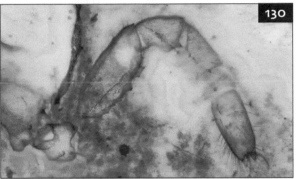

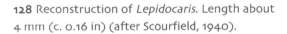

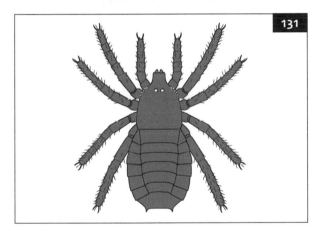

128 Reconstruction of *Lepidocaris*. Length about 4 mm (c. 0.16 in) (after Scourfield, 1940).

129 Juvenile trigonotarbid arachnid *Palaeocharinus* in approximately sagittal section (NHM). Whole animal was about 2 mm (c. 0.08 in) long.

130 Leg of *Palaeocharinus*, about 0.5 mm (c. 0.02 in) long. Note the claw arrangement.

131 Reconstruction of *Palaeocharinus*. Body length about 3 mm (c. 0.1 in).

plant stems in groups of up to 25 individuals. The chert texture around the animals indicates that they were preserved in water. Their morphology suggests affinity with the Branchiopoda, a group which includes fairy shrimps, brine shrimps, and water fleas. The presence of a univalve carapace suggests an affinity with the water fleas, but no specimens show much in the way of appendages, which would be necessary to identify them firmly. If, indeed, this animal does belong in this group, it would be their earliest known occurrence and would extend their fossil record back from the early Cretaceous.

Chelicerates. This group of arthropods is characterized by the presence of a pair of chelicerae (small claws or fangs) in front of the mouth, and no antennae. Trigonotarbids (**129–131**) are extinct arachnids similar in appearance to spiders but lacking the definitive features of venom glands and silk-producing organs, and show features which are primitive to spiders such as segmentation in the opisthosoma (abdomen). Their remains are common in the Rhynie Chert, and were first described as *Palaeocharinus rhyniensis* in the 1920s by Hirst and by Hirst and Maulik. The

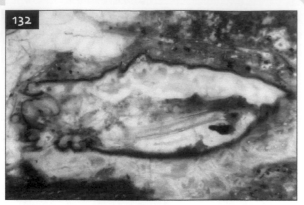

132 The oldest known harvestman (Opiliones), *Eophalangium sheari*, female; the banded tube inside the abdomen is the ovipositor (PBM). Length about 5 mm (c. 0.2 in).

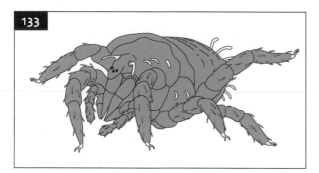

133 Drawing of a Rhynie mite. Body length 0.3 mm (c. 0.01 in) (after Hirst, 1923).

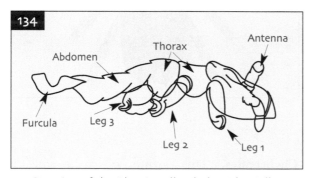

134 Drawing of the Rhynie collembolan *Rhyniella*. Body length 1 mm (c. 0.04 in) (after Scourfield, 1940).

discovery of well preserved book-lungs (internal air-breathing organs connected to the outside by small spiracles) in Rhynie trigonotarbids by Claridge and Lyon (1961) removed any possible doubt that these were truly terrestrial air-breathers (see also a modern study using 3D imaging by Kamenz *et al.*, 2008). *Palaeocteniza crassipes* was described as a spider by Hirst (1923), but restudy by Selden *et al.* (1991) concluded that this was erroneous and that *Palaeocteniza* is probably a moult of a juvenile trigonotarbid. Trigonotarbids were carnivores, as are all arachnids except some parasitic mites, presumably feeding on any animals they could catch. Like spiders, having caught their prey they would have poured digestive fluid into it through holes made by cheliceral fangs and then sucked out the liquified flesh.

The oldest known harvestman (Opiliones) was described by Dunlop *et al.* (2004) as *Eophalangium sheari* (**132**). Both males and a female are known, including (on the male) a penis and (on the female) an ovipositor. Both of these structures are of modern appearance, and the fossils also show air-breathing tracheae which are very similar to those of living harvestmen. The species was identified tentatively to the living suborder Eupnoi, which suggests remarkably little has changed among this group of arachnids since the Devonian. Another, somewhat doubtful, arachnid was described by Dunlop *et al.* (2006) as *Saccogulus seldeni*. This animal has a very thick, spongy cuticle, and the mouth has a unique filtering device which implies it was a terrestrial animal – a liquid feeder that digested its food before consuming it, just like spiders and trigonotarbids. Regrettably, little else of this animal is preserved, so we do not yet know if it really is the world's oldest spider.

The oldest known mites which can be placed unequivocally in that order (Acari) occur in the Rhynie Chert (**133**). Like spiders and trigonotarbids, mites are arachnids, but they are very small and their prosoma and opisthosoma are not clearly defined. Hirst thought that the Rhynie specimens all belonged to the same species, which he named *Protacarus crani* and placed, with some doubt, in the modern family Eupodidae. They were restudied by Dubinin (1962), who separated them into five species belonging to four modern families: *Protacarus crani* (Pachygnathidae), *Protospeleorchestes pseudoprotacarus* (Nanorchestidae), *Pseudoprotacarus*

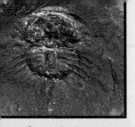

MAZON CREEK

BACKGROUND: THE COAL MEASURES

Once life became established on land it quickly developed a more complex habitat structure; as plants developed tree forms so forests evolved (by the Late Devonian), and alongside this there was a great diversification of animal life. The Late Devonian also saw the evolution of tetrapods (four-legged vertebrates), which emerged onto land in early Carboniferous (Mississippian) times and started preying on the abundant invertebrate life already established there. By the late Carboniferous (Pennsylvanian) there were extensive forests across the equatorial areas of the globe, which included the present areas of north-west and central Europe, eastern and central USA, and elsewhere in the world such as southern China and South America. These forests are represented in the fossil record by coal seams, which are preserved because in these places the forests developed on mires – permanently waterlogged ground – where anoxic conditions prevented complete decay of the forest litter and thus prompted the formation of peat which, when compressed under a great thickness of later sediment, turned to coal. It was these vast beds of coal, and associated ironstones, clays, and other natural resources which, in Britain in particular, provided the raw materials for the Industrial Revolution.

A Coal Measure sequence of rocks presents a more complex and interesting suite of environments than a simple swamp forest, and most Upper Carboniferous Coal Measure sequences represent deltas with a range of environments from marine bays through brackish lagoons to sand bars, freshwater lakes, levees, and swamp forests. Delta lobes are geologically short-lived. If their sediment supply is cut off, they rapidly sink and seawater transgresses the land surface, swamping the forests. Thus, in many sequences there is a band of mudstone or limestone bearing marine fossils immediately above a coal seam. The sea may persist in the area for many tens or hundreds of years before a new delta lobe builds out into the area, and relatively quickly establishes new silt and sand substrates upon which new forests can develop. Even in established swamp forests, floods are commonplace. As a result of these changing environments, Coal Measure sequences show a distinctive pattern of thin mudstone or shale layers, siltstones (often regularly laminated), coarse sandstones, and coal seams.

Some of the vascular plant groups discussed in Chapter 7 (The Rhynie Chert) continued with little morphological change into the Carboniferous and beyond, for example the bryophytes, while others, such as the psilophytes, gave rise to the horsetails, clubmosses, and ferns, which attained gigantic proportions in the Carboniferous. Many of these groups also formed the understorey vegetation, together with seed-ferns, cordaites, and early conifers. Animals, too, had diversified and moved into the new niches provided by the plants. Insects had appeared, evolved wings and, by the Permian, some griffenflies (early dragonfly relatives) had wingspans of 710 mm (c. 28 in) (though those found in the Mazon Creek beds were smaller). Myriapods became giant in the Carboniferous too, with large, armoured

136 Locality map of the Mazon Creek area. **A**: Location map, with area of marine rocks shown in blue; **B**: palaeoenvironmental reconstruction of the Mazon Creek area in relation to the strip mines; **C**: locations of the strip mines and dumps from shafts (after Baird *et al.*, 1986).

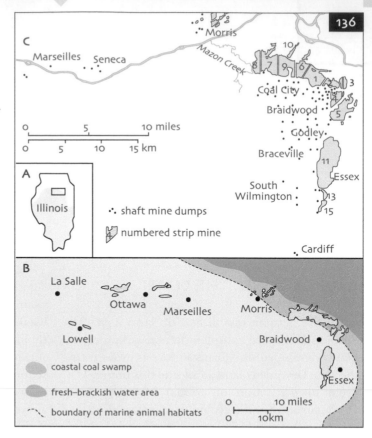

millipedes and the enormous (more than 2 m [c. 6 ft] long) arthropleurids – the largest known land arthropods. Vertebrates had not only followed the arthropods onto land but amphibious tetrapods had also attained large size (up to 1 m [c. 3 ft] long), and there were freshwater sharks with bizarre dorsal spines.

HISTORY OF DISCOVERY OF THE MAZON CREEK FOSSILS

Mazon Creek is a small tributary of the Illinois River, situated some 150 km (c. 90 miles) south-west of Chicago (**136**), and has given its name to this Lagerstätte, which actually comes principally from spoil heaps of the strip coal mines which have operated in the area over the last century. The importance of the Mazon Creek biota is that it has been so well collected, particularly by an army of keen amateurs, that it has yielded the most complete record of late Palaeozoic shallow marine, freshwater, and terrestrial life. More than 200 species of plants and 300 animal species have been described, including representatives of 11 animal phyla.

Plant fossils were collected and described from natural outcrops and small mine tips in the Mazon Creek area for many years before the large Pit 11 open-strip mine was opened in the 1950s. In the late years of that decade, Peabody Coal bought out the Northern Illinois Coal Company and allowed local fossil collectors to visit the pit and collect from the waste material. In strip mining, the overburden (in this case the Francis Creek Shale) is stripped off by giant buckets on drag lines to reveal the coal beneath, which is then simply dug out by smaller diggers and loaded into trucks for removal to the sorting plant. The coal is dug in long strips, so the overburden is used to back-fill the strip where coal was previously removed. It is the Francis Creek Shale which is the source of the exceptional fossils at Mazon Creek. Once news of the coal company's generosity in allowing access to its site spread, there was a regular stream of amateur fossil collectors to the site looking for the elusive special fossil.

The fossils at Mazon Creek occur in clay ironstone nodules (concretions). These usually require a winter or so of weathering before they will split easily with a single hammer-blow, usually along the weakest line – which is that of the fossil.

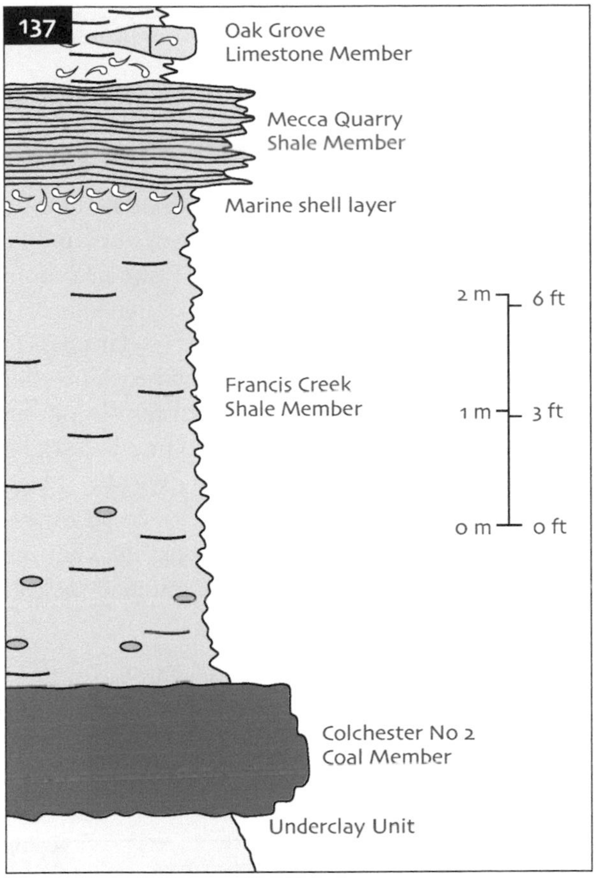
137
Oak Grove
Limestone Member
Mecca Quarry
Shale Member
Marine shell layer
2 m 6 ft
Francis Creek
Shale Member
1 m 3 ft
0 m 0 ft
Colchester No 2
Coal Member
Underclay Unit

137 Stratigraphic log of the Francis Creek Shale and associated Members of the Carbondale Formation, Illinois (after Baird *et al.*, 1986).

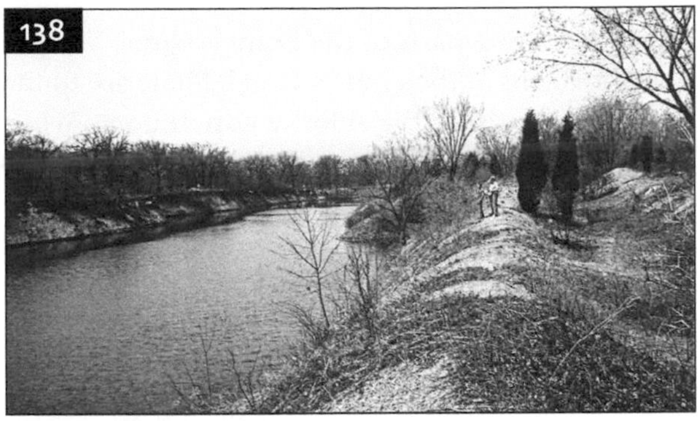

138 Disused Pit 8 in the Colchester N° 2 Coal. The spoil forming the raised banks is the source of the Mazon Creek nodules.

Some collectors artificially freeze and thaw the nodules to accelerate this process. Many of the concretions contain seed-fern fronds, some contain indeterminate shapes which were termed 'blobs' and consequently thrown away. Later research has shown that most of these blobs are actually fossil jellyfish, attesting to the unusual preservation of soft-bodied animals at Mazon Creek. The scramble of collectors at Pit 11 could have resulted in the loss of the best fossils to personal cabinets, never to be studied by experts were it not for the efforts of Dr E. S. ('Gene') Richardson, who encouraged the collectors to meet at regular intervals at the Field Museum in Chicago and show their finds. In this way they learned from both the scientists and each other about what the animals and plants were and the best ways of finding them. They could be swapped, and the best specimens presented to the museum for study.

Some 25 years ago the Peabody Coal Company sold Pit 11 for the construction of a nuclear power plant. While mining no longer goes on there, the tips remain and are still picked over. Moreover, during construction of the power station, boreholes were drilled which passed through the Francis Creek Shale and yielded some important information about its environment of deposition.

STRATIGRAPHIC SETTING AND TAPHONOMY OF MAZON CREEK

The Mazon Creek fossils occur in siderite (ironstone, $FeCO_3$) concretions in the Francis Creek Shale Member of the Carbondale Formation, of Westphalian D age (306 Ma). The Francis Creek Shale overlies the Colchester № 2 Coal Member, and is itself overlain by the Mecca Quarry Shale Member (**137, 138**). The Colchester Coal is generally about 1 m (c. 3 ft) thick; the Francis Creek Shale is typically a grey, muddy siltstone with minor sandstones and varies from complete absence up to 25 m (c. 80 ft) or more in thickness. The siderite concretions occur only in the lower 3–5 m (c. 10–15 ft) of the member, and only where the shale is more than 15 m (c. 50 ft) thick. The shale is coarser and bears sandstones near its top. The Mecca Quarry Shale is a typical Pennsylvanian black shale, which peels easily into sheets, and contains a rich fauna of sharks and their coprolites, which was exhaustively monographed by Zangerl and Richardson (1963). It is generally about 0.5 m (c. 1 ft 6 in) thick but is absent over areas where the Francis Creek Shale is more than about 10 m (c. 30 ft) thick, and thus also over the nodule-bearing parts of the Francis Creek Shale.

The fossils in the Francis Creek Shale are found almost only within the siderite concretions. When these are broken open they reveal a nearly three-dimensionally preserved organism, though the fossil becomes more flattened towards the edge of the nodule. Fossils are generally preserved as external moulds, commonly with a carbonaceous film (if a plant). There may be crystals such as pyrite, calcite, or sphalerite on the mould surfaces. The commonest mineral on these surfaces, however, is kaolinite – a white, soapy, clay mineral. This is often found completely infilling the space between the moulds (i.e. forms a cast). It is soft and therefore easy to remove mechanically. Fossils with few hard parts or little rigidity, such as jellyfish, may collapse completely and be preserved as composite moulds; even arthropods may show dorsal and ventral structures superimposed.

Most fossils show very little decay and, indeed, there are instances of bivalves preserved on the edge of the nodule at the end of their death trail (**139**)! The concretions do not normally extend far beyond the fossil, so the size and shape of the nodule correlate with the organism inside. Few concretions exceed 300 mm (c. 12 in), so large animals are rare in the biota. The evidence presented so far tends to indicate that the concretions formed very soon after the death and burial of the organisms – molluscs stopped in their tracks, seed-fern pinnules at right-angles to the bedding, and little decay (**140**). Large fish and amphibians, it is presumed, could escape this environment and not be preserved. That the organisms are preserved three-dimensionally, at least

139 Death trail of undescribed solemyid bivalve. The bivalve was still alive and attempting to escape while the siderite nodule was forming, resulting in a fossilized trail and the bivalve at the edge of the nodule (CFM). Scale bar is in cm.

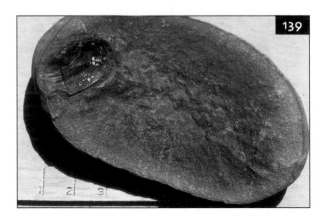

140 Diagram showing the rapid formation of a siderite nodule around a dead shrimp. **A**: Dead shrimp lands on sea floor; **B**: partial decay by bacteria, volatiles rise; **C**: siderite precipitation while compaction begins; **D**: further compaction of surrounding sediment while dewatering (syneresis) causes cracks to extend from the centre of nodule outwards. (after Baird *et al.*, 1986).

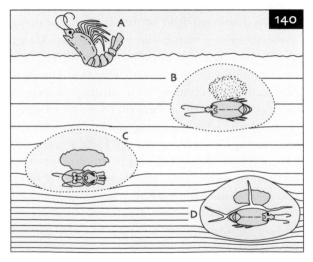

in the cores of the nodules, while the surrounding matrix, like all siltstones, is greatly compressed implies that the concretions formed before any appreciable compaction. Indeed, the siltstone laminae can be seen to widen progressively from the matrix towards the centre of the nodules (**141**), suggesting that the concretions grew during compaction. Also, cracks within the nodules, commonly infilled with kaolinite, can be related to dewatering of the sediment (synaeresis) during their formation.

Because the fossil-bearing concretions mirror the shape of the fossil, and the fossils are generally situated fairly centrally within the concretions, we can assume that the organisms contributed significantly to their formation. Moreover, barren nodules can usually be explained as containing rather flimsy fossils, unrecognizable organic matter, trace fossils, and the like. The nodules contain about 80% siderite cement, which implies that when the concretions formed there was at least 80% water by volume in the sediment before compaction. Iron would normally react with sulphur under the influence of anaerobic bacteria in the presence of decaying organic matter to form pyrite (FeS_2), as in the Hunsrück Slate (Chapter 6), in preference to siderite, but once any sulphate was used up by this process (some pyrite does occur in the nodules), then methanogenic bacteria would help to generate

siderite. Conditions at Mazon Creek which helped this process would have been an abundance of iron and a weak supply of sulphate. The Mazon Creek nodules are commonly asymmetrical, with flatter bottoms and more pointed tops. This is due to the effects of gravity: the weight of the carcass presses into the sediment beneath, the concretion can grow more easily upwards (where there is less compaction), and any light fluids resulting from decay would also rise preferentially.

DESCRIPTION OF THE MAZON CREEK BIOTA

The Mazon Creek biota actually consists of two biotas: the Braidwood and the Essex; the former occurs mainly in the north of the area, the latter in the south. Eighty-three percent of the Braidwood nodules contain plants, with the next commonest (7.8%) inclusions being coprolites. Following these (in descending order) can be found: freshwater bivalves (1.8%), freshwater shrimps (0.5%), other molluscs (0.4%), horseshoe crabs (0.3%), millipedes (0.1%), fish scales (0.1%), then insects, arachnids, fish, and centipedes form the remainder (< 0.1%). On the other hand, only 29% of the Essex biota consists of plants, the commonest animal being the 'blob' *Essexella* (42%). Following these (in descending order) are burrows and trails (5.9%), the marine solemyid bivalves (5.5%), coprolites (4.8%), worms (2.8%), miscellaneous molluscs (1.9%), the marine shrimp *Belotelson* (1.8%), the marine bivalve *Myalinella* (1.4%), miscellaneous shrimps (0.5%), the crustacean *Cyclus* (0.5%), the enigmatic Tully Monster (0.4%), the scallop *Pecten* (0.3%), the jellyfish *Octomedusa* (0.3%), miscellaneous fish (0.2%), then insects, millipedes and centipedes, hydroids, horseshoe crabs, arachnids, and amphibians form the remainder (< 0.1%). From these lists it can be seen that the Braidwood biota consists of terrestrial and freshwater organisms while the Essex biota consists of predominantly marine organisms, with some drifted plants and so on. No marine organisms can drift into freshwater but freshwater and terrestrial organisms can be washed down into the sea. Notice, too, that the marine Essex biota is not a typical marine biota (there are no brachiopods, corals, or crinoids), so it must represent a reduced-salinity, and perhaps muddy, environment which fully marine animals could not tolerate.

141 Section through the Francis Creek Shale laminites and part of a siderite nodule. Note the laminae widen towards the nodule, indicating greater compaction of the surrounding silt/clay laminites than the nodule, and that compaction had started during growth of the nodule. Thickest part of specimen (left) is 40 mm (c. 1.6 in).

142 The seed-fern *Neuropteris* (MU). Pinnule is 60 mm (c. 2.4 in) long.

143 The seed-fern *Pecopteris* (MU). Scale bar is in cm.

144 The seed-fern *Alethopteris* (MU). Nodule is 70 mm (c. 2.7 in) long.

Plants. The Mazon Creek nodules preserve a typical Coal Measure flora, exceptional only in that its preservation is better in the concretions than in other Upper Carboniferous clayrocks. Large fossils are not preserved in the Francis Creek Shale, but the presence of tree-sized clubmosses and horsetails can be inferred from pieces of bark (*Lepidodendron* and *Calamites*, respectively) and foliage (*Lepidophylloides* and *Annularia*, respectively). Seed-fern pinnules are the commonest fossils in all Mazon Creek nodules; examples include *Neuropteris* (**142**), *Pecopteris* (**143**), and *Alethopteris* (**144**). Comparison with modern coastal swamp forests suggests that much of the plant debris found in the Mazon Creek nodules originated not from the forest itself but was carried down by streams from far inland, so possibly represents a mixture of coastal and upland forest. Plant debris is allochthonous (i.e. drifted from its original life position) in both Braidwood and Essex biotas, but is more common in the former.

Cnidarians. This phylum includes the corals and anemones (Anthozoa), hydroids (Hydrozoa), Scyphozoa and Cubozoa (jellyfish), and a few other groups. Apart from the corals, they have no mineralized skeletons but some jellyfish have stiffened parts. Most 'blobs' in Mazon Creek nodules are jellyfish remains, and some show distinct tentacles and other structures. *Essexella*, for example (**145**), shows a bell with a cylindrical sheet hanging below it. *Essexella* belongs in the Scyphozoa, while *Anthracomedusa* (**146**), with four bunches of numerous tentacles, is a cubozoan.

Bivalves. Bivalves are common fossils on account of their hard, calcareous shells (valves), but those in the Mazon Creek nodules are important because they commonly preserve soft-part morphology. There is a high diversity of bivalves in the Mazon Creek, with 12 superfamilies represented. It is convenient to divide the fauna into freshwater and marine forms: i.e. those found in the Braidwood and Essex biotas, respectively. The most common marine bivalve occurs as articulated valves (called 'clam-clam' specimens by amateur collectors), and is occasionally found at the end of an unsuccessful escape trail (**139**). Previously misidentified as the marine pholadomyoid *Edmondia*, it has been shown recently to be a new genus of solemyid, now called *Mazonomya*. These solemyids are burrowers (infaunal benthos), but other common bivalves in the Essex biota are the thin-shelled swimmers (nekton) *Myalinella* and *Aviculopecten* (**147**). The family Myalinidae also includes freshwater forms such as *Anthraconaia*, found in the Braidwood biota.

Other molluscs. Three other classes of mollusc occur in the Mazon Creek biota: Polyplacophora, Gastropoda, and Cephalopoda. Though there are many freshwater gastropods (snails) today, the only gastropods at Mazon Creek are from the marine Essex biota. Polyplacophora (chitons) are exclusively marine animals which are rare in the fossil record

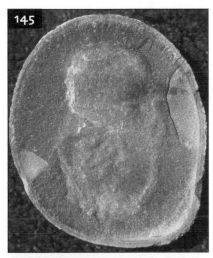

145 The jellyfish *Essexella asherae* (MU). Nodule is 60 mm (c. 2.4 in) long.

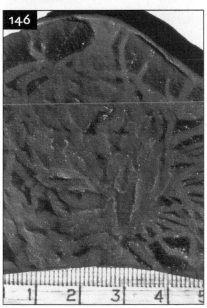

146 The jellyfish *Anthra-comedusa turnbulli* (MU). Scale bar is in cm.

147 The bivalve *Aviculopecten mazonensis* (MU). Scale bar is in cm.

because they usually inhabit rocky shores. However, one genus, *Glaphurochiton* (**148**), occurs in the Essex biota. Cephalopods are also wholly marine and, though normally common in fully marine Carboniferous rocks, they are rarer than chitons at Mazon Creek, but quite diverse in the Essex biota. In addition to the orthocone bactritoids, coiled ammonoids and nautiloids with external shells, coleoids (with internal hard parts) are also present. *Jeletzkya* is a small, squid-like coleoid with an internal shell similar to a cuttlebone.

Worms. Apart from their tiny jaws, called scolecodonts, polychaete annelids are rarely preserved as fossils because they are soft-bodied. Therefore, the great diversity of polychaetes found at Mazon Creek is an important contribution to the fossil record of this important group of marine animals. One of the commonest Mazon Creek polychaetes is *Astreptoscolex* (**149**), which shows a segmented body and short chaetae (spines) along each side.

Shrimps. A wide variety of Crustacea occur at Mazon Creek, many of which have a shrimp-like body shape. There are both freshwater and marine shrimps, found in the Braidwood and Essex biotas, respectively. *Belotelson magister*, a robust species, is by far the most common shrimp in the marine Essex biota. The second commonest marine shrimp is *Kallidecthes*. *Acanthotelson* (**150**) and *Palaeocaris* are freshwater shrimps which occur in the Braidwood biota and are also found rarely in the Essex nodules, where they presumably were washed down by currents.

Other crustaceans. Some squat, crayfish-like forms also occur at Mazon Creek: *Anthracaris* in the Braidwood biota and *Mamayocaris* in the Essex biota. There are also phyllocarids (*Dithyrocaris*) in the Essex biota, conchostracans (crustaceans almost completely enclosed in a bivalved carapace), which were probably fresh- or brackish-water forms, some ostracods, and barnacles. A crustacean which commonly occurs in Upper Carboniferous nodules, *Cyclus*, is also found in the Essex biota. As its name suggests, *Cyclus* has a round, dish-like carapace, and is most likely a fish parasite (see also Chapter 10, Grès à Voltzia).

148 The chiton *Glaphurochiton concinnus* (MU). Scale bar is in cm.

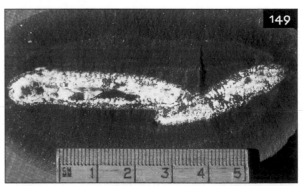

149 The polychaete annelid *Astreptoscolex anasillosus* (MU). Scale bar is in cm.

150 The shrimp *Acanthotelson stimpsoni* (MU). Scale bar is in cm.

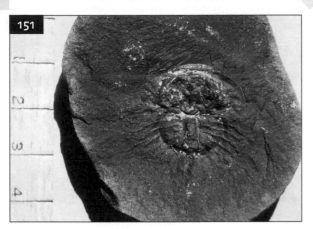

151 The horseshoe crab *Euproops danae* (MU). Scale bar is in cm.

152 Cast of uropygid arachnid *Geralinura carbonaria* (CFM). Nodule is 60 mm (c. 2.4 in) long.

Chelicerates. The arthropod subphylum Chelicerata includes horseshoe crabs (Xiphosura), scorpions, eurypterids, spiders, mites, and other arachnids. Mazon Creek has yielded some exceptionally fine fossils of these animals, which have provided a great deal of information on the evolutionary history of the chelicerates. *Euproops danae* (**151**) is one of the best known horseshoe crabs in the fossil record. Work by Dan Fisher (1979) of the University of Michigan revealed the amphibious mode of life of these animals, which are found predominantly in the Braidwood biota.

Eurypterids were discussed in Chapter 4 in connection with the Soom Shale. By late Carboniferous times they were mostly amphibious and represented largely by the genus *Adelophthalmus*, with numerous specimens in Mazon Creek nodules.

Terrestrial arachnids are well represented in the Braidwood biota, and the extinct group Phalangiotarbida has the most numerous representatives; trigonotarbids are also well represented. The latter are close to spiders in morphology but lack venom and silk glands and, as arachnids go, are relatively common in late Palaeozoic terrestrial ecosystems. Two living orders of arachnids, Uropygi (whip scorpions) and Amblypygi (whip spiders), have some well-preserved examples in the Braidwood biota (**152**). An interesting order of living arachnids, Ricinulei, is represented at Mazon Creek by three genera. Ricinulei are rarely encountered, even today, and are restricted to tropical forests and caves. They are only known from the Upper Carboniferous and Recent, but seem to have changed little in between. Three other arachnid orders occur at Mazon Creek: Opilionida (harvestmen), Solpugida (camel spiders), and Scorpionida. The scorpions are the most ancient of arachnids and, though exclusively terrestrial today, are found in aquatic environments in the Silurian. Scorpions are found only in terrestrial environments by the Upper Carboniferous, and are the second most abundant arachnid in the Braidwood biota.

Insects. Six orders of insects occur at Mazon Creek, all of which are now extinct. Much of our knowledge of Pennsylvanian insects comes from the 150 species found in Mazon Creek nodules. Palaeodictyoptera were medium to giant flying forms with patterned wings which comprised half of all known Palaeozoic insects. Both nymphs and adults were terrestrial and had sucking mouthparts. Megasecoptera were similar to Palaeodictyoptera but had more slender, often petiolated (stalked) wings. Diaphanopterodea resembled Megasecoptera with the one exception that they could fold their wings over their backs like modern butterflies and damselflies. Protodonata, as their name suggests, were distantly related to modern Odonata (dragonflies and damselflies). Some reached giant proportions in the late Carboniferous and Permian, with wingspans of 710 mm (c. 28 in). It is presumed that, like Odonata, the nymphs led an aquatic life, but none has been found.

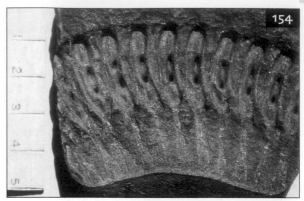

153 'Roachoid' (Blattodea) (MU). Scale bar is in cm.

154 The giant millipede *Myriacantherpestes* (CFM). Scale bar is in cm.

155 The millipede *Xyloiulus* (CFM). Nodule is 70 mm (c. 2.7 in) long.

'Protorthoptera' is a general name given to a large group of Palaeozoic insects which resemble the modern Orthoptera (grasshoppers, locusts, and crickets) but lacked jumping legs. They form the stem group which gave rise to modern Orthoptera, Dermaptera (earwigs), Phasmatodea (mantids), and some other extinct and living orders. Twelve families occur at Mazon Creek, and *Gerarus* is the commonest insect fossil found there. *Gerarus* belongs to the group which is thought to have given rise to the giant Triassic *Titanoptera*, now extinct. It was once thought that the familiar cockroaches occurred in rocks as old as Carboniferous because of the abundant remains of cockroach-like animals in Coal Measure nodules. However, it has now been shown that these animals possess some rather primitive traits, such as a large external ovipositor which evolved even before the origin of flight, so these stem-group Dictyoptera are better described as 'roachoids' or Blattodea (Grimaldi & Engel, 2005). In the nodules from Mazon Creek (**153**) their veined wings are often mistaken for seed-fern pinnules.

Myriapods. Myriapods are multilegged arthropods which include the centipedes (Chilopoda), millipedes (Diplopoda), two other living classes (Symphyla and Pauropoda), and the Palaeozoic Arthropleurida. Myriapods were among the earliest known land animals (Chapter 7, The Rhynie Chert), but by Upper Carboniferous times some had become gigantic, and many sported fierce spines,

presumably for defence against predators. Short millipedes which could roll up into a ball (Oniscomorpha: *Amynilyspes*) were present in the Braidwood biota, but the most dramatic forms belong to the extinct order Euphoberiida. *Myriacantherpestes* (**154**) probably reached 300 mm (c. 12 in) in length and had long, forked, lateral spines and shorter dorsal spines. *Xyloiulus* (**155**) was a more typical, cylindrical spirobolid millipede. Millipedes are generally detritus feeders, while centipedes are carnivorous. Centipedes at Mazon Creek include the scolopendromorph *Mazoscolopendra* and the fast-running scutigeromorph *Latzelia*. Arthropleurids range from tiny forms in the

Silurian to the Upper Carboniferous, when they became the largest known terrestrial arthropods, reaching about 2 m (c. 6–7 ft) in length. Nevertheless, like millipedes, they were probably detritus feeders. Isolated legs and plates of *Arthropleura* occur at Mazon Creek. Onychophora (velvet worms) should also be mentioned here. They are known from the Cambrian (e.g. *Aysheaia*, Chapter 2, The Burgess Shale), when they were marine, to the Recent, when they are wholly terrestrial. The Mazon Creek *Ilyodes* was collected from natural outcrops and it is not known whether it came from the terrestrial Braidwood or marine Essex biota.

Other arthropods. Euthycarcinoids are an odd group of apparently uniramous arthropods (those with a single leg-branch, like myriapods) which range from the Silurian to Triassic (see Chapter 7, The Ryhnie Chert). Mazon Creek has three species.

Another group of arthropods of unknown affinity is the Thylacocephala, which ranged from Cambrian to Cretaceous. Commonly called flea-shrimps, they may or may not belong to the Crustacea. They have a bivalved carapace which encloses most of the body, and large eyes. *Concavicaris* is quite common in Essex nodules.

Other invertebrates. Brachiopods are common in marine sediments of normal salinity, but they are rare in Mazon Creek nodules, being represented only by inarticulates such as *Lingula*, which is known to prefer brackish waters. *Lingula* is the only infaunal brachiopod, living in a vertical burrow into which it can retract by means of a long, fleshy pedicle. Numerous specimens from Pit 11 preserve *Lingula* in life position with burrow and pedicle intact.

Like brachiopods, echinoderms are usually found in fully marine waters and, apart from one crinoid specimen, the only echinoderm found at Mazon Creek is the holothurian (sea cucumber) *Achistrum*, which is actually quite common in Essex nodules. It can be distinguished from other worm-like creatures by the ring of calcareous plates which forms part of the sphincter at one end of the animal.

Perhaps the most interesting animal of all at Mazon Creek is one popularly known as 'Tully's Monster', named after its discoverer, the avid

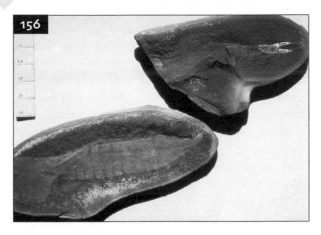

156 The 'Tully Monster', *Tullimonstrum gregarium* (MU). Scale bar is in cm.

157 Cycles of clay–silt pairs in the Francis Creek Shale laminites (MU). Scale bar is in mm.

collector Francis Tully. *Tullimonstrum gregarium*, to give it its scientific name, ranges up to 300 mm (c. 12 in) long. It has a segmented, sausage-shaped body with a long proboscis at the anterior, which terminates in a claw with up to 14 tiny teeth (**156**). Posteriorly, there is a diamond-shaped tail fin. Near the base of the proboscis is a crescentic structure, and just behind this is a transverse bar bearing an eye at each end. A number of ideas have been put forward as to the affinity of *Tullimonstrum*: conodont animal, annelid, nemertean, mollusc, or a group on its own. *Tullimonstrum* was clearly nektonic and predatory, and its overall appearance, proboscis, eyes, and tooth structure are all very reminiscent of a group of shell-less gastropods known as heteropodids. The fame of the Tully Monster was assured a few years ago when it was voted as State Fossil of Illinois.

Fish. More than 30 species of fish are known from Mazon Creek, but their identification is hampered by the fact that many of the fossils are small juveniles or isolated scales. Agnathans are represented by a hagfish and a lamprey, as well as two agnathans which cannot be assigned to known groups. Cartilaginous fish (chondrichthyans) are represented at Mazon Creek by a rare but diverse fauna of mainly juveniles. Interestingly, they do not appear to be the juveniles of the better known Mecca Quarry Shale sharks described by Zangerl and Richardson (1963), which were approximately coeval with the Francis Creek Shale forms, but seem to be the young of sharks which lived in a different habitat from the Mecca Quarry. *Palaeoxyris*, which is believed to be a shark egg-case, is a common fossil in the Braidwood biota and, to a lesser extent, the Essex biota. About 15 genera of bony fish (osteichthyans) occur at Mazon Creek. Most specimens are small and not easy to identify. However, there is a great variety of types from different habitats, from fresh through brackish to marine waters. Palaeoniscids, including deep-bodied forms as well as fusiform species generally referred to as '*Elonichthys*', are common in both Essex and Braidwood biotas. Among sarcopterygians, rhipidistians (which gave rise to tetrapods), coelacanths, and lungfish are all present at Mazon Creek.

Tetrapods. Tetrapods are rare at Mazon Creek, but are diverse and include 23 specimens of amphibian and one reptile. Temnospondyl amphibians are represented by larval *Saurerpeton*, both adult and larval *Amphibamus*, and a possible branchiosaurid. A single fragment (four vertebrae) of an anthracosaurid is known. Aïstopods – limbless, snake-like amphibians – are represented by two species and numerous specimens, and orders Nectridea, Lysorophia, and Microsauria are represented by single specimens. A single, immature specimen of a lizard-like, captorhinomorph reptile is known.

Trace fossils. Coprolites occur in both Essex and Braidwood biotas. While perhaps not as aesthetically pleasing as plants or animals, coprolites can tell us a great deal about what animals were eating. For example, spiral coprolites containing fish remains indicate that there were probably quite large sharks swimming in the Mazon Creek area, for which we have no body fossil evidence. At least 20 ichnogenera of other traces have been found, including worm burrows such as *Planolites*, *Arenicolites*, *Diplocraterion*, and *Skolithos*. The ichnofacies represented by these traces indicates a nearshore, shallow marine environment: the Cruziana ichnofacies.

PALAEOECOLOGY OF THE MAZON CREEK BIOTA

It is obvious from the evidence presented that the Mazon Creek area represents a variety of habitats – terrestrial, freshwater, brackish, and restricted marine – associated with a deltaic environment. The Colchester Coal represents an environment of swamp forest dominated by tree-sized clubmosses and horsetails with an understorey of seed-ferns, among other plants. The Francis Creek flora is dominated by fern, seed-fern and horsetail debris, which suggests it came from a more upland setting. The terrestrial animal fauna, such as myriapods, arachnids, and insects, presumably lived among these plants.

The Francis Creek Shale coarsens upwards, so the initial inundation of the swamp forest was rapid, and later delta-derived sediments filled the marine embayment. Many sedimentological and palaeontological features suggest rapid deposition: failed bivalve escape structures, *Lingula* buried in life position, and edgewise seed-fern pinnules, as well as the preservational features associated with rapid burial mentioned under taphonomy, above. Rapid sedimentation is characteristic of conditions adjacent to a delta. Cores resulting from boreholes drilled to test the foundation of the nuclear power plant in the vicinity of Pit 11 revealed complete sedimentation records. Moreover, the clay–silt laminae in these cores are paired, and the pairs widen and narrow in a cyclical fashion (**157**). Kuecher *et al.* (1990) studied the cores and the cyclicity of the paired laminae and interpreted the cyclicity as tidal in origin. The thin clay bands represent still-stands, either flood slack or ebb slack, i.e. when the tide is in the process of turning and the water is not flowing in either direction (**158**). The thicker silt layers represent periods of greater deposition (from the landward direction). The thicker silt bands were laid down during the ebb tide, when the outgoing tide allowed a high water flow into the basin; the narrower silt bands represent flood tides, when the incoming tide resists the flow to some extent. Thus, a single tidal cycle consists of two clay bands and two silt bands. The widest bands correspond to spring tides, with the highest tidal range, and the narrowest to neaps. Kuecher *et al.* (1990) found that there were 15–16 tides in a cycle from springs to next springs. This corresponds with half a lunar month; a complete lunar cycle has two springs (when the Moon and Sun align with the Earth) and two neaps (when the Moon–Earth axis is perpendicular to the Sun–Earth axis). Evidence from the cyclicity of coral growth (Johnson & Nudds, 1975) suggests that the lunar month consisted of 30 days in the Carboniferous period (i.e. the Earth–Moon system has accelerated over 350 million years to today's 28-day lunar cycle). Thus, the tidal cycle at Mazon Creek was the diurnal type, as found today in some parts of the world such as the Gulf of Mexico; tides around British coasts are semi-diurnal (i.e. there are two highs and two lows in a 24-hour period).

The tidal cyclicity provides a rare, direct measure of sedimentation rate. Each fortnightly cycle measures from 19 to 85 mm (c. 0.75 to 3.3 in) in thickness. This provides a deposition rate of about 0.5–2.0 m (c. 1.5–6.5 ft) per year of compacted sediment. The entire Francis Creek Shale was therefore deposited in 10–50 years. The tidal cycles provide independent, quantitative evidence of rapid sedimentation already concluded from qualitative evidence from sediments and fossils.

COMPARISON OF MAZON CREEK WITH OTHER UPPER PALAEOZOIC BIOTAS

Calver (1968) recognized a sequence of onshore–offshore communities in the hard-part fossil record of the Westphalian strata of northern England. His estheriid association roughly corresponds to the Braidwood biota, and his myalinid association, consisting principally of *Edmondia* and myalinid bivalves, corresponds to the Essex biota. It appears that the Mazon Creek biotas are common in Upper Carboniferous deltaic settings, but that Mazon Creek exceptionally preserves the soft-bodied biota which is normally lost by taphonomic processes at other localities.

Mazon Creek-type biotas have been found in ironstone concretions in other parts of the world, but nowhere have they been as well studied. For example, a number of localities in the British Coal Measures have yielded good nodule biotas, e.g. Sparth Bottoms (Rochdale), Coseley (West Midlands), and Bickershaw, Lancashire (Anderson *et al.*, 1997). A similar biota occurs at Montceau-les-Mines in France (Poplin & Heyler, 1994). Other Upper Carboniferous localities contribute additional information to our knowledge of Upper Carboniferous non-marine life. For example, the non-nodule locality of Nýřany (Czech Republic) is renowned for its exceptional tetrapod fossils. Schram (1979), in a systematic study of mainly crustaceans in Carboniferous non-marine biotas, argued that stable, predictable associations persisted throughout the period in a continuum. The Grès à Voltzia biota (Chapter 10) is an extension of Schram's continuum into the Triassic (Briggs & Gall, 1990). Thus, the marginal marine ecosystem seems to have been little affected by the great Permo–Triassic extinction event.

MUSEUMS AND SITE VISITS

Museums
1. Field Museum of Natural History, Chicago, Illinois, USA.
2. National Museum of Natural History, Smithsonian Institution, Washington DC, USA.
3. Illinois State Museum, Springfield, Illinois, USA (online exhibit: http://www.museum.state.il.us/exhibits/mazon_creek).
4. Burpee Museum of Natural History, Rockford, Illinois, USA.

Sites
Collection at the Mazon Creek sites is best done through field trips organized by the Public Programs of the Field Museum of Natural History (Tel. (+1-312-665-7400).

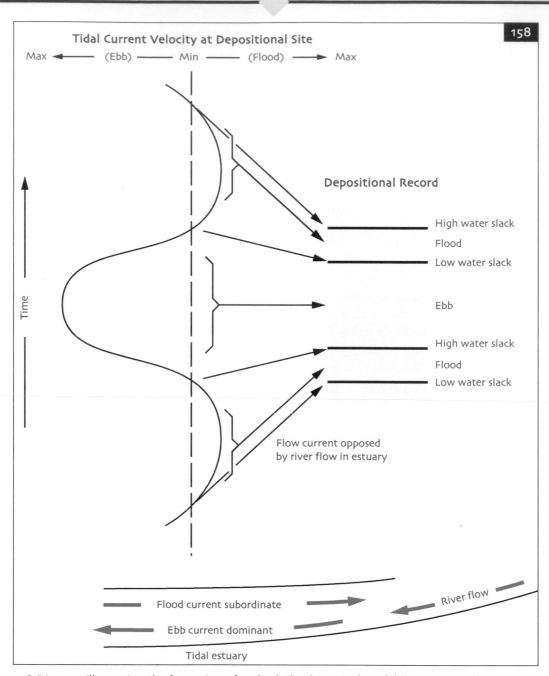

158

Tidal Current Velocity at Depositional Site

Max ◄──── (Ebb) ──── Min ──── (Flood) ──── ► Max

Time

Depositional Record

High water slack
Flood
Low water slack

Ebb

High water slack
Flood
Low water slack

Flow current opposed
by river flow in estuary

Flood current subordinate

River flow

Ebb current dominant

Tidal estuary

158 Diagram illustrating the formation of cyclical silt–clay pairs by tidal deposition (after Kuecher *et al.*, 1990).

KAROO

BACKGROUND: THE KAROO SUPERGROUP

Strictly speaking, the Karoo is not a Fossil-Lagerstätte but rather a supergroup of strata in which many important fossils and evidence for major evolutionary events can be found. In this chapter, we concentrate on three formations within the supergroup which span the Permo–Triassic mass extinction, and typify terrestrial ecosystems of Permian and Triassic times in Gondwana.

Rocks of the Karoo Supergroup (**159**) crop out over almost two-thirds of southern Africa, and the strata record an almost continuous sequence of deposition from late Carboniferous to middle Jurassic, a period of about 100 million years. The sedimentary sequence represents environments ranging from glacial to arid, and is capped by basaltic lavas of the Drakensberg Group. The Karoo sequence totals about 10 km (c. 6.2 miles) in thickness, and accumulated in numerous basins. The main Karoo Basin covers about three-fifths of the surface area of South Africa (about 600,000 sq. km [231,660 sq. miles]). The supergroup rests unconformably on the Cape Supergroup (see Chapter 4, The Soom Shale), and the basins developed following the uplift and folding of the Cape Fold Belt to the south during the Carboniferous Period. The main Karoo Basin developed as a foreland basin with the fold-thrust belt lying along its southern margin and the subduction zone along the palaeo-Pacific margin of Gondwana.

In this account, we concentrate on the main Karoo Basin in South Africa, which is the best studied. The oldest Karoo rocks, the Dwyka Formation, preserve evidence of the widespread Gondwana glaciation in the form of glacial sediments (diamictites) on top of beautiful striated pavements on older rocks (**160, 161**). Following the ice age, the Ecca Group records sedimentation in a shallow, landlocked sea. Erosion of the Cape Fold Belt to the south poured delta sands and turbidites into the Karoo Basin while, to the north, extensive deltas on more gentle slopes supported extensive forests. The Ecca Group is thus rich in coal deposits. The overlying Beaufort Group sediments mark a period of extensive flood plains in warmer and drier climates, with sedimentary input from all directions. We shall make a special study of fossil-rich horizons in the Lower Beaufort near Fraserburg, in the Teekloof Formation. The Permo–Triassic boundary occurs within the Beaufort Group, and is marked by a rapid change from fairly humid to rather arid conditions. We shall study this momentous event in Earth history in the Lootsberg Pass area, by looking in detail at the Balfour Formation.

The upper part of the Beaufort Group was laid down in the Triassic Period. Initial arid conditions following the Permo–Triassic extinction were eventually replaced by a more humid climate, and the late Triassic Molteno Formation, directly above the Beaufort Group, contains a remarkable flora and fauna which we shall look at in detail. To complete the story, following the Molteno, red marls of the Elliot Formation mark a return to semi-arid

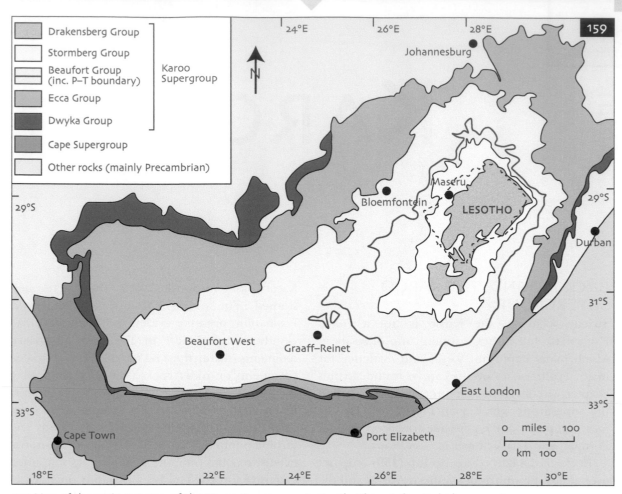

159 Map of the main outcrop of the Karoo Supergroup in South Africa (after Rubidge, 2005).

Legend:
- Drakensberg Group
- Stormberg Group
- Beaufort Group (inc. P–T boundary) — Karoo Supergroup
- Ecca Group
- Dwyka Group
- Cape Supergroup
- Other rocks (mainly Precambrian)

160 Dwyka glaciated pavement on Precambrian Mozaan Quartzite, Ithala Game Reserve, KwaZulu-Natal, South Africa.

161 Upper Carboniferous Dwyka glacial diamictite, near Matjiesfontein, South Africa.

162 Jurassic Clarens Formation dune sandstones, Golden Gate Highlands National Park, Free State, South Africa.

163 The Sentinel, Drakensberg basalts, overlying Clarens Formation sandstones, Royal Natal National Park, KwaZulu-Natal, South Africa.

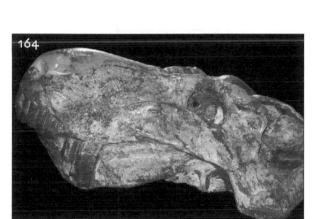

164 Skull of the gorgonopsian *Rubidgea* (RC). Length about 750 mm (c. 30 in).

conditions and then greater aridity still during the deposition of the Clarens Formation: spectacular dune sandstones of Jurassic age (**162**). By this time, break-up of Pangaea had proceeded to the extent that stretching of the Earth's crust allowed immense volumes of basaltic lava to pour out onto the Clarens Formation surface, fed by many dykes which can be seen criss-crossing the sedimentary layers beneath. The flood basalts can now be seen as the vast Drakensberg plateau in Lesotho and escarpment facing KwaZulu-Natal (**163**).

The major palaeontological importance of the Karoo Supergroup is its preservation of terrestrial biota through the Permian and Triassic periods, including excellent exposures of the Permo–Triassic boundary. In late Carboniferous times, a group of reptiles appeared which had a hole (fenestra) in the skull behind the eye socket (orbit) between the postorbital, squamosal, and jugal bones; i.e. in a lower position. This skull pattern is termed synapsid; other arrangements include two fenestrae behind the orbit (diapsid, seen in dinosaurs and birds, for example), no fenestrae (anapsid, seen in primitive reptiles and turtles), and an upper fenestra (euryapsid, found in the extinct aquatic reptiles: ichthyosaurs and plesiosaurs). The synapsid arrangement is found in a group called the Synapsida: the mammal-like reptiles and mammals. The mammal-like reptiles are so-called because they have certain features which are commoner in mammals rather than other reptiles; for example, their teeth are sometimes differentiated, with larger canine-like teeth and smaller chewing teeth, and their jaw action became more precise than that seen in most reptiles. As the group evolved, the dentary bone in the lower jaw enlarged at the expense of the other lower-jaw bones until, in the true mammals, which arose from the synapsids in the Triassic, the only bone in the lower jaw is the dentary.

In Permian times, early synapsids, such as the first herbivorous reptile, *Edaphosaurus*, and the carnivorous *Dimetrodon*, bore characteristic sails on their backs, probably to help regulate their body temperature. From among these, in the late Permian, arose the therapsids, which were more mammal-like than reptilian in many ways. The earliest of these were the Dinocephalia, which included some large (5 m [c. 16.5 ft] long) herbivores such as *Moschops*, known from South Africa. *Moschops* has a skull with a thick roof, possibly adapted for head-butting, as in goats today. The most diverse synapsids in the late Permian were the

165 Stratigraphy of the Karoo Supergroup, with vertebrate Assemblage Zones; yellow indicates sandstone-rich units, brown in western section marks hiatus (after Rubidge, 2005).

PERIOD	GROUP	FORMATION WEST OF 24°E	FORMATION EAST OF 24°E	ASSEMBLAGE ZONES
JURASSIC	STORMBERG	DRAKENSBERG	DRAKENSBERG	
JURASSIC	STORMBERG		CLARENS	
TRIASSIC	STORMBERG		ELLIOT	
TRIASSIC	STORMBERG		MOLTENO	
TRIASSIC	BEAUFORT		BURGERSDORP	Cynognathus
TRIASSIC	BEAUFORT		KATBERG	Lystrosaurus
PERMIAN	BEAUFORT		BALFOUR: Palingkloof Mbr / Elandsberg Mbr / Barberskrans Mbr / Daggaboersnek Mbr	Dicynodon
PERMIAN	BEAUFORT	Steerkampsvlakte Mbr (TEEKLOOF)		
PERMIAN	BEAUFORT	Oukloof Mbr (TEEKLOOF)	Oudeberg Mbr (BALFOUR)	Cistecephalus
PERMIAN	BEAUFORT	Hoedemaker Mbr (TEEKLOOF)	MIDDLETON	Tropidostoma
PERMIAN	BEAUFORT	Poortje Mbr (TEEKLOOF)		Pristerognathus
PERMIAN	BEAUFORT	ABRAHAMS-KRAAL	KOONAP	Tapinocephalus
PERMIAN	BEAUFORT	ABRAHAMS-KRAAL	KOONAP	Eodicynodon
PERMIAN	ECCA	WATERFORD	WATERFORD	
PERMIAN	ECCA	TIERBERG/FORT BROWN	FORT BROWN	
PERMIAN	ECCA	LAINSBURG/RIPON	RIPON	
PERMIAN	ECCA	COLLINGHAM	COLLINGHAM	
PERMIAN	ECCA	WHITEHILL	WHITEHILL	
PERMIAN	ECCA	PRINCE ALBERT	PRINCE ALBERT	
CARB.	DWYKA	ELANDSVLEI	ELANDSVLEI	

dicynodonts, which nearly died out in the Permo–Triassic mass extinction, but recovered again in the Triassic. *Diictodon* (p. 110) and *Lystrosaurus* (p. 117) belong to the dicynodonts. Alongside the dicynodonts in the late Permian were the predatory gorgonopsians, such as *Rubidgea* (**164**), and the therocephalians, also carnivores. The cynodonts also appeared at this time. They were the most mammal-like of the synapsids; indeed, mammals arose from among this group. Many features of cynodonts, such as the Triassic *Thrinaxodon* (p. 118) are mammalian, e.g. elaborate dentition, complex jaw mechanics, a double occipital condyle where the skull articulates with the spinal column, and differentiated vertebrae. Advanced cynodonts continued into the Jurassic Period. The Karoo is also important for its early mammal fossils, such as the Jurassic *Megazostrodon*, from the Jurassic upper Elliot Formation.

So abundant are the synapsid reptiles in Karoo sediments, in comparison with the rarity of most other biota, that they have been used in biostratigraphy. At present, eight assemblage biozones are recognized in the mid-Permian to mid-Triassic Beaufort Group: *Eodicynodon*, *Tapinocephalus*, *Pristerognathus*, *Tropidostoma*, *Cistecephalus*, *Dicynodon*, *Lystrosaurus*, and *Cynognathus* (Hancox & Rubidge, 2001) (**165**). The Teekloof Formation covers the upper part of the *Pristerognathus* and the whole of the *Tropidostoma* and *Cistecephalus* zones; the Balfour Formation includes all of the *Dicynodon* zone and the basal part of the *Lystrosaurus* zone; the Molteno Formation contains no synapsid remains and lies outside of this scheme.

HISTORY OF DISCOVERY OF THE KAROO FOSSILS

The first fossil reptiles from the Karoo were brought to the world's attention in the 1830s by Andrew Geddes Bain. Bain emigrated to the Cape from Scotland in 1816 where he eventually became a highly respected civil engineer, building many road passes across the Karoo, which led him to an interest in geology. He discovered many fossils, including the herbivorous dicynodont *Oudenodon bainii*. These were the first specimens of mammal-like reptiles to be discovered.

Another Scot, Robert Broom, born in 1866 and educated in medicine, settled in South Africa in 1897. In 1903 he became Professor of Geology and Zoology at Victoria College, Stellenbosch, and spent all his vacations collecting fossils in the Karoo. Many of his specimens went to the South African Museum, Cape Town, which appointed him curator of fossil reptiles. Broom was privileged with a free railway pass so that he could travel on his collecting expeditions, but a new government in 1909 took away the pass, throwing Broom into such depression that he retired from his chair and from the museum and went to London to examine the Karoo collections there. On his return in 1910 he published a paper about mammal-like reptiles which revolutionized knowledge of their structure and relationships and laid the foundations of therapsid classification, which has stood the test of time. Later, he became interested in anthropology and is, perhaps, better known for his studies of fossil hominids and particularly for excavations at Sterkfontein Cave in Gauteng. Nevertheless, he continued work on mammal-like reptiles until his death in 1951.

During his fossil-collecting visits to the Karoo, Broom would call on the Kitching family in the village of Nieu-Bethesda, close to his base at Graaff-Reinet, and encourage them to look for fossils on their land. The entire family became dedicated and skilful fossil collectors, and none more so than the eldest son, James, then aged six. At the age of seven he discovered the type specimen of *Youngopsis kitchingi*. When the University of the Witwatersrand set up the Bernard Price Institute for Palaeontological Research in 1945, Kitching was appointed as the first staff member, and immediately set off on a collecting trip to Graaff-Reinet.

One of the farms Robert Broom visited during his travels in the Karoo was Wellwood, which has been run by the Rubidge family since 1838 (**166**). In 1934, in answer to a question about fossils from his 10-year-old daughter, Sidney Rubidge organized a family picnic, which led to the discovery of a fossil skull with large teeth 'resembling a horse's snout'. Robert Broom visited the farm in 1935 to look at the specimen and described it as the holotype of the gorgonopsian *Dinogorgon rubidgei*. He thus encouraged Sidney Rubidge to continue collecting. Between 1935 and 1940, James Kitching and his four sons accompanied Rubidge on numerous collecting expeditions and spent many weekends collecting in the Graaff-Reinet area. So, a large collection was amassed at Wellwood. Rubidge (1956) wrote 'Kitching would convey his wife and entire family in his second-hand Buick car to the far-off fossil-bearing ridges where he initiated his four sons, from five to 13 in age, into the craft, for craft it is, of fossil-hunting'.

Sidney Rubidge encouraged his son to enjoy

166 Wellwood, home of the Rubidge family since 1838.

fossil collecting, and Richard continued the day-to-day running of the Merino sheep farm and stud. In 1965 a building on the farm was restored specifically to house the collection. Special cabinets were built to display the fossils to best advantage and the Rubidge collection of Karoo fossils now includes over 800 catalogued specimens, of which 118 are named types. Bruce, grandson of Sidney, has made a career in palaeontology and is the Director of the Bernard Price Palaeontological Institute in Johannesburg.

First discovered by Carruthers in 1871, the Molteno Formation was mapped in great detail by Alexander du Toit in the early years of the 20th Century. He had no reliable base-maps on which to mark geological boundaries; he travelled on foot or by bicycle, carrying a plane-table on his back and explored every inch of the country south and east of Aliwal North in Eastern Cape. Alexander du Toit was an enthusiastic supporter of Wegener's theory of Continental Drift, in support of which he could see evidence in the fossil plants of what we now call the Gondwana continents. He was also the first person to recognize the glacial nature of the Dwyka sedimentary sequence (du Toit, 1921). The Molteno Formation became the focus of research interest by Heidi and John Anderson of the National Botanical Institute in Pretoria in the 1960s. Concerted collecting over nearly 40 field seasons at 69 localities has resulted in the accumulation of 30,000 catalogued slabs, and an enormous collection of mostly plants and insects from the Molteno outcrop. The Andersons amassed a vast quantity of data on the Molteno Formation, from which it is possible to make a plausible reconstruction of the ancient ecosystem.

STRATIGRAPHIC SETTING AND TAPHONOMY OF THE KAROO BIOTA

The Teekloof Formation

The oldest of the sequences to be studied in more detail here comes from the lower Beaufort Group, of late Permian age. The Beaufort Group consists of alternating mudstone and sandstone units with mud cracks and caliche horizons which accumulated on a vast, semi-arid flood plain. Within this sequence, the Teekloof Formation consists of 1,000 m (c. 3,300 ft) of green, grey, and maroon mudstones, with channel sandstones such as the Poortje Sandstone. A number of cycles can be recognized locally (**167**); most of the important fossils come from the Hoedemaker Member, which lies above the Poortje Sandstone.

A taphonomic study of the therapsid fossils in the Hoedemaker Member (c. 257 Ma) by Smith (1993a) classified the remains into eight taphonomic classes, ranging from completely articulated skeletons (sometimes found in pairs in a burrow), through isolated skulls (with or without lower jaw), to isolated and/or fragmented bones such as ribs, limb bones, and vertebrae. The distance and duration of post-mortem transport are likely to have ranged from none in the first of these classes, to far and long in the last. Study of the distribution of these taphonomic classes among the various sedimentological settings (channel, channel bank, proximal flood plain, distal flood plain) revealed six taphonomic pathways: 1) disarticulated skeletons embedded at the site of death and buried by alluvium; these remains formed the bulk of the finds in the study and

167 Cyclicity in the Teekloof Formation as shown by sandstone scarps in this view of the Karoo near Putfontein, Beaufort West, Western Cape, South Africa.

168 Burrow of *Diictodon* in the Teekloof Formation, Putfontein, Beaufort West, Western Cape, South Africa. Hand lens is 30 mm (c. 1.2 in) long.

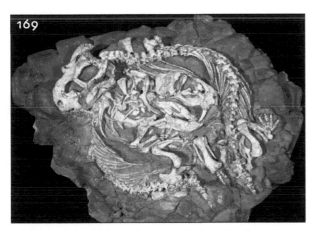

169 Two *Diictodon* skeletons curled up in a burrow. *Diictodon* is about 450 mm (c. 17.7 in) in length (SAM).

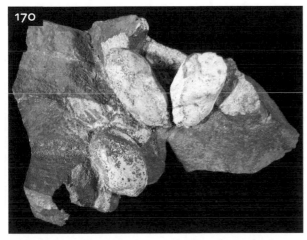

170 Cluster of bone-bearing coprolites from the Late Permian Teekloof Formation (SAM). The coprolites are 25 mm (c. 1 in) long.

occurred abundantly in the proximal flood plain environment, but also in the channel bank and distal flood plain. They represent carcasses of animals which died naturally on the flood plains and over time were scavenged and disarticulated by periodic floods; isolated *Diictodon* skulls are common in this setting. 2) *Diictodon* skeletons preserved in underground burrows (**168, 169**); these occurred primarily in the proximal flood plain, but also in the channel bank deposits where bone-bearing coprolites (**170**) were commoner. 3) Articulated, disarticulated and transported small postcranial elements accumulated in embayments of the low-stand lake margin; not only is this as far as small bones could be transported but also it is where animals could find water during the dry season, and where they would also die, and be predated and trampled. 4) Carcasses of animals that died in the channel during floods, became disarticulated during transport downstream and their bones added to the river gravels. 5) Caliche-encrusted bones reworked from the river banks when the channel migrated. 6) Numerous *Diictodon* skeletons accumulated in the muddy sediments in and around shrinking waterholes along the river channel, and were probably disarticulated through predation; burial may have been accelerated by larger animals trampling the skeletons into the soft substrate.

The Balfour Formation

The second sequence studied in this chapter occurs mainly around the Lootsberg Pass, between Graaff-Reinet and Middelberg (**171**), and around Bethulie in Free State, in the Balfour Formation of the upper part of the Beaufort Group. Pinning the Permo–Triassic boundary and the associated mass extinction event is not easy; a number of changes happened within a short space of geological time, but not exactly at the same instant (c. 250 Ma). The base of the Triassic in the Karoo was traditionally defined by the first appearance of the dicynodont *Lystrosaurus*, and placed at the base of the Katberg Formation, a thick sandstone overlying the muddy Palingkloof Member of the Balfour Formation. More recently Smith (1995) and Ward *et al.* (2000) showed that the transition between the shale-dominated Balfour Formation and the sandstone-dominated Katberg Formation is gradational and shows no evidence of disconformity other than those normally found within fluvial sequences. Moreover, *Lystrosaurus* fossils appear below the Katberg sandstone at Lootsberg Pass and near Bethulie, and the range of *Lystrosaurus* thus overlaps the last *Dicynodon* zone fossils by about 41 m (c. 135 ft) (Smith & Ward, 2001). So, the base of the Triassic cannot be defined easily on vertebrate fossils.

Smith and Ward (2001) collected vertebrate fossils from 52 stratigraphic levels at seven localities in the Bethulie and Lootsberg Pass areas (**172**). They recorded eight species in Permian strata, seven in Triassic strata, and one (*Lystrosaurus*) which spans the boundary. The greatest diversity of late Permian animals occurs in grey mudrocks

171 View across the Karoo, looking south from Lootsberg Pass, between Graaff-Reinet and Middelberg. The P-T boundary is just below this spot.

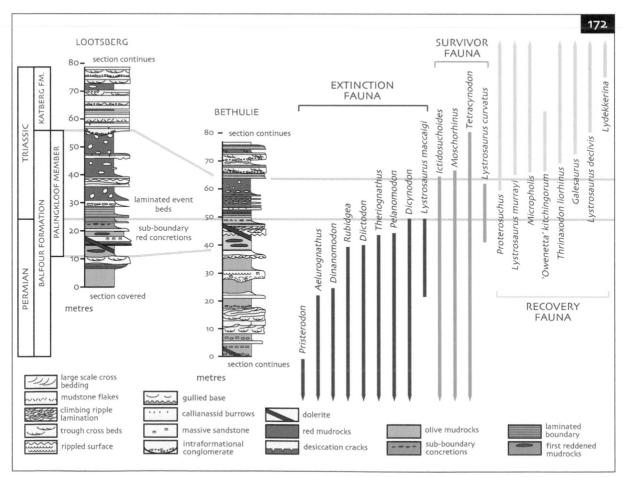

172 Sedimentary logs through the P-T boundary at Lootsberg Pass (**171**) and Bethulie, showing the stratigraphic position and elements of the Extinction, Survivor and Recovery faunas (after Smith & Ward, 2001; Smith & Botha, 2005).

typical of the Palingkloof Member of the Balfour Formation. The traditional Triassic indicator fossil *Lystrosaurus* first appears in this rock type some 41 m (c. 135 ft) below the P-T boundary. Most of the late Permian taxa (e.g. *Diictodon*, *Rubidgea*, *Theriognathus*) disappear in the succeeding reddish mudrocks. This rock sequence is terminated by a 1 m (c. 3 ft) thick red mudrock capped by a single layer of large, brown-weathering calcareous nodules which coincides with the last occurrence of *Dicynodon lacerticeps*, the index fossil of the latest Permian biozone. This layer of brown-weathering nodules is taken as the P-T boundary: the event-bed. The boundary layer is immediately overlain by 3–5 m (c. 10–15 ft) of distinctively laminated red mudrock made up of thinly bedded dark reddish-brown and olive-gray siltstone–mudstone couplets. This rock type is unique in all the sections, and intensive prospecting failed to find any vertebrate fossils in it. A complete lack of late Permian gymnosperm pollen, an increase in woody debris, and abundant remains of fungi were reported from this layer at a locality at Carlton Heights (Steiner *et al.*, 2003) but it is likely that, particularly since biostratigraphy is poor at this section, the fungal spike might not correspond to the immediate post-extinction event but be higher in the section, at the base of the Katberg Formation (Ward *et al.*, 2005). The laminated layer does, however, coincide with a carbon isotope anomaly that has been identified as spanning the P-T boundary in marine sections (Ward *et al.*, 2005). The first appearances of Triassic species (as well as the reappearance of *Lystrosaurus*) are in a red siltstone with minor sandstone sheets. These thin sandstone bodies have distinctive gullied basal scours and there are abundant curled-up *Lystrosaurus* skeletons, thickly enveloped in dark reddish brown nodular material, in these beds.

The Molteno Formation

The Molteno Formation (**173**) crops out in a roughly oval belt covering an area of about 25,000 sq. km (c. 9,650 sq. miles) and attains a maximum thickness of about 460 m (c. 1,500 ft) in the south-east, thinning to the north. The formation is highly unconformable with the underlying Burgersdorp Formation of the Permo–Triassic Beaufort Group and interdigitates with the overlying upper Triassic Elliot Formation. The Molteno Formation was deposited during upper Triassic (Carnian, c. 230

173 The Birds River meanders through outcrops of the Molteno Formation in this view from the Birds River fossil locality.

Ma) times when uplift of the southern margin of the Karoo Basin resulted in the accumulation of a thick wedge of terrigenous sediment. The sediments consist mainly of coarse-grained pebbly sandstones, interbedded with occasional fine sandstones, shales, and rare coals. At its thickest, six fining-upward cycles can be seen. The environments represented range from river channel-fills (including abandoned channel-fill deposits), through flood plains with crevasse-splay and sheetflood (overbank deposits), lacustrine, marshes, and peat-swamp deposits.

DESCRIPTION OF THE KAROO BIOTA

Plants. Permian vegetation in Gondwana was dominated by the *Glossopteris* flora. The distinctive, tongue-shaped leaves of the gymnospermous tree *Glossopteris* have been found on all of the continents which make up Gondwana, and this distribution was one of the main pieces of evidence put forward to support the idea of continental drift. *Glossopteris* forest thrived in the cool, moist conditions which prevailed in the southern hemisphere following the Dwyka glaciation. *Glossopteris* and the woodlands it characterized were casualties of the end-Permian mass extinction, though other seed-ferns continued into the Mesozoic Era. Immediately after the end-Permian extinction, there was a rise in diversity of smaller herbaceous plants including lycophytes such as

174 Petrified log of the gymnosperm tree *Agathoxylon*, Karoo National Park, Beaufort West, Western Cape, South Africa. The widely spaced annual rings indicate rapid growth of this tree in a seasonal climate.

175 The distinctive seed-fern *Dicroidium* from the Molteno Formation (NBI). Length of frond 250 mm (c. 9.8 in).

176 A different species of *Dicroidium* (left), together with a finely divided leaf of the ginkgoalean *Sphenobaiera* from the Molteno Formation (NBI). Width of view 185 mm (c. 7.3 in).

177 Leaf of the gymnosperm tree *Ginkgo waldeckensis* from the Molteno Formation (NBI). Length of longest lobe 70 mm (c. 2.8 in).

Selaginellales (clubmosses) and Isoetales (quill-worts). Later on, other groups of gymnosperms again become dominant but suffer periodic die-offs. The recovery of gymnosperm forests took about 4–5 million years. Fossil woods are common in Permian strata of the Karoo (**174**). While it is notoriously difficult to identify the tree to which they belong without attached leaves or cones, something can be deduced about the environment in which the tree grew. For example, well-developed tree rings imply a seasonal (wet–dry or warm–cool) climate.

We know much about Triassic Gondwana floras from studies of the Molteno Formation plants. The flora includes 56 genera and 204 species of macroplants. Diversity among the gymnosperms and non-gymnosperms is approximately equal. The seed-fern *Dicroidium* (**175**), with 19 species, characterizes the flora as it does throughout the Gondwana Triassic. While some *Dicroidium* fronds have been found attached to branches, we do not yet know what the rest of the bush or tree might have looked like. Several other gymnosperms, most notably *Lepidopteris*, *Sphenobaiera* (**176**), and *Ginkgo* (**177**), are also prominent. The cycads are diverse, with five genera and 21 species, but generally uncommon. The conifers are far less diverse, with three genera and five species, but one genus, *Heidiphyllum*, occurs frequently and abundantly. The non-gymnosperms include some 20 species of bryophyte (mosses and liverworts), six species of possible lycopod (clubmosses), 21 species of horsetail in five genera, and some 50 species of ferns. The horsetail *Equisetum* (nine species) dominates a number of assemblages.

Invertebrates. No body fossils of invertebrates have been found in the Permian Teekloof Formation but trace fossils record their presence. *Planolites* is likely to have been made by worms, the trackway *Umfolozia* is evidence of freshwater crustaceans, and gently meandering grooves (**178**) may reflect the furrowing activity of freshwater gastropods.

In the Palingkloof Member of the Balfour Formation, which straddles the P-T boundary, some inclined burrows with scratch marks have been described as *Katbergia* and attributed possibly to crustaceans (Gastaldo & Rolerson, 2008).

Insects are by far the most important component of the Molteno fauna. They are known from 43 of the 100 plant assemblages and the number of specimens collected exceeds 3,000. Provisional taxonomy of the full collection reveals 117 genera including 333 species. These represent 18 orders with the roachoids by far the most abundant (956 specimens), the beetles second (453 specimens), and the bugs third (229 specimens). The beetles (Coleoptera) are the most

diverse order with nearly half of the described species (161). The bugs (Hemiptera) come second in diversity, with 69 species, and the dragonflies (Odonata) (**179**) are third with 22 species (Cairncross *et al.*, 1995).

The bivalved crustaceans Spinicaudata (conchostracans) are the next most significant faunal element in the Molteno fauna, with about three genera and eight species from 20 plant assemblages, in some of which they occur in abundance on bedding planes. Two genera and species of bivalves have been reported, and a spider species is known from two specimens in two assemblages. The spider (**180**) was identified as the oldest member of the Araneomorphae ('true spiders') (Selden *et al.*, 2009).

Fish. Fossil fish occur throughout the Karoo Supergroup. The majority of these are palaeoniscids, but hybodontid sharks, coelacanths, and lungfish have also been found. Palaeoniscids are a diverse group of freshwater fish which are represented by three species from the Dwyka, one from the Ecca, 25 from the Beaufort groups, and one from the Clarens Formation (Rubidge, 2005). A locality at the base of the Burgersdorp Formation (Beaufort Group) has yielded a fish fauna from the

179 Dragonfly wing from the Molteno Formation, Birds River locality (BPI). Maximum length of wing preserved is 30 mm (c. 1.2 in).

178 Trail, probably made by furrowing snail, crossing ladder ripples, Teekloof Formation, Putfontein, Beaufort West, Western Cape, South Africa. Hand lens is 30 mm (c. 1.2 in) long.

early Triassic (Scythian), including lungfish, saurichthyids, and many isolated chondrichthyan teeth, fin spine fragments, and actinopterygian scales and teeth (Bender & Hancox, 2004). A fish fauna from a mid-Triassic locality at Bekkerskraal farm near Rouxville, higher in the Burgersdorp Formation, has yielded 14 lower actinopterygian fish, several specimens of *Lissodus africanus*, a freshwater hybodontid shark, and the coelacanth *Whiteia africana* (Jubb & Gardiner, 1975). Fish, however, are rare in the three formations described here: fish fin trails (*Undichnus*) are common in the Teekloof Formation (Smith, 1993b), and three species of fish, including *Semionotus* (**181**), have been reported from the Molteno Formation (Cairncross *et al.*, 1995).

Amphibians. The Beaufort Group preserves one of the most diverse Permo–Triassic temnospondyl faunas in the world. However, only the family Rhinesuchidae is present in the Permian part of the Beaufort Group. Although a number of rhinesuchid genera have been described, currently only three taxa are considered valid (Rubidge, 2005). The Rhinesuchidae appear to be endemic to Gondwana and have been recorded from Brazil, India, and Malawi, as well as South Africa. An explosion of amphibian taxa in the Triassic came in the wake of the end-Permian extinction when a renewed adaptive radiation of temnospondyl amphibians occurred with a broad range of new superfamilies and families coming into prominence (Rubidge, 2005).

Anapsids. The oldest reptiles from Gondwana are the freshwater mesosaurs from the Whitehill Formation of the Ecca Group (early Permian). Mesosaurs are the oldest known anapsid reptiles and the first reptiles to adapt to an aquatic lifestyle. Mesosaurs are also known from Namibia and Brazil, and this distribution was one of the pieces of evidence for continental drift used by Alex du Toit. Recent phylogenetic studies have identified southern Africa as the place of origin of anapsids (Modesto, 2000). *Eunotosaurus* and the pareiasaurs are restricted to the Permian Beaufort Group, while the procolophonids extend into the Triassic. Pareiasaurs are found throughout Pangaea, including southern Africa, Brazil, China, Germany, Scotland, and Russia. They appear to be restricted

to the Permian. Ten of the 16 recognized species are from the Karoo. There are essentially two hypotheses for the relationship of the turtles (Chelonia) to the other anapsid reptiles: either they are closely related, which makes turtles rather primitive, or they derived from somewhere in the diapsids by closing off the two post-orbital fenestrae. The most primitive representative of the turtles from Africa is *Australochelys* from the Jurassic part of the Elliot Formation, though the oldest turtle known is from the Upper Triassic of China.

180 The oldest araneomorph spider *Triassaraneus andersonorum*, from the Molteno Formation, Upper Umkomaas locality, KwaZulu-Natal, South Africa (NBI). Leg span 6 mm (c. 0.24 in).

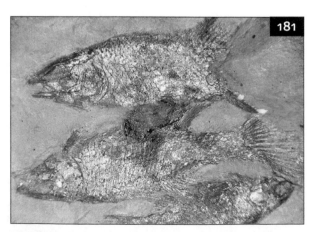

181 The holostean fish *Semionotus* from the Molteno Formation (BPI). Length 150 mm (c. 6 in).

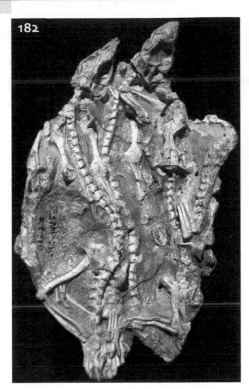

183 The late Triassic–early Jurassic Elliot Formation exposed in the Rooidraai roadcut, Golden Gate Highlands National Park, Free State, South Africa. The Elliot Formation is the red mudstone in the lower part of the cliff, the Clarens Formation sandstone sits on top. Slanting basalt dykes feeding the overlying Drakensberg lavas (seen in the distant hills) cut the sedimentary sequence at either end of the cutting.

182 The oldest diapsid in South Africa, *Youngina*, seen here as an assemblage of five juvenile skeletons preserved in a den (SAM). Length 110 mm (c. 4.3 in).

Diapsids. This group includes lizards, snakes, crocodiles, pterosaurs, dinosaurs, and birds. The oldest diapsid in South Africa is *Youngina* from the Teekloof Formation (Smith & Evans, 1995; **182**). The specimen consists of five juvenile skeletons preserved in life position in a shallow depression, suggesting that they were aestivating in a burrow when flood water filled the burrow and drowned them. The skull of a small captorhinid *Saurorictus* was found at the same locality, the first to be discovered in South Africa (Modesto & Smith, 2001).

Archosaurs include dinosaurs, crocodiles, birds, and pterosaurs, and one of the earliest is *Proterosuchus* from South Africa, which first appears at the beginning of the Triassic. *Euparkeria* occurs in slightly younger Anisian rocks of the Burgersdorp Formation. The position of this genus near the base of the archosaurs is still the subject of much debate among vertebrate palaeontologists. Dinosaurs first appeared in the Carnian, which in South Africa equates to the Molteno Formation. Footprints of sauropod dinosaurs have been found in the Molteno Formation, which provides evidence that they were there.

By the time of the late Triassic–early Jurassic Elliot Formation, dinosaurs were well established in the Karoo, and the fauna includes several prosauropods and sauropods. During the Jurassic part of the deposition of the Elliot Formation, the large early sauropods disappeared and were replaced by smaller forms such as *Massospondylus*. Clutches of dinosaur eggs from the Lower Jurassic part of the Elliot Formation at Rooidraai (**183**) have been ascribed to *Massospondylus*, and are some of the oldest dinosaur eggs preserved with embryos yet discovered.

Synapsids. Non-mammalian synapsids are the most abundant reptiles in the Beaufort Group, and it is for these fossils that the Karoo is internationally renowned. Karoo rocks preserve some of the oldest

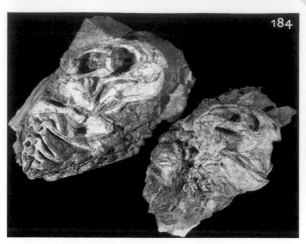

184 Female (left) and male (right, with tusks) skulls of the dicynodont *Diictodon* (SAM). Skulls are about 110 mm (c. 4.3 in) long.

185 Skull of the large, carnivorous gorgonopsian *Aleurognathus* from the Teekloof Formation (SAM). Canine tooth is about 50 mm (c. 2 in) long.

therapsid fossils in the world, and a Gondwanan origin for several therapsid groups is possible, although new material from China lends support to a Laurasian origin for the group (Liu *et al.*, 2009). The most primitive synapsids following the sail-back forms were the Biarmosuchia, most of which are from South Africa. Biarmosuchians include most of the ictidorhinids and the more derived burnetia-morphs. The latter are rare, carnivorous therapsids with large canine teeth and bony protruberances on their skulls. They are known from the Teekloof Formation, in which they form a rare component of the fauna (Sidor & Smith, 2007).

Dinocephalians are a diverse group of basal therapsids most abundantly represented in Middle Permian rocks of South Africa. There are two types: carnivorous anteosaurids and herbivorous tapino-cephalids. *Australosyodon* and *Tapinocaninus* from the lowermost Beaufort Group represent the most basal forms of these two groups. Despite being fairly common in the lowermost Beaufort, dino-cephalians became extinct before deposition of the Teekloof Formation began.

The herbivorous anomodonts, which include the dicynodonts, are the most abundant therapsids in the Beaufort Group. Anomodonts appeared in the middle Permian and survived through to the late Triassic, and occur throughout the world. The earliest anomodonts occur in China, so it is supposed that they originated in Laurasia, but

Eodicynodon, the most primitive dicynodont, is known exclusively from the lowermost Beaufort Group, which suggests that the dicynodonts evolved in Gondwana. The lowermost Beaufort Group has a rich diversity of dicynodonts, includ-ing *Robertia*, *Pristerodon*, *Chelydontops*, *Diictodon*, *Colobodectes*, and *Lanthanostega*. *Diictodon* is one of the commonest fossils, especially in the Teekloof Formation, where it has been found in helical burrows (Smith, 1987; **169**). *Diictodon* was the size and shape of a small Dachshund dog. Like most dicynodonts, it lacked teeth at the front of the jaw but had a sharp, horny beak to slice off vegetation. Study of variation in the large canine teeth showed that there was sexual dimorphism between males and females (Sullivan *et al.*, 2003; **184**).

Dicynodont diversity crashed at the end-Permian mass extinction, and in the Triassic only two dicynodonts, *Lystrosaurus* and *Myosaurus*, are found in South Africa. *Lystrosaurus* occurs nearly world-wide, and was originally considered to herald the start of the Triassic, but *Lystrosaurus* has been found in the late Permian at Lootsberg Pass (Smith & Ward, 2001) and thus it is the only dicynodont known to have survived the mass extinction. Dicynodonts underwent a second radiation during the middle to late Triassic, with taxa such as *Kannemeyeria* becoming relatively common. Gorgonopsians (**185**) were the dominant large carnivores of the late Permian and are abundantly

186 Two skeletons of the cynodont *Thrinaxodon* curled up in a burrow (SAM). Skulls about 50 mm (c. 2 in) long.

187 The cynodont *Galesaurus*; the differential rib morphology could be evidence that the animal had a mammal-like, muscular diaphragm (SAM). Skull about 80 mm (c. 3 in) long.

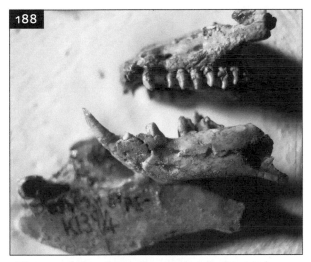

188 Jaws of the advanced cynodont *Pachygenelus* (SAM). Lower jaw length 16 mm (c. 0.6 in).

preserved in the lower Beaufort Group, e.g. the Teekloof Formation, but they became extinct at the end of the Permian. The first therocephalians are found with the primitive therapsids in the lowermost Beaufort Group. Diversity of therocephalians increased through the late Permian, by which time they were all carnivorous. In contrast, Triassic bauriid therocephalians were herbivorous. Advanced therocephalians parallel the development of several cynodont features, but became extinct by middle Triassic times (Rubidge, 2005).

The most advanced mammal-like reptiles are the cynodonts, which are considered to have given rise to the mammals. The oldest cynodont known is *Charassognathus* from the late Permian Teekloof Formation (Botha *et al.*, 2007). *Thrinaxodon* (**186**) and *Galesaurus* (**187**) are well-known representatives from the early Triassic, which belong to groups thought to have crossed the P-T boundary (Sidor & Smith, 2004), and a number of genera, such as *Cynognathus*, *Diademodon*, and *Trirachodon*, are known from the middle Triassic Burgersdorp Formation. *Trirachodon* and *Thrinaxodon* burrows indicate that cynodont burrows have a design different from those of earlier synapsids. The structure of the burrow suggests that *Thrinaxodon* adopted a mammalian posture inside its burrow (Damiani *et al.*, 2003).

Mammals. There are two mammals known from the Upper Elliot Formation: *Erythrotherium* and *Megazostrodon*. A number of advanced cynodonts, such as *Tritylodon* and *Pachygenelus* (**188**), occur together with these, but there is still debate as to which group of cynodonts the mammals evolved from.

PALAEOECOLOGY OF THE KAROO BIOTA

The Teekloof Formation

A reconstruction of the palaeoenvironment and ecosystem of the Teekloof Formation can be made following the work of Smith (1993a,b) and others. The sedimentary environment is a river system and its flood plain (**189**). The major rivers were Mississippi-sized, draining into a shallow lake, and annual snowmelt in the Gondwanide Mountains to the south probably maintained year-round flow in the main channels. Elsewhere, however, there were likely to have been ephemeral streams and playa lakes. A semi-arid climate is confirmed by gypsum deposits and calcretes. Evidence from fossil soils suggests that riparian vegetation flourished along the river banks and abandoned channels, and consisted of *Glossopteris* trees, horsetails, ferns, and clubmosses. Out on the flood plains vegetation was sparser and probably consisted mainly of semi-arid adapted shrubs and cyanobacterial mats (**190**).

Far and away the commonest fossil collected in the Teekloof Formation is the herbivorous dicynodont *Diictodon* (**184**). Collections made in the Teekloof Pass area in the 1980s show that this animal makes up 69% of the fossil finds, followed by other dicynodonts (27%) and the carnivorous gorgonopsians, therocephalians, and burnetiamorphs making up a meagre 4% of the vertebrate fauna (Sidor & Smith, 2007). Calculations based on museum records confirm that the herbivore: carnivore ratio in the Teekloof is about 24:1 (Nicolas & Rubidge, 2010). Herbivores generally outnumber carnivores in typical food chains. *Diictodon* was a common animal in this ecosystem, nipping the vegetation with its horny beak. Most likely they, like their food, lived close to the water courses. The burrows containing *Diictodon* are not sufficiently common to be considered normal dwelling burrows, and more likely represent aestivation structures dug during the dry season (Smith, 1987). Invertebrates are represented only by their traces, and we can only glimpse the occasional crayfish and snail moving across the wet, muddy surface (**178**).

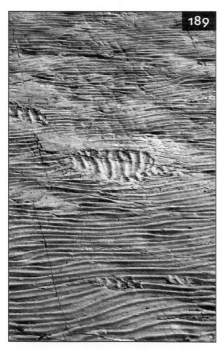

189 Patches of wave ripples interfering with current ripples in the Teekloof Formation sandstone at Putfontein, Beaufort West, Western Cape, South Africa. This is evidence for shallowing water following a flood on the flood plain.

190 Wrinkled sandstone surface caused by microbial mat, Teekloof Formation sandstone at Putfontein, Beaufort West, Western Cape, South Africa. Hand lens is 30 mm (c. 1.2 in) long.

191 Bone bed accumulation of *Lystrosaurus*. Note the rack of ribs, centre top, which indicates desiccation of the carcass during drought conditions (SAM). Width of picture about 700 mm (c. 28 in).

The Balfour Formation

Smith and Ward (2001) interpreted the results of concerted collecting of fossils around the P-T boundary as showing the mass extinction of vertebrates occurring within a single sedimentary facies, overlain by a distinctly lifeless bed immediately above the last occurrence of *Dicynodon* (**172**). The sequence represents a rapid transition from meandering to braided rivers in response to the climate becoming arid. Evidence for drought conditions includes bone beds consisting of *Lystrosaurus* with clues such as racks of ribs which had been held together by mummified skin rather than scattered by erosion or scavengers (**191**). The extinction occurred during a period of environmental change from wet flood plain to dry flood plain environments, which caused extinction of the Permian fauna and its replacement by drought-tolerant taxa in the early Triassic. Ward *et al.* (2005) showed that the extinction of vertebrates started gradually in the late Permian, a number of species had become extinct in the grey mudrocks of the lower part of the Palingkloof Member below the P-T boundary, but there was nevertheless a markedly higher level of extinction at the boundary. Moreover, members of the typical Triassic fauna were already appearing in the late Permian (*Lystrosaurus*), and an enhanced origination also typified the P-T boundary. Thus, the extinction reflects fairly rapid environmental change, but not a severe and instantaneous event as the end-Cretaceous bolide impact (p. 219). Nevertheless,

some 69% of late Permian terrestrial vertebrates died out in a period of about 300,000 years preceding the P-T boundary, and a lesser extinction wiped out 31% of the survivor taxa about 160,000 years later (Smith & Botha, 2005). The second extinction coincides with the base of the Katberg Formation, and also the disappearance of most Late Permian gymnosperm pollen in a layer containing solely the abundant remains of fungi. This fungal spike apparently represents widespread devastation of the world's forests and the subsequent bloom of fungal decomposers (Steiner *et al.*, 2003).

Three faunas can be distinguished in the Palingkloof Member of the Balfour Formation: an extinction fauna, consisting of those animals which died out before or at the P-T boundary; a survivor fauna, being those animals which survived the main extinction but died out soon after; and a recovery fauna, consisting of those animals which originated in the early Triassic recovery phase and persisted later into the Triassic (Smith & Botha, 2005). The extinction fauna consisted essentially of those animals we saw in the Teekloof Formation: common, small, herbivorous dicynodonts such as *Dicynodon* and *Diictodon*, which cropped riverside vegetation with their horny beaks, and were preyed upon by rather rare, larger carnivores such as *Rubidgea*. Only four genera out of 13 in the late Permian survived the extinction: the dicynodont *Lystrosaurus* (represented by only one species: *L. curvatus*) and the therocephalians *Tetracynodon*,

192 The skull of the cynodont
Progalesaurus lootsbergensis (SAM).
Length 98 mm (c. 4 in).

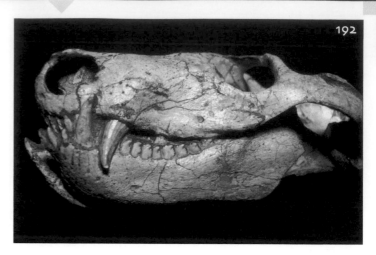

Moschorhinus, and *Ictidosuchoides*. None of the thero-cephalian genera survived into the succeeding early Triassic Katberg Formation because they died out in the upper Palingkloof and are the survivor fauna. The end-Permian extinction clearly affected the dicynodonts most, whereas the carnivorous gorgonopsians were already well in decline before the end of the Permian (Smith & Botha, 2005).

The following early Triassic recovery fauna comprises proterosuchian archosauromorphs, small amphibians, small procolophonoids (e.g. '*Owenetta*'), medium-sized dicynodonts (e.g. *Lystrosaurus*), and small carnivorous cynodonts (e.g. *Progalesaurus*, **192**, *Galesaurus*, **187**, and *Thrinaxodon*, **186**) (Smith & Botha, 2005). These animals had adaptations that enabled them to survive arid conditions. For example, many aspects of the skull of *Lystrosaurus* show that it is adapted to eat tough vegetation, and palaeobotanical evidence shows that Isoetales (quillworts) and Equisetales (horsetails) survived the extinction (Retallack *et al.*, 2003). Moreover, *Lystrosaurus* and the cynodonts *Thrinaxodon*, *Galesaurus*, and *Progalesaurus* were burrowers, either for dwelling or aestivation, which would help them survive the periodic droughts in the new, more arid environment. Shelter sharing by different taxa has been reported, and considered to be an activity brought on by adverse environmental conditions. In one instance, the meerkat-sized cynodont *Galesaurus* was found sharing its burrow with the gecko-like procolophonoid '*Owenetta*' and a large millipede (Abdala *et al.*, 2006).

The Molteno Formation

The environment of deposition of the Molteno Formation was that of a sand-dominated braidplain which prograded north from the newly uplifted source area. The Molteno flora can be classified into a number of distinct habitats, including two types of *Dicroidium* riparian forest, *Dicroidium* woodland, *Sphenobaiera* woodland, *Heidiphyllum* thicket, *Equisetum* marsh, and fern/*Ginkgophytopsis* meadow (Cairncross *et al.*, 1995).

The riparian forest types occupied the river-sides: a multi-storeyed climax forest bordered mature river channels and abandoned channels, it had a medium-diversity insect fauna, dominated in approximately equal numbers by roachoids and beetles, with a minor component of bugs, and other orders very scarce or absent. The spider occurs in this habitat. A second type of *Dicroidium* forest was an immature, single-storey, medium-diversity association lining shifting braided channels of the flood plain. It had a medium-diversity insect fauna, also dominated by roachoids and beetles, but with a much greater representation of other orders, including bugs (Hemiptera), dragonflies (Odonata), Paraple-coptera (extinct), Orthoptera (grasshoppers, crickets), and Mecoptera (scorpionflies). *Dicroidium* woodland was a low–medium-diversity woodland of open flood plains. It had a medium-diversity insect fauna dominated by bugs and beetles, with roachoids less common, and with other orders rare.

Sphenobaiera woodland was a medium-diversity woodland association lining lakes on the flood plain. It had a high-diversity insect fauna with many orders represented, roachoids being dominant in numbers, beetles, bugs, and dragonflies also common, and mayflies (Ephemeroptera), Protodonata, Paraplecoptera, and Plecoptera (stoneflies), like Odonata, all with aquatic nymphs and flying adults associated with water bodies, well represented. The Birds River locality represents this habitat.

Heidiphyllum thicket consists of shrubby coniferous stands of *Heidiphyllum* associated with areas of higher water table in the flood plain, or on sand bars in the river channels. It had a low–medium-diversity insect fauna dominated by roachoids, with less common beetles and bugs; mantids are commoner in this habitat than in other associations, but the remaining orders are rare. *Equisetum* marsh consists of monospecific stands of horsetail in flood plain marshes, lake margins, and river sand bars. It had a very low-diversity insect fauna characterized by beetles and bugs, and with the otherwise ubiquitous cockroaches conspicuously rare or absent. The fern/*Ginkgophytopsis* meadow is a low-diversity herbaceous association found colonizing the sand bars of braided rivers, and no insects have been found in this habitat.

Herbivorous insects dominated over carnivores, in both abundance and diversity. Insect damage to leaves is widespread in the Molteno Formation collections on many plant species at numerous localities. The damage includes feeding traces (predominantly continuous marginal traces), leaf mines (including linear and possible blotch varieties), and probable leaf galls. Damaged taxa include a wide variety of gymnosperms including the conifer *Heidiphyllum*, the ginkgoaleans *Ginkgo* and *Sphenobaiera*, the seed-ferns *Dicroidium* and *Dejerseya*, the pentoxyalean *Taeniopteris*, and the gnetopsid *Yabeiella* (Scott *et al.*, 2004).

COMPARISON OF THE KAROO WITH OTHER PERMO–TRIASSIC LAGERSTÄTTE

No other Fossil-Lagerstätte can yet rival the Karoo for its preservation of evidence of the evolution of mammal-like reptiles or terrestrial life in Gondwana. However, recent studies in Antarctica and China have provided exciting new insights into Permian–Triassic vertebrates (Liu *et al.*, 2009) and the end-Permian extinction (Retallack *et al.*, 2006).

MUSEUMS AND SITE VISITS

Museums
1. Iziko Museums of Cape Town, 25 Queen Victoria Street, Cape Town, South Africa.
2. Bloemfontein National Museum, 36 Aliwal Street, Bloemfontein, South Africa.
3. Bernard Price Institute for Palaeontology, University of the Witwatersrand, Johannesburg, South Africa.
4. Rubidge Collection, Wellwood Farm, Graaff-Reinet, South Africa.
5. National Botanical Institute, 2 Cussonia Road, Pretoria, South Africa.

Sites
Among the best place to see fossils in the field is to visit the Karoo National Park, Beaufort West, which has a short (400 m [c. 440 yd]) Fossil Trail with some fossils on display, set in magnificent Karoo scenery. The Kitching Fossil Exploration Centre is situated next to the famous Owl House in Nieu Bethesda; half-hour guided tours take you onto the Gats river bed where you can see fossils exposed in the rock. Best of all, stay with Marion and Robert Rubidge at Wellwood's guest accommodation, see the Rubidge Collection, and visit Nieu Bethesda, which is very close by. Also nearby is Lootsberg Pass (**171**) and the historic town of Graaff-Reinet.

Chapter Ten

GRÈS À VOLTZIA

BACKGROUND: THE PERMO–TRIASSIC TRANSITION

In the last chapter we witnessed the end of both the Permian Period and the Palaeozoic Era, which was defined by the sudden change in fauna and flora that resulted from a major mass extinction event – the greatest the Earth has ever witnessed. As we now enter the Triassic Period, and the Mesozoic Era, life on Earth is quite different from what it was before. Trilobites, eurypterids, and graptolites are extinct; bivalved molluscs, not brachiopods, are now the dominant shelled animals on the sea floor; there are new types of corals forming reefs; and gymnospermous trees now dominate the flora on land.

In this chapter the famous Lagerstätte of Grès à Voltzia, in the northern Vosges mountains of north-eastern France (**193**), is described. As mentioned in Chapter 8, there are certain similarities with the Late Carboniferous Mazon Creek Lagerstätte (the Grès à Voltzia was laid down in a deltaic environment, for example), but there are preservational differences and, of course, the living biota of the Triassic was different from that of the Carboniferous.

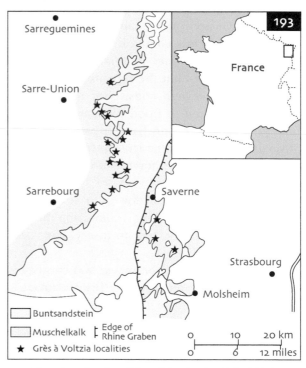

193 Map showing fossil localities in the Grès à Voltzia sandstone in the northern Vosges Mountains, northwest of Strasbourg, France (after Gall, 1985).

194 The magnificent Gothic Cathedral of Notre-Dame in Strasbourg, Alsace, France, built from Grès à Voltzia sandstone from the northern Vosges mountains.

HISTORY OF DISCOVERY OF THE GRÈS À VOLTZIA FOSSILS

Triassic sandstone has been quarried in the northern Vosges for centuries, for building (e.g. the magnificent Gothic cathedral in Strasbourg, **194**) and for millstones. Associated with the sandstones (the Grès à meules or millstone grit) are clayrocks, which are a nuisance to the quarrymen, but are rich in fossils. In Britain, Triassic sandstones are also extensively quarried as building stones, but few make good freestones (stones which lack a prominent 'grain' and so can be carved in any direction into intricate decorative tracery). Moreover, the clay wayboards (lenses of clay and silt intercalated in the sandstones) in the British Triassic are strongly oxidized so preserve fossils poorly, while those in the Vosges are not oxidized and special conditions prevailed which allowed fine preservation. Systematic collection of the fossils began in the middle of the twentieth century by Louis Grauvogel of the Louis Pasteur University, Strasbourg. Léa Grauvogel-Stamm, Louis's daughter, joined in the research with her interest in palaeobotany, and in 1971 Jean-Claude Gall became another member of the research team on the Grès à Voltzia biota and its palaeoecology, which continues today.

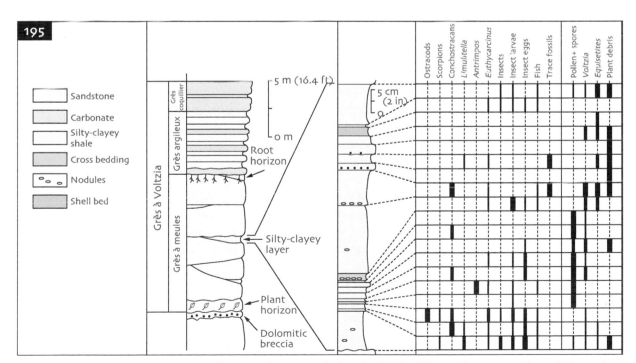

195 Stratigraphic log of the Grès à Voltzia Sandstone, with details of a fossiliferous horizon in the Grès à meules and relative abundances of the biota (after Gall, 1971).

STRATIGRAPHIC SETTING AND TAPHONOMY OF GRÈS À VOLTZIA

The Triassic Period was named by von Alberti in 1834 because of its three-fold division in Central Europe into a lower Buntsandstein (sandstone), a middle Muschelkalk ('mussel-chalk' – a marine limestone), and an upper Keuper Sandstein (sandstone). In Britain the Muschelkalk is missing but the name Triassic has been accepted into international stratigraphical usage. The Grès à Voltzia is in the upper part of the Buntsandstein (c. 243 Ma). The lower unit of the Grès à Voltzia, the Grès à meules, is a fine sandstone, while the upper unit, the Grès argileux, is silty and marks the beginning of the marine transgression as the Muschelkalk sea spread across the continent. In this chapter, we are concerned only with the Grès à meules, in which three facies have been recognized (Gall, 1971, 1983, 1985; **195**): a) thick lenses of fine-grained sandstone, sometimes grey or pink but most often multicoloured, containing land-plant debris and amphibian bone fragments; b) green or red silt/clay lenses, generally composed of a succession of laminae each a few millimetres thick, with well-preserved fossils of aquatic and terrestrial organisms; and c) beds of calcareous sandstone (or carbonate breccia) with a sparse marine fauna. It is in the greenish (rarely red) silt/clay laminites (b) that the beautifully preserved biota occurs.

Evidence from the sediments and fossils points to a deltaic sedimentary environment (Gall, 1971, 1983). The sandstones represent point bars deposited in strongly sinuous channels, the clay lenses represent the settling of fine material in brackish ponds, and the calcareous sandstone results from brief incursions of seawater during storms. The palaeogeographical position, the red beds, and the xeromorphic nature of the land flora together suggest a semi-arid climate, although the low-lying situation suggests that the delta itself was not too arid. The climate was probably seasonal, with the pools filling during the wet season and evaporating in the dry season. The presence of desiccation cracks, reptile footprints, salt pseudomorphs, and land plants in life position at the top of each clay lens indicates complete drying-out of the pools. Moving upwards through each clay lens a transition from aquatic to terrestrial biota is also observable (Gall, 1983).

Drying-up of the pools led to the death of the aquatic fauna. The abundance of conchostracans is significant in that these crustaceans are adapted to swift completion of their life cycle in temporary water bodies. Regular high evaporation rates of the water bodies also favoured deoxygenation, consequent mass mortality of the aquatic fauna, and the rapid proliferation of microbial films. It seems that microbial films ('veils', Gall, 1990) shielded the carcasses from scavenging activity and created, by the production of mucus, a closed environment which inhibited the decomposition of organic material. Later deposition of a new detrital load (clay, silt) buried the microbial films and the organisms (Gall, 1990).

Sediment compaction resulted in some squashing of the fossils, but in others there is three-dimensional preservation by casting with calcium phosphate, which is a rare casting material in invertebrates. While it is present in organic tissues, when liberated it is swiftly recycled by other organisms. However, because of the exceptional taphonomic conditions present in the Grès à meules, rapid phosphatization occurred. Phosphatization requires a low-oxygen environment and abundant organic matter. The microbial film would have sealed the phosphates being released from the organic matter in the decaying animals, thus preventing its re-use by other organisms. Acidic conditions produced by decaying organic matter would have released free calcium ions, which combined with the phosphate to form apatite. Once a phosphatic nodule had formed, it would have prevented further flattening of the body during sediment compaction.

Thus, the taphonomy of the Grès à Voltzia is quite different from those of other Fossil-Lagerstätten. While there are similarities – for example, in the rapid burial, reducing conditions, and fine sediment – between Grès à Voltzia and Mazon Creek, no microbial mat or phosphatization has been reported for the latter. Phosphatization occurs in many Lagerstätten, e.g. Santana (Chapter 16), but the process is different to that seen in the Grès à Voltzia.

196 The conifer *Voltzia heterophylla* (GGUS). Scale bar is 10 mm (c. 0.4 in).

197 The gymnosperm *Albertia* (GGUS). Width of specimen 25 cm (c. 9.8 in).

198 The fern *Anomopteris* (GGUS). Frond is about 140 mm (c. 5.5 in) across.

DESCRIPTION OF THE GRÈS À VOLTZIA BIOTA

Plants. The Grès à Voltzia is named for the abundant remains of the conifer *Voltzia heterophylla* (**196**). This is a bushy conifer which formed thickets between the delta distributaries together with other gymnosperms such as *Albertia* (**197**), *Aethophyllum*, and *Yuccites*. On the banks of the waterways were probably thick stands of vegetation which was adapted to rooting in shifting sand substrates and frequent flooding, such as horsetails (*Equisetum*, *Schizoneura*). Ferns such as *Anomopteris* (**198**) and *Neuropteridium* were also present, and cycads and ginkgos have also been reported. Many of the macroplant remains occur in the coloured sandstone lenses (facies (a), above) together with amphibian remains, although drifted plant debris also occurs in the silt-clay laminites.

Cnidarians. A medusoid, *Progonionemus vogesiacus*, is known from 10 juvenile to adult specimens (Grauvogel & Gall, 1962). It consists of a bell-shaped umbrella about 8–40 mm (c. 0.3–1.6 in) across, and many tentacles of 9–40 mm (c. 0.4–1.6 in) in length. Both primary and secondary tentacles can be recognized, and gonads can be seen in the adult forms. *Progonionemus* was identified as close to *Gonionemus*, belonging to the group Limnomedusae, which includes fresh- and brackish-water forms.

Brachiopods. The inarticulate brachiopod *Lingula tenuissima* occurs in the Grès à meules. *Lingula* is a burrowing brachiopod which is generally found in shallow, brackish-water settings that most articulate brachiopods cannot tolerate. Most interestingly, *Lingula* is preserved in its upright, life position, indicating that it lived exactly where it died (autochthonous) rather than being drifted in by currents.

Annelids. A few specimens of annelids have been described by Gall and Grauvogel (1967), including the polychaetes *Eunicites* and *Homaphrodite*. These marine worms also occur in waters of variable salinity.

Molluscs. A variety of bivalves and gastropods has been recorded from the Grès à meules, including pectens (free-swimming forms) and the trigonioid *Myophoria*. In general, these animals are rare in the brackish-water pools, but indicate marine water nearby in time and space.

Arthropods. These are the most abundant animals found in the laminites and, with their brown cuticles (probably not the original coloration), their remains look remarkably fresh. The commonest arthropods are those associated with water sometime in their life cycle. The horseshoe crab (limulid) *Limulitella bronni* (**199**) is a common component of the fauna of the Grès à meules; its trackways (*Kouphichnium*) occur as well as remains of the dead animals and moulted exoskeletons. Horseshoe crabs are predominantly marine, but can be found a great distance up rivers, especially when they come ashore in hordes to mate, and in late Carboniferous times (e.g. Mazon Creek, Chapter 8) they appear to have tolerated both fresh and saline water, and were possibly amphibious.

Horseshoe crabs are the only primarily aquatic members of the arthropod subphylum Chelicerata that are still alive today. In the middle of the Palaeozoic Era another chelicerate group, the scorpions, were also living in water, but by Carboniferous times they had left the water and become terrestrial. Scorpions occur in the Grès à meules (**200**) and form part of the terrestrial fauna. Three genera have been described, in two new families (Lourenço & Gall, 2004). One of these fossil families is in the modern superfamily Buthoidea; the other fossil family was tentatively placed in the superfamily Mesophonoidea and exhibits several features inherited from Palaeozoic lineages. They give evidence for the recovery of terrestrial ecosystems severely reduced during the

199 The horseshoe crab *Limulitella bronni* (GGUS). Length of animal (including tail spine) is 55 mm (c. 2.2 in).

200 The scorpion *Gallioscorpio voltzi* from the Grès à meules (GGUS). Length of body is 60 mm (c. 2.4 in).

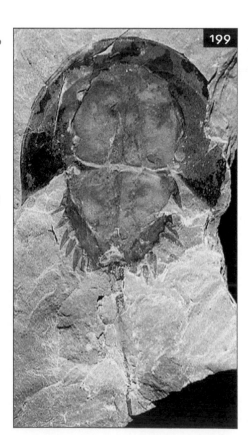

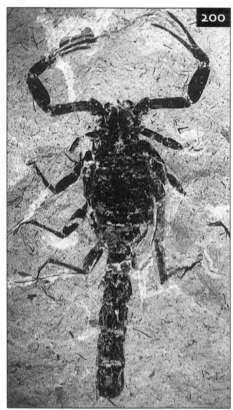

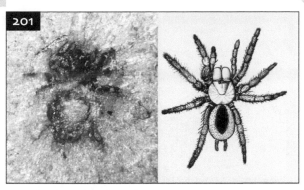

201 The funnel-web spider *Rosamygale grauvogeli*. Photograph (left) and reconstruction (right) (GGUS). Length about 6 mm (c. 0.2 in) (after Selden & Gall, 1992).

late Permian mass extinction. Another, more familiar, group of Chelicerata is the Araneae: spiders (**201**). About a dozen specimens are known from the Grès à meules, and these were described by Selden and Gall (1992) as the oldest representative of the mygalomorph spiders (trapdoor, funnel-web, and tarantula spiders). The single species, *Rosamygale grauvogeli*, was placed in the modern family Hexathelidac, a mainly Gondwanan family today, although some species occur in the Mediterranean region.

Among Crustacea the notostracan *Triops cancriformis* occurs in the Grès à meules. Today, these animals are characteristic of ephemeral pools, and so their presence in the temporary water bodies of the Grès à Voltzia environment is to be expected. Spinicaudata (conchostracans) are also common, and these small, bivalved crustaceans are frequently found in lacustrine and other non-marine environments. Other crustaceans found in the Grès à meules are the mysid shrimp *Schimperella*, the isopod *Palaega pumila* (Gall & Grauvogel, 1971), and some crayfish: *Antrimpos* (**202**), a predominantly nektonic form, and the benthic *Clytiopsis*. These crustaceans are a common component of the fauna of the temporary pools, and many were originally described by Bill (1914).

A group of rather strange crustaceans, now extinct, is the Cycloidea. These range from the early Carboniferous to the Cretaceous, and are found worldwide. The genus *Halicyne* (**203**) occurs in the Grès à meules. Cycloids resemble modern fish-lice and, indeed, are thought by some palaeontologists to be just that, i.e. branchiuran crustaceans, while there have been many other suggestions, ranging from trilobites and xiphosurans through numerous crustacean orders, in the past. They are characterized by a nearly circular carapace with only small legs protruding from beneath the rim. *Halicyne* from the Grès à Voltzia is particularly well preserved with its appendages. Similar forms have come to light from the Triassic of Poland and Madagascar (Dzik, 2008), which allowed for a new investigation of the group. This study confirmed their identity as branchiurans.

Myriapods are represented by some rather beautifully preserved specimens of the millipede *Hannibaliulus wilsonae* (**204**). Millipedes are important feeders on detrital plant material at the present day. *Hannibaliulus* has around 44 diplosegments, but they are females so cannot be easily identified to order. Nevertheless, it appears to belong to the order Callipodida (Shear *et al.*, 2009).

A group of arthropods which ranged from possibly as old as the Cambrian (as evidenced by possible trackways) to the Triassic Period and which, though crustacean-like in appearance, have some affinities with insects, is the Euthycarcinoidea (see **135**). These animals have a head with antennae, a long body with many legs, a short 'abdomen' and a spine-like tail. A few specimens of *Euthycarcinus* are known from the Grès à meules. These were the first euthycarcinoids to be described, and are among the youngest known examples of the group (Wilson & Almond, 2001).

Insects in the Grès à meules include both aquatic and terrestrial forms. Many terrestrial insects have aquatic larvae, and these are well represented in the beds (**205, 206**). Their presence indicates fresh, rather than brackish, water. The insects present include: Ephemeroptera (mayflies), Odonata (dragonflies, including a large species), Blattodea (roachoids), Coleoptera (beetles), Mecoptera (scorpion flies), Diptera (true flies), and Hemiptera (bugs).

In addition to the adults and larvae of insects and crustaceans, masses of eggs of these arthropods have been found in the Grès à meules. Gall and Grauvogel (1966) described a variety of different types of egg (*Monilipartus*, *Clavapartus*,

203 The cycloid crustacean *Halicyne ornata* (GGUS). Width of body 15 mm (c. 0.6 in).

202 The crayfish *Antrimpos* (GGUS). Length (excluding antennae) 50 mm (c. 2 in).

204 The millipede *Hannibaliulus wilsonae* (GGUS). Length about 50 mm (c. 2 in).

205 Insect larva (GGUS). Body length about 17 mm (c. 0.7 in).

206 Insect larva (GGUS). Length (including tails) about 15 mm (c. 0.6 in).

Furcapartus) which seem most similar to the eggs of chironomid midges. These eggs look like small, dark, circular or oval spots and occur in chains (*Monilipartus*) or discrete packages (*Clavapartus*, *Furcapartus*) as if they were enclosed in mucus in life. Some eggs also occur within or associated with the shells of conchostracans and so could represent the eggs of these animals.

Fish. A few fish occur in the Grès à meules; juveniles are especially abundant. Actinopterygians are represented by a single specimen of *Saurichthys*, many specimens (mainly juveniles) of *Dipteronotus* (**207**), and a few dozen examples of the holostean *Pericentrophorus*. Coelacanth scales also occur. All of these fish are nektonic (i.e. swimming rather than bottom-living) forms. The shark egg-case *Palaeoxyris*, which also occurs at Mazon Creek (Chapter 8), is common in the Grès à meules, as are clutches of eggs of other fish.

Tetrapods. Amphibian remains are quite common in the sandstone facies. They have been attributed to the capitosauroid temnospondyls *Odontosaurus* and *Eocyclotosaurus*. These were aquatic animals with flattened, triangular heads, and the group was entirely Triassic. Reptile remains are rare in the Grès à Voltzia, but *Chirotherium* trackways, which are common in the Triassic of Britain, provide evidence for the presence of fairly large animals.

PALAEOECOLOGY OF THE GRÈS À VOLTZIA BIOTA

Much of the biota of the Grès à Voltzia has yet to be described systematically but, in contrast, the palaeoecology of the Lagerstätte has been well studied and the whole scenario is well worked out, particularly by many years of research by Jean-Claude Gall of the University of Strasbourg (Gall, 1971). The interdigitation of the three facies in the lower Grès à Voltzia indicates deposition in a complex deltaic setting adjacent to the sea. Following the facies changes upwards, we witness the gradual change from an alluvial plain to a delta, and finally a marine transgression by the Muschelkalk Sea.

Facies (a) of the Grès à meules represents fluvial distributary channels on the delta top. Lenses a few metres thick of fine-grained pink or grey sandstone with an erosive base represent channel bars, and coarser-grained, poorly sorted sandstones with plant and amphibian debris and mud-flake conglomerates are interpreted as crevasse splay deposits. Proximal sandstones contain coarser bone and vegetation debris than do distal.

Facies (b) provides evidence of temporary pools. Lenses of green or red silty clay occur as intercalations within the sandstones. They consist of a series of graded laminations, each layer usually only a few millimetres thick. The sediment was deposited by flood water spilling over from river channels or as

207 Juvenile fish *Dipteronotus* (GGUS). Length 35 mm (c. 1.4 in).

a result of exceptionally high tides. The occurrence of salt pseudomorphs and mud cracks is evidence for desiccation (i.e. periodic emergence); increased salinity is reflected in the high boron content of the clay minerals (Gall, 1985). Individual pools probably existed for only a short period of time – a few weeks to several seasons. One 600 mm (c. 2 ft) lens preserves what appears to be a single annual cycle in gymnosperms, from inflorescence to seed (Gall, 1971). The tops of the fine-grained lenses usually show signs of emergence, such as mud cracks and plant-root traces (Gall, 1971, 1985). Aquatic organisms are concentrated near the bases of the lenses, which presumably reflects periods when the water was deeper and more permanent. Dramatic changes in the proportions of different faunal elements from lamina to lamina within the lenses (Gall, 1971) indicate that a new assemblage was established with each initial influx of water (and accompanying sediment). It is clear that the aquatic fauna is autochthonous and that the terrestrial biota was derived from the immediate vicinity of the stagnating pools. The evidence includes the preservation of organisms in life position (e.g. *Lingula*), the absence of any current directional orientation, the presence of both larval and adult forms in the same assemblage, and the concentration of organisms in former residual ponds. The fauna, which was dominated by euryhaline forms (those tolerant of a wide range of salinities), was subsequently killed (sometimes with clear evidence of mass mortality), probably by shifts in oxygen levels associated with a decrease in the size of the water body through evaporation.

Facies (c), calcareous sandstones or brecciated carbonate, represents temporary marine incursions. The carbonate breccias are storm deposits and the rarer lenses of calcareous sandstone containing gastropods suggest occasional, minor marine transgressions.

The palaeoecological scenario that can be envisaged for the Grès à Voltzia delta is one of low-lying sandy substrates, with thickets of gymnosperms (e.g. *Voltzia, Yuccites*) and lycopods (e.g. *Pleuromeia*), with horsetails and ferns (e.g. *Neuropteridium, Anomopteris*) between. The vegetation appears to have been restricted to just a few main species: those which were able to root in relatively unstable sand and could survive (or quickly recolonize after) flooding. Among this vegetation there was a similarly species-poor fauna of insects (mostly those with aquatic larvae), millipedes, scorpions, spiders (*Rosamygale*), amphibians, and reptiles. In the brackish-water pools there was an abundance of life, but each taxon exhibited low species diversity and many animals were euryhaline forms. The fauna included medusae (*Progonionemus*), polychaete annelids, *Lingula*, some bivalves, *Limulitella*, crustaceans (e.g. *Triops*, conchostracans, *Schimperella, Antrimpos, Clytiopsis, Halicyne*), euthycarcinoids, fish (e.g. *Dipteronotus*), and insect eggs (e.g. *Monilipartus*) and larvae. Gall (1985, Figures 8, 9) illustrated reconstructions of the palaeoecology of the Grès à meules.

Low diversity is characteristic of both semi-arid terrestrial and brackish-water communities, and it would appear that, thanks to the exceptional preservation of soft-bodied organisms, there is little biota absent from the reconstruction. Moreover, the low diversity, sandy substrate, semi-arid (seasonal) ecosystem was well established by the Triassic Period and continues, albeit with somewhat different taxic composition, today.

COMPARISON OF GRÈS À VOLTZIA WITH OTHER BIOTAS

Briggs and Gall (1990) made a quantitative comparison of the Grès à Voltzia biota with four major Carboniferous Lagerstätten, to test and extend Schram's (1979) concept of a faunal continuum among these Carboniferous Lagerstätten. A particularly important aspect of this work was the realization that not only should comparison be made between the taxic composition of the biotas but also the stratigraphic position (especially when comparing biotas either side of an extinction event), palaeoenvironmental differences, and preservational effects should be evaluated. Using a new coefficient of similarity, these authors compared the Grès à Voltzia with the Namurian Bear Gulch Member of the Heath Formation of Montana, the Stephanian Montceau-les-Mines Lagerstätte of France, the Lower Carboniferous Glencartholm Volcanic Beds of Scotland, and the Westphalian Mazon Creek biota of Illinois. The results showed that the Grès à Voltzia was much closer to Mazon Creek than to the other biotas, which ranked (in decreasing similarity): Glencartholm, Bear Gulch, and Montceau-les-Mines.

Stratigraphic age clearly had little effect on the

results: Glencartholm is the furthest, stratigraphically, from the Grès à Voltzia and yet came second-most similar, while the youngest biota (and closest geographically), Montceau-les-Mines, was least similar. Taphonomy was important in that with better preservation more taxa are preserved. So, for example, the taphonomy of Glencartholm is not good enough to preserve organims without some sort of cuticle, and medusae, polychaetes, and many terrestrial forms, such as spiders and insects, are absent from this Lagerstätte, even though they were probably around at the time of deposition. The factor which came out as most influential in determining similarity was palaeo-environment. Sedimentological evidence from both Mazon Creek (Chapter 8) and Grès à Voltzia suggests palaeoenvironmental settings transitional between terrestrial and a marine-influenced delta. Taxonomic groups common to both Lagerstätten include medusoids, brachiopods, polychaete annelids, bivalve and gastropod molluscs, horse-shoe crabs, scorpions, spiders, ostracods, shrimps, cycloids, euthycarcinoids, millipedes, insects, fish, and tetrapods. Organisms adapted to fluctuating conditions (e.g. euryhaline species, plants adapted to unstable substrates) show strong congruence at the family and lower levels between the Lagerstätten, and these groups were little affected by the end-Permian extinction event. The main taxonomic differences between the Grès à Voltzia and Mazon Creek lie in the higher crustaceans and the insects. Many of those represented in the Grès à Voltzia appeared in the Permian and radiated across the Permo–Triassic boundary as Palaeozoic forms became extinct. Thus, Briggs and Gall (1990) concluded that there is a striking continuity between the biotas of Carboniferous and Triassic transitional sedimentary environments.

Other Triassic biotas include the Molteno Formation of late Triassic age (see Chapter 9, Karoo), which is dominated by Gondwanan floras and insects, and the almost coeval Solite Lager-stätte from eastern North America (Fraser *et al.*, 1996). A slightly older biota which is rather similar in some respects to the Grès à Voltzia in its constituent biota (*Voltzia* occurs there, for example), though rather richer in insects, is known from Madygen, Kyrgyzstan (Shcherbakov, 2008). The palaeoenvironment of Madygen was recon-structed as an intermontane river valley with a seasonally arid climate, with oxbow lakes and ephemeral ponds on the flood plain in which the biota is mainly preserved. The famous Lagerstätte of Monte San Giorgio, Ticino, Switzerland, has a quite different setting to the Grès à Voltzia. It records life in a tropical lagoon environment, sheltered and partially separated from the open sea by an offshore reef, so is more comparable in this respect to Solnhofen (Chapter 13). Monte San Giorgio preserves diverse marine life, including reptiles, fish, bivalves, ammonites, echinoderms, and crustaceans, and also some terrestrial fossils including reptiles, insects, and plants.

Biotic recovery from the Permo–Triassic mass extinction was discussed in relation to the Karoo therapsids in Chapter 9. It has been estimated that global recovery from the extinction event lasted a long time, some 8–10 Ma at least, and repopulation of ecosystems proceeded from refugia. The Grès à Voltzia may have acted as a refugium (Gall & Grauvogel-Stamm, 2005). Three biotas can be identified: 1) Palaeozoic survivors of crustaceans, amphibians, insects, and plants similar to those found at Mazon Creek (Chapter 8), as well as the long-ranging 'living fossils' *Lingula* and *Triops*, which are still alive today; 2) a recovery biota, akin to the Modern Fauna of Sepkoski (1984), con-sisting of new crustacean, arachnid, and insect groups; and 3) pioneering species ('weeds') which rapidly invaded the disturbed ecospaces, such as the herbaceous conifer *Aethophyllum* and the lycopsid *Pleuromeia* (Grauvogel-Stamm & Ash, 2005).

MUSEUMS AND SITE VISITS

Museums
The Grès à Voltzia fossils are held in the Université Louis Pasteur de Strasbourg and are not on display.

Sites
The Grès à Voltzia sandstone is quarried extensively in the northern Vosges Mountains (**193**). Permission from the quarry owner should be sought before entering any of these quarries, working or not.

Chapter Eleven

THE HOLZMADEN SHALE

BACKGROUND: THE MESOZOIC MARINE REVOLUTION

From the late Triassic onwards, while dinosaurs were dominating the land and pterosaurs were taking to the air, a revolution in marine life was taking place beneath the waves. As the super-continent of Pangaea began to break up, sea levels rose, and vast areas of low-lying land were flooded by epicontinental seas which supported coral reefs teeming with life. In such environments, full of potential prey, marine reptiles quickly diversified and dominated the seas; they included ichthyosaurs, plesiosaurs, crocodiles, and turtles.

The ichthyosaurs and plesiosaurs are not closely related, but both are euryapsid reptiles of uncertain affinities. The crocodiles are diapsid archosaurs (along with pterosaurs and dinosaurs), while the turtles belong to the more primitive anapsid reptiles (see p. 106).

Ichthyosaurs were fish-shaped reptiles (see **213**), entirely adapted to life in the sea, with streamlined bodies and paddle-like limbs, looking very much like modern sharks (which are fish) or dolphins (which are mammals). They had long, thin snouts lined with sharp, conical teeth and fed on fish and cephalopods. Their large eye sockets suggest that they had good eyesight – perhaps needed in the deep and muddy waters of the Jurassic seas; in fact ichthyosaurs had the largest eyes ever recorded in the animal kingdom. They swam by beating the flexible body and powerful tail from side to side, like a shark, and they used the small front paddles for steering and a dorsal fin for balance. They were so well adapted to life in the sea that unlike most marine reptiles they could not come ashore to lay eggs, but gave birth to live young (like whales and most sharks).

The plesiosaurs were the true giants of the Mesozoic seas. Their bodies were broad and flat (see **216**) and they swam by rowing the body along by means of two pairs of paddle-like limbs, or alternatively by subaqueous 'flight', in the manner of turtles and penguins. The tail served merely as a rudder.

There are two groups; the true plesiosaurs had very long necks (up to 72 cervical vertebrae) and tiny heads. Contrary to popular reconstructions, however, their necks were rather inflexible with only limited lateral and dorsal movement, and they probably held the neck straight out in front, or possibly bent downwards probing the seabed for food, feeding mainly on small fish and soft-bodied prey. Their long, slim, pointed teeth were homodontous (all the same type). A second group, the pliosaurs, had larger heads (up to 4 m [13 ft] long) and shorter necks (as few as 13 cervical vertebrae). But the real difference was in their dentition, which was heterodontous (with different types of teeth serving a variety of functions). The giant pliosaurs, up to 13 m (c. 43 ft) long, were the top predators of the Jurassic seas and preyed on other marine reptiles in the open sea, just as killer whales today feed on smaller whales and seals. With their more powerful paddles it is possible that the plesiosaurs could have struggled ashore to lay eggs.

The only invasion of the sea by the archosaurs was the development in the Jurassic of the short-lived group of marine crocodiles known as the thalattosuchians (literally 'sea crocodiles'). Crocodiles and alligators are arguably the most successful four-footed hunters that have ever lived on Earth; they can equally well walk and swim and catch prey on land and in water. Most of the Jurassic thalattosuchians belong to the teleosaurids, a semi-aquatic group with slender, streamlined bodies, similar to the gavials of the River Ganges in India today. Evolving from these, however, was a highly specialized, fully marine group, the metriorhynchids, which flourished in the later Jurassic and persisted into the early Cretaceous. They were so well adapted for life in the sea that their limbs were modified into swimming paddles and their tail developed into a fish-like fin, akin to that of the ichthyosaurs. The metriorhynchids were open-sea hunters, and stomach contents include indigestible pterosaur bones and hooks from belemnoid or squid tentacles.

In the Jurassic seas dominated by ichthyosaurs, plesiosaurs, and crocodiles were also turtles and a variety of fish which were going through their own revolution. They included actinopterygians (ray-finned fish), sarcopterygians (lobe-finned fish), and chondrichthyans (cartilaginous fish) (see Chapter 6, The Hunsrück Slate). While most of the early Jurassic actinopterygians were of the more primitive groups, either holosteans (i.e. the bony ganoid fish) or chondrosteans (i.e. the sturgeons), a new group of advanced fish, the teleosteans (or modern bony fish), became dominant by the end of the period.

Lower Jurassic marine ecosystems are known from many locations across Europe, but around the small village of Holzmaden in the Schwäbische Alb of Baden-Württemberg, southern Germany (**208**), the Posidonienschiefer (*Posidonia* Shale) contains an abundant and sometimes completely preserved biota within black bituminous marls. All the major groups of marine reptiles and fish are exquisitely preserved, often with the outline of their skin and soft body tissue clearly visible. In addition there are rare pterosaurs and dinosaurs and a variety of marine invertebrates, dominated by coleoid cephalopods such as squids and belemnoids, sometimes complete with ink sacs and tentacles.

HISTORY OF DISCOVERY AND EXPLOITATION OF THE HOLZMADEN SHALE

Shale has been quarried around the villages of Holzmaden, Ohmden, Zell, and Boll to the south-east of Stuttgart since the end of the sixteenth century. The shale, known as 'Fleins', was initially used for roofing and paving, but because of its poor resistance to weathering it was later confined to internal use, such as oven bases, hearths, window sills, wall cladding, flooring, wash stones, tanner's slates, and laboratory tables. At Holzmaden the Fleins is a regular, 180 mm (c. 7 in) thick bed which splits into four equal layers. Within the same sequence limestone was also quarried as a building stone for cellars.

The shale itself is bituminous and contains up to 15% organic matter. This has resulted in the past in several serious shale fires when excavations had been neglected. In 1668 a shale digging at Boll burned for 6 years during which time oil flowed from the burnt shale and was sold locally. The last big shale fire was at Holzmaden from 1937 to 1939.

During times of emergency and especially during the First World War, this shale oil has been utilized as an alternative energy source by companies such as the Jura Oil Shale Works at Göppingen; the shale yields a maximum of 8%. After the war production ceased, but the bituminous shale continued to be used for a short time as a heat source for the manufacture of cement from the White Jura Limestone and Liassic Marl. Just prior to the Second World War, the Portland cement works in the Balingen area produced shale oil as a by-product, and during the war a crude oil plant was again established for the production of shale oil. Oil and tar extracted from the *Posidonia* Shale are now used in the pharmaceutical industry, and at Bad Boll Spa the shale is finely ground and marketed as a medicinal mud!

The manual quarrying of former years brought to light many remarkable fossils from the Holzmaden Shale. In 1939 about 30 quarries employed 100 workmen, but unfortunately this came almost to a complete stop during the Second World War. In 1950 about 20 quarries reopened, but since then the industry has suffered from competition from imported slate, marble, and synthetic material; nowadays only a handful of quarries remain (**209**) and because their

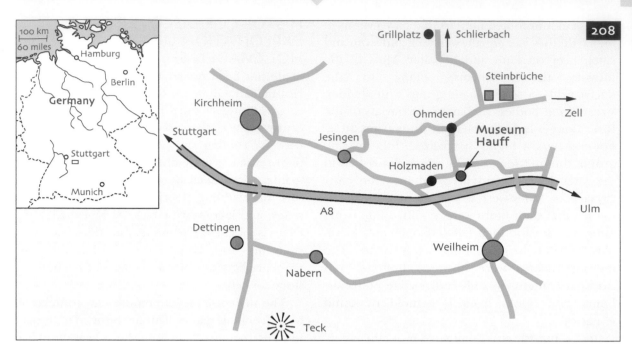

208 Locality map to show the Holzmaden Shale region in the Schwäbische Alb of southern Germany.

209 Quarry in Holzmaden Shale (Schieferbruch Kromer) at Ohmden in the Schwäbische Alb of southern Germany.

operations are largely mechanized, fewer fossils are found. The area is now protected by law, but some quarries are open to collectors.

Fossils were first discovered in this area during excavations at the spa town of Boll in 1595, but it was a sensational discovery in 1892 which brought them to the attention of the wider scientific world. Bernhard Hauff (1866–1950), the son of a chemist who had come to Holzmaden in 1862 to pursue his interest in extracting oil from shale, found many new fossils in his father's quarry which he prepared meticulously. In 1892 he succeeded in exposing the outline of the body of a fossil ichthyosaur complete with its skin. Previous reconstructions of ichthyosaurs had shown them without dorsal fins and with a long, thin tail, but Hauff's specimen proved the presence of a triangular dorsal fin and a fleshy upper lobe to the tail, neither of which had skeletal support.

Bernhard Hauff and his own son, Bernhard Hauff junior (1912–1990), built the Urwelt-Museum Hauff at Holzmaden to exhibit the finest fossils from this unusual Fossil-Lagerstätte (see Hauff & Hauff, 1981).

STRATIGRAPHIC SETTING AND TAPHONOMY OF THE HOLZMADEN SHALE

The Holzmaden Shale is the informal name given to the Posidonienschiefer at its outcrop in the Holzmaden area. Here it consists of 6–8 m (c. 20–26 ft) of black, bituminous marls (**210**) with intercalated limestones of Lower Toarcian age (Lower Jurassic; zones of *Dactylioceras tenuicostatum*, *Harpoceras falcifer*, and *Hildoceras bifrons*; approximately 180 million years ago). The Tübingen geologist August Quenstedt subdivided the Jurassic of the Schwäbische Alb into six divisions termed alpha to zeta, and the Holzmaden Shale falls within the 'Lias epsilon (ε)'.

It has been further subdivided (see Hauff & Hauff, 1981, p. 10) into the lower (εI), middle (εII), and upper (εIII) divisions (**211**). The Fleins (see p. 134) occurs in the middle epsilon (εII 3). Above the Fleins is the 'Untere Schiefer' (εII 4), from which the shale oil was extracted, and it is from the lower part of this layer that the best preserved fossils occur, including the ichthyosaurs with soft-tissue preservation of skin and muscles. Above this is the 'Untere Stein' (εII 5), a hard and weather-resistant limestone formerly used as a building stone (p. 134), which has yielded uncompressed fossil fish. The 'Oberer Stein' (εII 8) is another limestone horizon and between these two is the 'Schieferklotz' (εII 6), which has yielded most of the crocodiles. Above this the layers become more irregular; the 'Wilder Schiefer' (εIII) consist of soft, dark, blue-grey shale increasing in thickness from west to east up to 7 m (c. 23 ft). In its lower part large numbers of compressed ammonites are found, but vertebrates are very rare.

At the beginning of the Jurassic the Schwäbische Alb lay under a wide epicontinental sea which was eventually to gain connections with the Tethys Ocean to the south (see also p. 159). The extensive sea, which covered much of northern Europe, was divided into a number of basins separated by submarine highs and islands; the Posidonienschiefer in the vicinity of Holzmaden was deposited in the South German Basin between Ardennes Island to the west and the Vindelicisch Land/Bohemian High to the south and east (Hauff & Hauff, 1981, Figure 4).

The dark colour of the fine-grained marls is caused partly by diffused pyrite and partly by the high percentage (up to 15%) of solid organic matter (kerogen and poly-bitumen). Together these suggest deposition within a stagnant basin, starved of oxygen and rich in H_2S. As with the Burgess Shale (Chapter 2) and the Solnhofen Limestone (Chapter 13) the marls are finely laminated and individual laminae can be traced over large distances, also suggesting deposition in still waters. Generally there is no evidence of bioturbation of the sediment, and benthic organisms are extremely rare, being restricted to a few echinoids, crustaceans, and burrowing bivalves.

Bottom conditions were clearly hostile to life, the absence of bottom currents leading to an anoxic seabed with insufficient oxygen to support life – other than anaerobic bacteria – or to enable aerobic bacterial decay of dead organic tissue.

210 Laminated bituminous marls with intercalated limestones at Schieferbruch Kromer, Ohmden.

Fresh water flowed into the South German Basin in the upper water zones only, and these well-oxygenated and nutritious surface waters supported a rich nektonic (swimming) and planktonic (floating) fauna. Occasional bio-turbated horizons (εI 3, εI 4, top of εIII) do suggest temporary periods favourable to epibenthic life and alignment of fossils at certain horizons similarly suggests occasional bottom currents, but generally there was no exchange of water near the seabed.

Micro-organisms falling to the seabed would have begun to decay, but this would soon have used up the available oxygen so that organic particles were instead incorporated into the sediment. Carcasses of macro-organisms, such as marine reptiles and fish, were thus buried in anoxic mud and decay of their soft tissue was arrested. Scavengers would also have been excluded by the noxious environment so that the carcasses were preserved intact. Such a stagnation depositional model for the Holzmaden Shale compares with conditions today in the Black Sea, an epicontinental basin with a restricted aperture.

Kauffman (1979) contested this model and suggested that some of the presumed pseudoplanktonic fauna (especially bivalves and crinoids, see pp. 141–142) was in fact benthic and that benthic communities characterized much of the depositional history of the shale. He suggested that only the sediment was anoxic and that the water body just above the sediment/water interface was hospitable. This has not been generally accepted, but it did lead to a modification of the stagnation model (Brenner & Seilacher, 1979; Seilacher, 1982) in which stagnant conditions were sometimes interrupted by high-energy events related to severe storms which had brief oxygenating effects.

One aspect of Kauffman's model that is interesting is his proposal of the development of an algal mat just above the sediment/water interface, which if nothing else would inhibit grazers and scavengers. The presence of such a cyanobacterial mat has also been suggested in the case of Ediacara (Chapter 1), Chengjiang (Chapter 3), Grès à Voltzia (Chapter 10), the Solnhofen Limestone (Chapter 13), and the Santana/Crato formations (Chapter 16) and is an important contributory factor in the preservation of soft tissue.

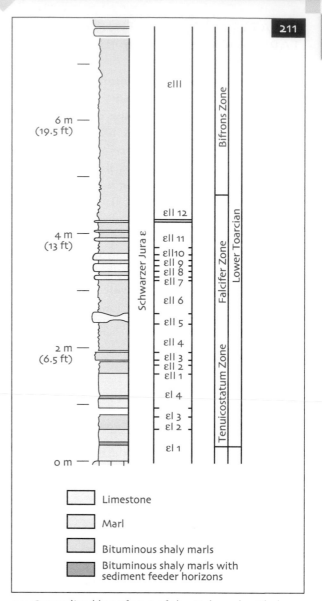

211 Generalized log of part of the Holzmaden Shale sequence in the Holzmaden area (after Wild, 1990).

212 The ichthyosaur *Stenopterygius quadriscissus*, with the body outline preserved as a black organic film (UMH). Length 1.2 m (c. 4 ft).

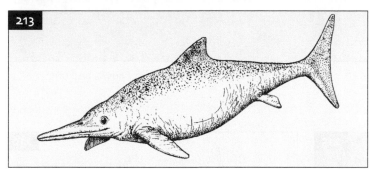

213 Reconstruction of *Stenopterygius*.

DESCRIPTION OF THE HOLZMADEN SHALE BIOTA

Ichthyosaurs. The Holzmaden Shale is famous for its complete vertebrate skeletons, which are often preserved with the skin of the animals marking the outline of the body as a black film. This is seen in sharks, in bony fish, in crocodiles and pterosaurs, but most spectacularly in the various species of the ichthyosaur *Stenopterygius* (**212**). Bernhard Hauff's initial discovery of this phenomenon (p. 135) proved that ichthyosaurs had a dorsal fin and that their tail had an upper lobe (**213**), neither of which have skeletal support and are therefore normally not fossilized. Many individuals of *Stenopterygius* are preserved with stomach contents (such as the indigestible hooks of cephalopods and the thick ganoid scales of fish), and more remarkable are the numerous females preserved either in the process of giving birth or with embryos within their bodies (**214**). The high percentage of pregnant females and juveniles suggests that this area was a spawning ground to which animals periodically migrated. More than 350 specimens were recorded by Hauff (1921).

Plesiosaurs and crocodiles. These are much more rare; only 13 complete skeletons and scattered remains of plesiosaurs have ever been found. They include the long-necked plesiosaur *Plesiosaurus* (**215, 216**), shorter-necked pliosaurs *Peloneustes* and *Rhomaleosaurus*, and also the enigmatic *Hauffiosaurus*, which is either a pliosaur or a basal plesiosaur. Teleosaur crocodiles are somewhat more numerous; Hauff (1921) recorded 70 specimens, the most common being *Steneosaurus* (**217, 218**), which had a long, narrow snout with many teeth and probably caught fish by quickly snapping shut its narrow jaws. Its eyes point upwards and outwards so it probably dived under shoals of prey, coming up to attack. The powerful legs and tail suggest that it could walk on land and was a strong swimmer. The much smaller *Pelagosaurus* had its eyes arranged laterally on the cranium and was a more skilful swimmer; some authorities consider it to be a basal metriorhynchid. The more heavily armoured *Platysuchus* is much rarer, being known from only four specimens worldwide.

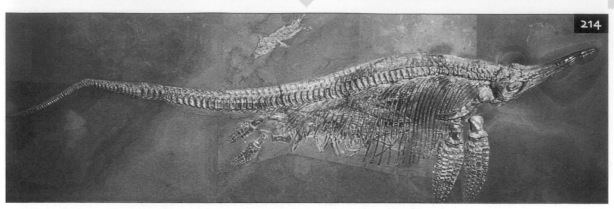

214 The ichthyosaur *Stenopterygius crassicostatus,* with five embryos preserved inside an adult, and one juvenile (UMH). Length 3 m (c. 10 ft).

215 The plesiosaur *Plesiosaurus brachypterygius* (UMH). Length 2.8 m (c. 9 ft).

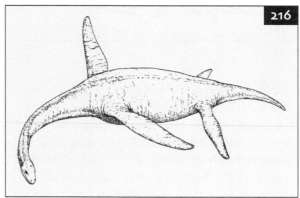

216 Reconstruction of *Plesiosaurus.*

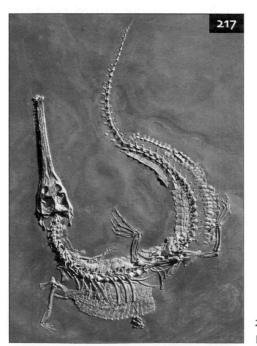

218 Reconstruction of *Steneosaurus.*

217 The teleosaur crocodile *Steneosaurus bollensis* (UMH). Length 2.7 m (c. 9 ft).

Pterosaurs and dinosaurs. Two genera of pterosaurs, *Dorygnathus* (**219**) and *Campylognathoides*, with wingspans of 1 m (c. 3 ft) and 1.75 m (c. 5 ft 9 in) respectively, are known from the Holzmaden Shale, both from complete skeletons; Hauff (1921) recorded about 10 specimens in total. The only dinosaur to be recorded is the cetiosaurid *Ohmdenosaurus*, a 4 m (c. 13 ft) long sauropod, known from a single tibia, and named after the local village of Ohmden.

Fish. Primitive holostean fish (ganoids) are represented by well-known Liassic genera such as *Lepidotes*, a heavily-built fish up to 1 m (c. 3 ft) in length (**220, 221**) with thick, shiny scales of ganoine which have been found in the stomach of *Baryonyx*, the fish-eating dinosaur; *Dapedium* (the 'Moonfish'), with a flat, round body, ganoid scales, and peg-like crushing teeth (**222, 223**), and the giant predator *Caturus*. *Ohmdenia* was originally considered to be a basal chondrostean, but is now also regarded as a holostean, and a member of the pachycormid group; it was a large suspension feeder, comparable to the baleen whales today. Much rarer teleosts include the sprat-like *Leptolepis*. Occasional sharks, such as *Hybodus* (**224**) and *Palaeospinax*, are often preserved with the black outline of their skin and

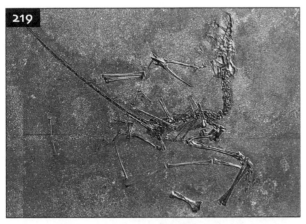

219 The pterosaur *Dorygnathus banthensis* (UMH). Height of block 420 mm (c. 16.5 in).

220 The holostean fish *Lepidotes elvensis* (UMH). Length 600 mm (c. 24 in).

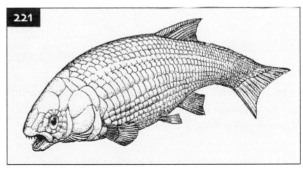

221 Reconstruction of *Lepidotes*.

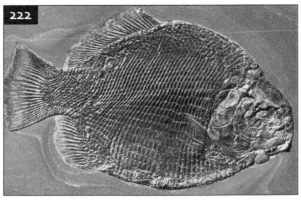

222 The holostean fish *Dapedium punctatum* (UMH). Length 335 mm (c. 13 in).

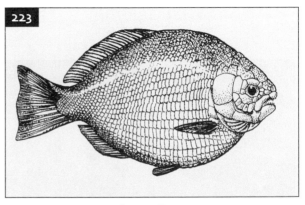

223 Reconstruction of *Dapedium*.

reached lengths of 2.5 m (c. 8 ft), but the largest fish, up to 3 m (c. 10 ft), were the sturgeons such as *Chondrosteus*. A single specimen of coelacanth, *Trachymetopon*, is also known.

Cephalopods. Many of the well-known Liassic ammonites, genera such as *Harpoceras* (**225**), *Hildoceras*, and *Dactylioceras*, are common consti-tuents of the Holzmaden fauna, but the most spectacular cephalopods are squids and belemnoids (e.g. *Passaloteuthis*), often with soft body tissues preserved including ink sacs, tentacles, and their hooks (**226**).

Crinoids. The crinoids *Seirocrinus* and *Pentacrinus*, which lived in colonies attached to floating logs (**227**), are common members of the Holzmaden fauna; a spectacular specimen of *Seirocrinus* displayed in the Urwelt-Museum Hauff in Holzmaden is over 18 m (c. 60 ft) long, attached to a 12 m (c. 40 ft) log.

224 Reconstruction of *Hybodus*.

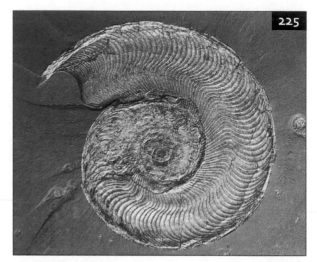

225 The ammonite *Harpoceras falcifer* (UMH). Diameter 200 mm (c. 8 in).

226 The belemnoid *Passaloteuthis paxillosa*, showing preservation of tentacles with their hooks (UMH). Length 230 mm (c. 9 in).

227 The crinoid *Pentacrinus subangularis*, attached to a floating log (UMH). Height 1.7 m (c. 5 ft 6 in).

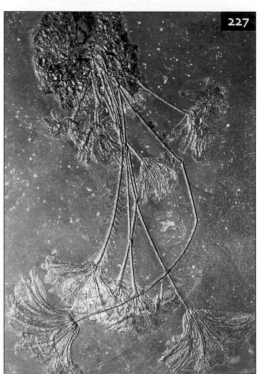

Bivalves. Most of the common Liassic bivalves, such as *Gervillia*, *Oxytoma*, *Exogyra*, and *Liostrea*, are also often found fixed by their byssal threads to floating logs or to the shells of ammonites and only occasionally to a temporarily hardened seabed. Some, such as *Bositra* (previously known as *Posidonia*), were nektoplanktonic and a few, such as *Goniomya*, were burrowers.

Plants. The Holzmaden flora comprises horsetails and gymnosperms, the latter including ginkgos, conifers, and cycads.

Trace fossils. Bioturbated horizons (εI 3, εI 4, top of εIII) consist mainly of the trace fossils *Chondrites* and *Fucoides*.

The fauna and flora from the Holzmaden Shale were well documented and illustrated by Hauff and Hauff (1981).

PALAEOECOLOGY OF THE HOLZMADEN SHALE BIOTA

The Holzmaden Shale represents an epicontinental marine basin community living at a depth that varied between 100 and 600 m (c. 300 and 2,000 ft), depending on basinal subsidence (Hauff & Hauff, 1981), and which was situated within the subtropical zone at about 30°N.

For much of the time stagnant and anoxic bottom waters severely limited benthic life. Occasional horizons are intensely bioturbated (for example the 'Seegrasschiefer' (εI 3), which is riddled with trace fossils such as *Chondrites* and *Fucoides*) and reflect periodic improved aeration of the seabed. Generally, however, benthic infauna (living in the sediment) was restricted to a few burrowing bivalves, such as *Solemya* and *Goniomya* (Riegraf, 1977), while the vagrant benthic epifauna (living on the sediment surface) included only minute diademoid echinoids, ophiuroids, the gastropod *Coelodiscus*, and possibly the crustacean *Proeryon*. Kauffman (1979) included some bottom-feeding fish, such as *Dapedium*, in the nektobenthos.

Well-aerated surface waters, however, supported a thriving planktonic and nektonic life. Most of the crinoids, bivalves, and inarticulate brachiopods had a pseudoplanktonic lifestyle attached either to floating logs or, in the case of the latter two groups, to the shells of living ammonites (Seilacher, 1982). Kauffman (1979), however, disputed this,

suggesting that the encrustations occurred on dead ammonite shells and sunken logs that were lying on the seabed and maintaining that these three groups were benthic in habit.

The bivalve *Bositra* (formerly *Posidonia*) has been interpreted as a nektoplanktonic (passively swimming) bivalve. True nektonic swimmers include the numerous ammonites and smaller fish, while the strongest swimmers were squids, belemnoids, large fish such as the sharks and sturgeons and, of course, the marine reptiles. Of these, the plesiosaurs and ichthyosaurs were inhabitants of the open sea, while the crocodiles would have lived near the coast.

The Vindelicisch landmass, some 100 km (c. 60 miles) to the south, supported an abundance of vegetation including horsetails and a variety of gymnosperms, some of which grew into large trees (ginkgos, conifers, and cycads). Leafy twigs and large logs were washed into the basin, where they were deposited as allochthonous elements of the biota along with the occasional disarticulated remains of small sauropod dinosaurs. Pterosaurs flying over the basin in search of fish were sometimes overcome by storms and drowned; their carcasses are still articulated and clearly were not transported any distance.

Trophic analysis identifies the filter feeders (crinoids and bivalves) and deposit feeders (gastropods, echinoids, and ophiuroids) as the primary consumers, being preyed on by primary predators such as the bony fish and the various cephalopods. Top predators were the ichthyosaurs, plesiosaurs, crocodiles, and sharks, with the large plesiosaurs at the top of the food web.

COMPARISON OF THE HOLZMADEN SHALE WITH OTHER JURASSIC MARINE SITES

Yorkshire coast, United Kingdom

The coastal outcrops of Lower Jurassic (Lias) marine sediments of the United Kingdom have been well known for many years. Those of the Dorset coast in southern England have been made famous by the celebrated nineteenth-century collector Mary Anning of Lyme Regis, who with her brother discovered one of the first ichthyosaurs and the first complete plesiosaur in rocks of Lower Lias age (Cadbury, 2000). The Liassic outcrops of the Yorkshire coast in the north-east of the country are, however, more comparable with the Holzmaden Shale, since in this area it is the Upper Lias sequence that has yielded most of the marine reptiles.

Benton and Taylor (1984) listed 55 crocodiles, 69 ichthyosaurs, 33 plesiosaurs, and one pterosaur from the Yorkshire sequence which, like the Holzmaden Shale, is of Lower Toarcian age (zones of *Dactylioceras tenuicostatum*, *Harpoceras falcifer*, and *Hildoceras bifrons*). There are, however, significant differences in the two faunas. The majority of the specimens come from the Jet Rock and Alum Shale formations, the former consisting of hard grey bituminous shales, the latter of soft grey micaceous shales, both with bands of calcareous concretions. Typical ammonites are *Dactylioceras*, *Harpoceras*, *Hildoceras*, and *Phylloceras*.

The first fossil ichthyosaur to be found in Yorkshire was a skull and partial skeleton collected in 1819, and a more complete skeleton was found in 1821. It is interesting that the original illustration of this second specimen shows the tail straightened out, as was common practice at the time, prior to Hauff's demonstration from Holzmaden specimens (p. 135) that ichthyosaurs had a bend in the vertebral column to accommodate the large tail fin. The Yorkshire ichthyosaurs belong to *Temnodontosaurus* and *Stenopterygius*.

A variety of plesiosaurs are known from Yorkshire including pliosaurs (*Rhomaleosaurus*), true plesiosaurs (*Microcleidus* and *Sthenarosaurus*), and also the enigmatic *Hauffiosaurus*, just as at Holzmaden (p.138) (Benson *et al.*, 2011). Yorkshire crocodiles are conspecific with the Holzmaden teleosaurs; most belong to *Steneosaurus*, with *Pelagosaurus* also present, but again rare.

A single specimen of the partial skull of a pterosaur, *Parapsicephalus*, was found in 1888, and a single dinosaur bone, identified as a theropod femur, was recorded in 1926. This specimen is unfortunately lost, but if verified would be the only known record of an Upper Liassic theropod; indeed, the only other dinosaur known from the Upper Lias is that of the sauropod *Ohmdenosaurus*, from Holzmaden (p. 140).

The Yorkshire marine reptiles are slightly younger than those from Holzmaden. Plesiosaurs and crocodiles are relatively more abundant, and ichthyosaurs are much less common in the Yorkshire Lias than at Holzmaden. Although the species of crocodile are shared between the two sites, the ichthyosaurs and plesiosaurs are generally not conspecific, with the notable exception of the ichthyosaur *Stenopterygius acutirostris*.

A large number of Yorkshire specimens are well preserved in an articulated state with little evidence of scavenging. The Jet Rock Formation, in particular, is rich in bitumen (kerogen), and Benton and Taylor (1984) suggested that bottom conditions were again anoxic.

MUSEUMS AND SITE VISITS

Museums
1. Urwelt-Museum Hauff, Holzmaden, Germany.
2. Urweltsteinbruch Museum, Holzmaden, Germany.
3. Staatliches Museum für Naturkunde, Stuttgart, Germany.

Sites
Some of the quarries around Holzmaden and Ohmden are open to collectors on payment of a small fee. These villages are approximately 40 km (c. 25 miles) south-east of Stuttgart and are accessed via the A8 Stuttgart–Munich autobahn. Leave the autobahn at the Aichelberg exit and follow the clear signs (with the crocodile *Steneosaurus*!) to the Urwelt-Museum Hauff in Holzmaden. Immediately opposite this museum is the Urweltsteinbruch Museum, which has a small quarry attached. Hammers and chisels are provided, but the beds here are not particularly fossiliferous. More productive is Schieferbruch Kromer at Ohmden (**209**), which is open on Mondays to Saturdays from April to October (telephone/fax +49-7023-4703). The Urwelt-Museum Hauff will arrange visits to the latter.

Chapter Twelve

THE MORRISON FORMATION

BACKGROUND: TERRESTRIAL LIFE IN THE MID-MESOZOIC

From the end of the Triassic to the close of the Cretaceous, life on land was, of course, dominated by the hugely successful dinosaurs. Since different periods of the Mesozoic were characterized by different assemblages of dinosaurs, it is useful to understand something of their classification. Dinosaurs are classified into two main groups, the reptile-hipped saurischians and the bird-hipped ornithischians.

In saurischians the hip bones are arranged similarly to those of most other reptiles. The blade-like upper bone, the ilium, is connected to the backbone by a row of strong ribs and its lower edge forms the upper part of the hip socket. Beneath the ilium is the pubis, which points downward and slightly forward, and behind this is the backwardly extending ischium.

The saurischians are further divided into two groups. Theropods include all of the carnivorous (meat-eating) dinosaurs; most have powerful hindlimbs ending in sharply-clawed, bird-like feet, lightly-built forelimbs, a long, muscular tail, and dagger-like teeth. Examples include the giant *Tyrannosaurus* ('tyrant lizard') and *Albertosaurus*, the small and agile *Velociraptor* ('fast-thief'), and some toothless types, such as *Oviraptor* ('egg-thief') and *Struthiomimus* ('ostrich-mimic'), all from the late Cretaceous. The group also includes the classically huge predators of the Jurassic, such as *Allosaurus* and *Megalosaurus* and the tiny *Compsognathus* from the same period (Chapter 13, The Solnhofen Limestone).

The second group of saurischians, the sauro-podomorphs, were all herbivorous (plant-eating) dinosaurs. They ranged in size from diminutive forms (the prosauropods, such as *Massospondylus*) from the late Triassic and early Jurassic, to the gigantic true sauropods of the late Jurassic, such as *Diplodocus*, *Apatosaurus* (previously known as *Brontosaurus*), *Brachiosaurus*, and *Camarasaurus*. They tend to have long, slender bodies, whip-like tails, long, shallow faces and thin, pencil-shaped teeth.

In the bird-hipped ornithischians the arrangement of the hip bones is similar to that of living birds (although confusingly the ornithischians did not give rise to birds; see Chapter 14, The Jehol Group). While the ilium and ischium are arranged in a similar manner to the saurischians, the pubis is a narrow, rod-shaped bone which lies alongside the ischium. In addition, all ornitihiscians seem to possess a small, horn-covered beak perched on the tip of the lower jaw.

Ornithischians were entirely herbivorous and are classified into five major groups: the ornithopods (medium-sized animals such as the early Cretaceous *Iguanodon* and *Tenontosaurus*, and the hadrosaurs, or duck-billed dinosaurs, such as the late Cretaceous *Edmontosaurus*); the ceratopsians (horned and frilled dinosaurs of the late Cretaceous, such as *Triceratops*); the stegosaurs (plated dinosaurs of the Jurassic, such as *Stegosaurus*); the pachycephalosaurs (with domed and reinforced heads, such as *Pachycephalosaurus*); and the ankylosaurs (armoured dinosaurs, covered in thick bony plates embedded in the skin, such as *Ankylosaurus*).

Fossil evidence for terrestrial life during much of the Jurassic is quite poor, but towards the later

228 Locality map to show the extent of the Morrison Formation outcrop in North America.

1   Dinosaur National Monument, Utah

2   Canyon City, Colorado

3   Morrison, Colorado

4   Como Bluff, Wyoming

5   Cleveland-Lloyd Quarry, Utah

6   Dry Mesa Quarry, Colorado

Montana

South Dakota

Wyoming

Nebraska

Utah

Colorado

Kansas

New Mexico

Arizona

Edge of Morrison basin

500 km
300 miles

229 Fossil Cabin Museum, Como Bluff, Wyoming. Dinosaur bones are so common that they are used as a building material.

part of the period there are some exceptionally rich deposits, especially in China, Tanzania, and North America. This chapter is based on the fossils of the Morrison Formation, a vast and highly productive sequence, long known for its spectacular dinosaur skeletons, which outcrops along the Front Range of the Rockies in the USA, from Montana in the north, to Arizona and New Mexico in the south (**228**).

Over such a huge area the Morrison Formation represents a variety of terrestrial conditions from wet swamps (with coal deposits) in the north, to desert conditions in the south. It is in the mid-west states of Colorado, Utah, and Wyoming, where the Morrison Formation represents mostly fluviatile and lacustrine deposits, that the richest finds have been made. Here, flash floods deposited literally tons of bones (**229**) in a Concentration Lagerstätte which gives a detailed insight into a late Jurassic terrestrial ecosystem that includes not only some of the largest dinosaurs known, but also some of the other land animals which coexisted alongside the dinosaurs, including the most diverse Mesozoic mammal assemblage yet known.

HISTORY OF DISCOVERY OF THE MORRISON FORMATION

The story of the discovery of the Morrison Formation dinosaurs has become known in American palaeontological folklore as 'The Bone Wars' and is the story of a bitter rivalry between two of America's leading palaeontologists of the late nineteenth century. It began in 1877 when two schoolteachers, Arthur Lakes and O. W. Lucas, independently discovered rich remains of dinosaur bones in Colorado. Lakes sent his fossils, which he had found near the town of Morrison, to Professor Othniel Charles Marsh of Yale's Peabody Museum, who was well known for his work on hadrosaurs from Kansas. In the same year Lucas found bones from the same horizon near Canyon City, which he sent to Edward Drinker Cope at Philadelphia, who had described some of the first ceratopsian dinosaurs from Montana.

Marsh and Cope were already bitter enemies due to an earlier dispute over Cope's description of a fossil reptile which Marsh had shown to be erroneous. Immediately, a frantic race began between the two men to describe the numerous new dinosaurs that were being collected. Cope's specimens from Canyon City were larger and much more complete and at first he had the upper hand, but later the same year (1877) new discoveries were made in equivalent beds at Como Bluff, Wyoming, and this time Marsh was first on the scene.

Como Bluff, near Medicine Bow, Wyoming, is a low ridge, approximately 16 km (c. 10 miles) long and 1.6 km (c. 1 mile) wide formed by a north-east–south-west trending anticline with a gently dipping southern limb and a steeply dipping northern limb. The southern limb is capped by the highly resistant Cloverly Formation of the lowermost Cretaceous, while the northern limb exposes the underlying beds of the uppermost Jurassic, which have since been named the Morrison Formation after the classic locality near Denver. It was in these latter beds that the rich dinosaur fauna, mainly of giant sauropods, was preserved.

The discovery at Como Bluff further fuelled the 'Bone Wars'. It was made by two workers on the transcontinental Union Pacific Railroad, which was then being driven through southern Wyoming to exploit the region's extensive deposits of coal (Breithaupt, 1998). In July 1877 William Edward Carlin, the station agent at nearby Carbon Station, and the section foreman, William Harlow Reed, wrote to Marsh informing him that they had discovered some gigantic bones of what they thought to be *Megatherium* (the Pleistocene ground-sloth), and offered to sell the fossils to Marsh and also to collect further specimens if required. They signed the secret letter with their middle names, Harlow and Edwards, to cover up their identities. Four months later Marsh's representative, Samuel Wendell Williston, arrived at Como Bluff to survey the site and to pay Carlin and Reed their costs. (A previous cheque made out by Marsh to 'Harlow and Edwards' could not be cashed since these were not their real names!)

Williston immediately informed Marsh of the richness of Como Bluff and transferred his collecting crews from Colorado to the new site. Carlin and Reed continued to work for Marsh, the latter eventually becoming the curator of the Geological Museum of the University of Wyoming in Laramie, and a respected palaeontologist. Not wishing to be outdone, Cope moved his crews to Como Bluff, eventually persuading Carlin to work for him. The feud extended into field operations and skirmishes between rival camps often broke out. Breithaupt (1998) reports spying on rival quarries, smashing of bones to prevent the opposition collecting them, and sometimes even 'fisticuffs'!

Over the ensuing years, however, a huge number of new dinosaurs were discovered including the carnivore *Allosaurus*, the strange, plated dinosaur *Stegosaurus*, and the largest sauropods then known, such as *Diplodocus* and the notorious *Brontosaurus* (now known as *Apatosaurus*). Alongside the dinosaurs the Morrison also yielded the most important fauna of Mesozoic mammals yet discovered.

The rivalry continued until Cope's death in 1897; Marsh died in 1899. By the end of their careers Marsh had described 75 new species of dinosaur, of which 19 are valid today, while Cope had described 55 species of which 9 are valid today. The Morrison Formation, now known from twelve different states in North America (**228**), remains one of the world's most prolific dinosaur 'graveyards' and its fossils can be seen on display in museums all over the world.

STRATIGRAPHIC SETTING AND TAPHONOMY OF THE MORRISON FORMATION

The Morrison Formation, traditionally divided into four members (Gregory, 1938), was restricted to a two-member division by Anderson and Lucas (1998): an upper Brushy Basin Member and a lower Salt Wash Member (**230**). It has been dated radiometrically and on microfossil evidence as Kimmeridgian to early Tithonian (Upper Jurassic; approximately 150 million years ago) and its outcrop along the Front Range of the Rocky Mountains (**231**) covers an area of more than 1.5 million sq. km (c. 0.6 million sq. miles). The thickness is highly variable, but at Dinosaur National Monument in Utah it is approximately 188 m (c. 620 ft).

In many areas the Morrison is capped and

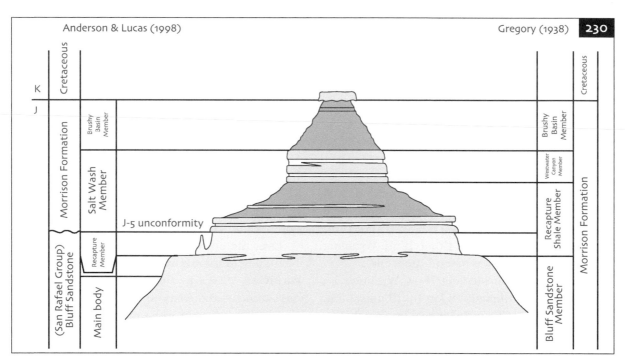

230 Diagram to show the stratigraphy of the Morrison Formation in the region of Bluff, Utah (after Anderson & Lucas, 1998).

231 Morrison Formation (foreground) with underlying red beds of the Permo–Triassic resting on Palaeozoic rocks of the Bighorn Mountains (the Front Range of the Rockies); near Buffalo, Wyoming.

232

232 The Lower Cretaceous Cloverly Formation (top) overlays alternating sandstones and shales of the Morrison Formation bone beds; Wyoming Dinosaur Center, Thermopolis, Wyoming.

protected by the highly resistant Cloverly Formation of the Lower Cretaceous (**232**), and is underlain by the Middle Jurassic Sundance Formation representing the marine deposits of the Sundance Sea. This succession, at both Dinosaur National Monument and Cleveland–Lloyd Dinosaur Quarry (Utah), two of the most important Jurassic dinosaur sites in the world, illustrates a regressional sequence coincidental with the final northward withdrawal of this vast, shallow sea during the mid Jurassic. At the latter location, for example, the intertidal beds of the Summerville Formation (equivalent to the Sundance) pass upwards into the supratidal Tidwell Member, then into fluvial and lacustrine deposits of the Salt Wash Member, and finally into overbank deposits with meandering rivers of the Brushy Basin Member (Bilbey, 1998; **230**).

Over such a vast area there is clearly enormous variation in the environment of deposition of the Morrison Formation, but in many of the richest beds the bone accumulations were deposited in poorly sorted sandstones that are thought to have resulted from cataclysmic flash floods. The wide, open plains left by the retreat of the Sundance Sea and traversed by meandering rivers, were home to herds of herbivorous dinosaurs roaming the rivers and lakes in search of vegetation. Cyclic periods of severe drought, perhaps with a periodicity of 5, 10, 40, or 50 years as in Kenya today, concentrated the dinosaur herds (and other vertebrates) around remnant water holes where they eventually died of dehydration. Following such mass mortality, subsequent flash floods swept the disarticulated bones a short distance before burial in sandbodies representing the channel fill of streams. This situation was observed at the Dry Mesa Dinosaur Quarry in Colorado by Richmond and Morris (1998), where a diverse assemblage, including 23 different dinosaurs, plus pterosaurs, crocodiles, turtles, mammals, amphibians, and lungfish is preserved. The assemblage has gained notoriety for its large sauropods, especially *Supersaurus* and *Ultrasauros*.

Such mass death assemblages, represented by bone beds in the fossil record, may be either catastrophic or non-catastrophic accumulations. The former is defined as sudden death (within a few hours at most, e.g. a poisonous ash fall) and includes all age and sexual groups. Non-catastrophic mass mortality occurs over a longer time span of hours up to months (e.g. starvation). The killing agent can be selective with regard to age, health, gender, and social ranking of individuals so that juveniles, females, and old adults dominate numerically. Most of the bone accumulations in the Morrison Formation are considered to represent non-catastrophic mass mortalities (Evanoff & Carpenter, 1998), which are characterized by a greater degree of disarticulation due to the longer time span.

The preservation of the disarticulated bones is favoured by an arid climate. Dodson *et al.* (1980) suggested that Morrison Formation dinosaur carcasses decomposed on dry open land or in channel beds prior to deposition. After death the arid climate dehydrated muscle tissue, ligaments, and skin, but there is little evidence of scavenging and little sign of cracking or exfoliation of the bones. Richmond and Morris (1998) suggested that they were exposed for no more than 10 years before burial by the flash flood.

DESCRIPTION OF THE MORRISON FORMATION BIOTA

Allosaurus. This was one of the largest predatory theropod dinosaurs of the Jurassic, up to 12 m (c. 40 ft) in length and weighing up to 1,500 kg (c. 1.5 tons) (**233–235**). The skull was almost 1 m (c. 3 ft) long, with jaws supporting 70 curved and sharp teeth. It is thought to have hunted in groups when attacking herds of sauropods. The feet supported three ferociously large claws used for tearing flesh.

233 The theropod dinosaur *Allosaurus fragilis* (AMNH). Length 12 m (c. 40 ft).

234 The skull of *Allosaurus fragilis*, with sharp serrated teeth for cutting meat (SMA). Length of skull 1 m (c. 3 ft).

235 Reconstruction of *Allosaurus*.

Diplodocus. A plant-eating sauropod described by Marsh – one of the longest dinosaurs known, up to 27 m (c. 90 ft), but slimly built and weighing only around 10,000–12,000 kg (c. 10–12 tons) (**236–238**). Both the neck and the tail were held more or less horizontally. The 6 m (c. 20 ft) long neck supported a tiny skull compared to the size of the animal, while the long tail with 73 vertebrae could be used like a whiplash. Recent fossil evidence suggests that it possessed a dorsal crest of triangular spikes, like a modern *Iguana*.

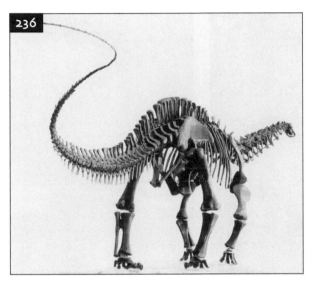

236 The sauropod dinosaur *Diplodocus*; note the whip-like tail (AMNH). Length up to 27 m (c. 90 ft).

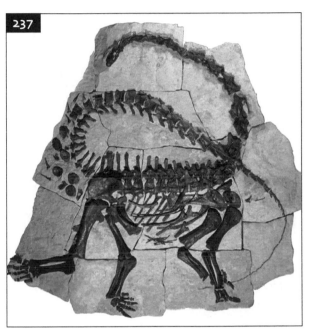

237 The sauropod dinosaur *Diplodocus* in matrix (SMA). Length, head to tail, 10.5 m (c. 35 ft).

238 Reconstruction of *Diplodocus*.

Apatosaurus. Another massive plant-eating sauropod related to *Diplodocus* and formerly known as *Brontosaurus*. Around 20 m (c. 65 ft) in length, but with a massive skeleton and weighing more than 20,000 kg (c. 20 tons), it was the largest dinosaur known after its discovery by Marsh in the Morrison Formation (**239–241**).

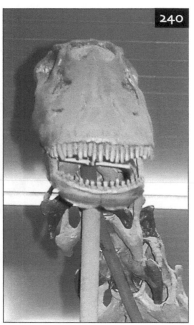

239 The massive sauropod dinosaur *Apatosaurus* (FPM). Length 23 m (c. 75 ft).

240 The skull of *Apatosaurus* (UWGM).

241 Reconstruction of *Apatosaurus*.

Camarasaurus. Described by Cope, this is the most common herbivorous sauropod and is nicknamed the 'Jurassic cow' (**242–245**). It is related to *Brachiosaurus* (see p. 154), and both had a more upright (giraffe-like) posture than the two previous genera, with the forelimbs longer than the hindlimbs. It was about 18 m (c. 60 ft) long, but had a massive skeleton and weighed up to 18,000 kg (c. 18 tons). The skull was much larger than that of *Diplodocus*, supporting 52 cone-shaped teeth for tearing vegetation (**243**).

242 The sauropod dinosaur *Camarasaurus* (WDC). Length c. 15 m (c. 50 ft).

243 The skull of *Camarasaurus*, with pencil-shaped teeth for tearing vegetation (WDC). Length of skull c. 550 mm (c. 22 in).

244 Vertebrae, skull, ribs, and limb bones of *Camarasaurus* in Morrison Formation bone bed; Dinosaur National Monument, Utah.

245 Reconstruction of *Camarasaurus*.

Stegosaurus. An unusual ornithischian dinosaur with large dorsal plates running in two rows down the length of the body. Their function is still not certain, but they may have been used as a temperature-regulating device; they consist of only a thin layer of bone and so could not have been used in defence. It is also characterized by a tiny skull, forelimbs shorter than the hindlimbs, and a massive tail with four sharp spikes – a defensive weapon against large predators (**246–248**).

Other dinosaurs. Other carnivores recorded from the Morrison Formation include *Torvosaurus, Ceratosaurus, Coelurus, Ornitholestes, Elaphrosaurus,* and *Tanycolagreus,* while herbivores include *Amphicoelias, Othnielosaurus, Camptosaurus,* and *Hesperosaurus.* The latter is a stegosaur from the lower part of the Morrison Formation and so is a little older than other Morrison stegosaurs.

Dinosaur trace fossils. Ornithopod eggshells, herbivore coprolites, and a variety of trackways are all known from the Morrison Formation.

Other reptiles and amphibians. These are never common, but do include some rare records of frogs (the oldest anuran is Lower Jurassic), the lizard-like sphenodons, some true lizards, crocodiles, and turtles, and a few records of pterosaurs, including pterodactyloids and rhamphorhynchoids. Records of birds have all been later refuted (Padian, 1998).

Mammals. The Morrison Formation mammals comprise one of the most important Jurassic mammal faunas ever discovered as they provide a rare window on the long early history of mammals in the Mesozoic. Known mainly from isolated jaw bones and their distinctive teeth are the primitive triconodonts, docodonts, symmetrodonts, and dryolestoids, while the multituberculates are a more developed group of rodent-like omnivores which, unlike the other groups, survived the K/T extinction and persisted into the Eocene (Engelmann & Callison, 1998).

Fish. Lungfish (sarcopterygians) were first reported by Marsh and for many years were the only known Morrison fish. More recently a variety of actinopterygians (ray-finned fish) have been reported, including a primitive teleostean (modern

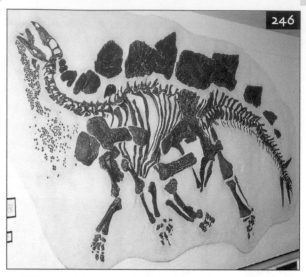

246 The ornithischian dinosaur *Stegosaurus* in matrix (UWGM). Length 4.5 m (c. 15 ft).

247 The ornithischian dinosaur *Stegosaurus,* with dorsal plates and tail spikes (SMA). Length 4.8 m (c. 16 ft).

248 Reconstruction of *Stegosaurus.*

bony fish), a variety of holosteans (bony ganoid fish), and a new chondrostean palaeoniscid, *Morrolepis*, the 'Morrison fish' (Kirkland, 1998).

Invertebrates. These include freshwater molluscs (gastropods and bivalves), ostracods, conchostracans, crayfish, and caddisfly cases.

Plants. Flora from the Brushy Basin Member includes bryophytes, horsetails, ferns, cycads, ginkgos, and conifers (Ash & Tidwell, 1998).

PALAEOECOLOGY OF THE MORRISON FORMATION BIOTA

The Morrison Formation was deposited in a terrestrial basin near the western margin of Laurasia (following the break-up of Pangaea), situated in the low mid-latitudes between 30° and 40° N. The climate is interpreted as having been arid to semi-arid, but with some seasonal rainfall (Demko & Parrish, 1998). A mountainous region to the west probably had a rain-shadowing effect and low annual rainfall is supported by the presence of evaporites, aeolian sandstones, and saline lake facies. However, the presence of various freshwater invertebrates and fish suggests that there were perennial streams and lakes present on the wide, open plains of the Morrison Basin, and the flora of horsetails, ferns, cycads, ginkgos, and various gymnosperms suggests at least short periods of a more humid, tropical climate (Ash & Tidwell, 1998). It seems that this fluvial–lacustrine environment was strongly influenced by repeated cycles of drought and flood.

The lush lake margins and swampy river courses were home to huge herds of herbivorous dinosaurs, which roamed the plains in search of food. Smaller quadrupedal herbivores, such as stegosaurs, browsed on low-level horsetails, ferns, cycads, and small conifers, while the giant sauropods with their long necks were eating the tops of the tallest trees, mainly conifers, ginkgos, and tree ferns. Meat-eating carnivores (such as *Allosaurus*) followed the herbivores and by pack-hunting were able to overcome and kill even the largest sauropod.

Frogs, sphenodons and lizards made their home in and around the lakes and streams, which were also inhabited by turtles and crocodiles, the latter being the top predator of these aquatic habitats. Pterosaurs, probably living on lake margins, also scanned the

lakes in search of fish. Meanwhile another group of small animals was keeping a low profile in caves and in trees, waiting for their day; these small, primitive mammals were mainly rat-like in appearance. Their food consisted mostly of insects; although some were true carnivores, their prey would of necessity have been small. Most were probably nocturnal (and arboreal) in habit in order to survive the threat of the great carnivorous dinosaurs.

COMPARISON OF THE MORRISON FORMATION WITH OTHER DINOSAUR SITES

Tendaguru Formation, Tanzania

The Upper Jurassic Tendaguru Beds of Tanzania outcrop about 75 km (c. 47 miles) north-west of Lindi, and are the richest deposits of Late Jurassic strata in Africa. This area was formerly part of German East Africa, and German expeditions from 1909 to 1913, led by Werner Janensch and Edwin Hennig, discovered huge accumulations of dinosaur bones comparable in their numbers, age, and taxa to those of the Morrison Formation. Approximately 100 articulated skeletons and many tons of bones were collected and sent to the Museum für Naturkunde in Berlin for study, where many remain today still in their plaster jackets!

The Tendaguru Formation of the Somali Basin differs from the Morrison in having marine horizons. Three members of terrestrial marls are separated by marine sandstones containing Kimmeridgian/Tithonian ammonites. At this time the Tendaguru region was situated between 30° and 40° S. The complex is approximately 140 m (c. 460 ft) thick in total and the depositional regime is interpreted as lagoonal or estuarine within the margin of a warm, epicontinental sea. Russell *et al.* (1980) considered that, as with the Morrison Formation, the dinosaur bones accumulated following mass mortality during periodic regional drought.

The fauna is similar to that of the Morrison in being dominated by giant sauropods, especially the huge *Brachiosaurus*, which was up to 25 m (c. 80 ft) in length and weighed 50,000–80,000 kg (c. 50–80 tons). The forelimbs were longer than the hindlimbs so that it had an upright, giraffe-like posture and stood up to 16 m (c. 50 ft) tall. Although *Brachiosaurus* was first described from fragmentary remains in the Morrison Formation, it is rare in North America and is better known from more complete skeletons from Tendaguru. Other dinosaurs include *Barosaurus* and *Dicraeosaurus* (both diplodocids), *Kentrosaurus* (a stegosaur), and the small theropod *Elaphrosaurus*, which also occurs in the Morrison Formation.

There are some notable differences between the Morrison and Tendaguru faunas, the most obvious being the rarity of large theropod dinosaurs such as *Allosaurus* in the latter. In addition to dinosaurs, the vertebrates include crocodiles, bony fish, sharks, pterosaurs, and mammals. Invertebrates include cephalopods, corals, bivalves, gastropods, brachiopods, arthropods, and echinoderms, all inhabitants of the shallow epicontinental sea. A flora of silicified wood plus a microflora of dinoflagellates, spores, and pollen may yield new palaeoecological data.

MUSEUMS AND SITE VISITS

Museums

1. American Museum of Natural History, New York, USA.
2. Museum of the Rockies, Montana State University, Bozeman, Montana, USA.
3. Field Museum of Natural History, Chicago, Illinois, USA.
4. Carnegie Museum of Natural History, Pittsburgh, Pennsylvania, USA.
5. Geological Museum, University of Wyoming, Laramie, Wyoming, USA.
6. Black Hills Institute of Geological Research, Hill City, South Dakota, USA.
7. Dinosaur National Monument, Vernal, Utah, USA.
8. The Wyoming Dinosaur Center, Thermopolis, Wyoming, USA.
9. Fossil Cabin Museum, Como Bluff, Medicine Bow, Wyoming, USA.
10. The Natural History Museum, London, UK.
11. Saurier Museum, Aathal, Switzerland.
12. Museum für Naturkunde der Humboldt-Universität zu Berlin, Germany.
13. Science Museum of Minnesota, St. Paul, Minnesota, USA.

Sites

There are many places to view the Morrison Formation, covering as it does such a vast area of the western United States. Most spectacular is Dinosaur National Monument in Utah, while for organized collecting the Wyoming Dinosaur Center and Dig Sites is excellent and accessible. Dinosaur National Monument was originally established to protect a quarry containing 1,600 exposed dinosaur bones from 11 different species of dinosaur (**244**). The Quarry Visitor Center is 11 km (c. 7 miles) north of Jensen, which is 21 km (c. 13 miles) east of Vernal, Utah on Highway 40. The main quarry face is currently closed to the public pending redevelopment, but a fossil trail through the rock sequence is accessible by shuttle bus. The short trail includes a rock face in the same bed as in the closed building where abundant dinosaur remains can be seen *in situ* in the rock. It is open daily except Thanksgiving, December 25, and January 1. The Wyoming Dinosaur Center and Dig Sites at Thermopolis, in the Big Horn Basin of Wyoming, consists of a new museum exhibiting many specimens collected on the adjacent land. Interpretive dig site tours visit several active collecting sites on an adjacent 6,000 ha (15,000 acre) ranch, while the 'dig-for-a-day' program allows visitors to work alongside professional palaeontologists in the field (website www.wyodino.org). The small town of Morrison, 25 km (c. 15 miles) west of Denver on Highway 70 is also worth visiting. The nearby Dinosaur Ridge, site of the original discoveries, now includes a geological nature trail which follows the sequence through the overlying Cretaceous, down into the Morrison Formation. Dinosaur footprints and bones can be observed *in situ*, including some spectacular dinosaur trackways in the Lower Cretaceous Dakota Sandstones.

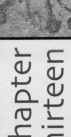

Chapter Thirteen

THE SOLNHOFEN LIMESTONE

BACKGROUND: MESOZOIC LITHOGRAPHIC LIMESTONES (PLATTENKALKS)

Both marine and terrestrial Jurassic ecosystems have already been analysed in detail (Chapter 11, The Holzmaden Shale and Chapter 12, The Morrison Formation) from the Lower and Upper Jurassic, respectively. However, the geological record of the mid to late Mesozoic includes a disproportionately high number of Fossil-Lagerstätten, due mainly to palaeogeographic conditions during the late Jurassic and early Cretaceous which resulted in a concentration of restricted marine basins at that time.

Many of these basins display a typical facies of very finely laminated, micritic limestones, known as 'lithographic limestones' (some horizons are ideal for lithographic printing) or, more accurately, as 'Plattenkalks', a German word meaning 'platy limestones', which conveys the idea of the lateral continuity of these beds for many kilometres. For a variety of reasons Plattenkalks often display exquisite preservation of soft tissues. Moreover, Plattenkalks often preserve a more complete ecosystem, including aquatic and terrestrial animals and plants, and thus give a fuller picture than is portrayed, for example, in the more restricted environments of the Holzmaden Shale and the Morrison Formation.

Most celebrated and most important of the various Mesozoic Plattenkalk Lagerstätten is the Solnhofen Limestone of Bavaria in southern Germany (249). Although fossils are by no means common, this limestone has produced over the years a range of spectacular specimens which illustrate the richness of life at the very end of the Jurassic. They include the delicate remains of vascular and non-vascular plants, a whole range of marine and terrestrial invertebrates (including preservation of the soft tentacles of squids and the fragile wings of dragonflies), fish and marine reptiles, rare dinosaurs, flying reptiles (sometimes complete with wing membranes), and most famous of all, the only known examples of *Archaeopteryx*, the world's earliest known bird, complete with its feathers.

HISTORY OF DISCOVERY AND EXPLOITATION OF THE SOLNHOFEN LIMESTONE

There is a long history of exploitation of the Solnhofen Limestone, first as a building stone and later as a lithographic printing stone. The regular bedding and the ease with which it splits along bedding planes into flat blocks or thin sheets have meant that it has been used at least since Roman times for building and for floor and roof tiles. It is still worked today, almost entirely by hand, and produces beautifully coloured floor and wall tiles for domestic use. At the end of the eighteenth century it was also discovered that certain fine-grained, porous, but hard layers of the Solnhofen Plattenkalk were ideal for lithographic printing, a process that originally required writing in oily ink onto a polished surface of limestone and then etching the exposed limestone in weak acid prior to printing. Hence its common name of 'lithographic limestone'.

The area of outcrop of the Solnhofen Limestone is a high plateau known as the Southern

249 Locality map showing the Plattenkalk basins in the Solnhofen Limestone region of the Southern Franconian Alb, Bavaria in southern Germany (after Barthel *et al.*, 1990).

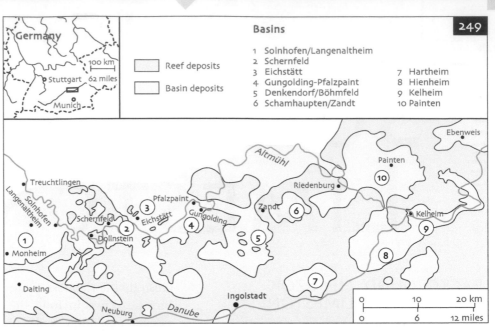

Basins

1 Solnhofen/Langenaltheim
2 Schernfeld
3 Eichstätt
4 Gungolding-Pfalzpaint
5 Denkendorf/Böhmfeld
6 Schamhaupten/Zandt
7 Hartheim
8 Hienheim
9 Kelheim
10 Painten

Reef deposits

Basin deposits

250 Manual working of the Solnhofen Limestone at Haardt Quarry, Solnhofen, in the Southern Franconian Alb of southern Germany.

251 Quarrying of domestic tiles in the Solnhofen Limestone at Haardt Quarry (see **250**), illustrating the fine lamination of the Plattenkalks.

252 Fine lamination of Plattenkalk beds in the Solnhofen Limestone at Haardt Quarry (see **250**), showing thicker beds of 'Flinz' and thinner beds of 'Fäule'.

Franconian Alb, which lies to the north of Munich in Bavaria (**249**). Outcrop is patchy, with massive biohermal limestones surrounding several distinct Plattenkalk basins. The main quarries are concentrated in the western part of the area, where the limestone is purer, especially around the small village of Solnhofen and the old Baroque town of Eichstätt (**250–252**).

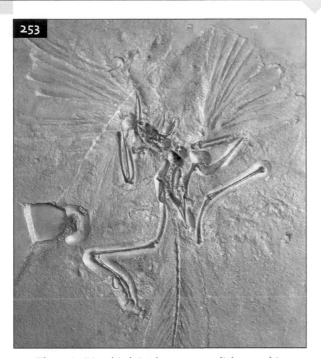

253 The primitive bird *Archaeopteryx lithographica* – the 'London specimen' (cast in JM; original in NHM). Wingspan 390 mm (c. 15 in).

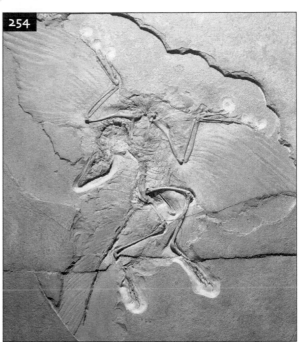

254 The primitive bird *Archaeopteryx lithographica* – the 'Berlin specimen' (HMB). Wingspan 430 mm (c. 17 in).

Fossils must have been known from these limestones throughout the period of its exploitation, but the real interest began in 1861 when a single fossil feather was discovered near Solnhofen. Further sensation quickly followed when an almost complete skeleton with a fan-like tail and feathered wings, and lacking only the skull, was found later the same year (**253**). Although portraying some reptilian features, this was clearly a fossil bird. Its discovery was timely – just 2 years after Darwin's publication of *The Origin of Species*, this appeared to be his predicted 'missing link' between reptiles and birds. The specimen, later described as *Archaeopteryx lithographica* (von Meyer, 1861), was given to the local doctor, Carl Häberlein (in lieu of medical fees), who sold it to the British Museum in London (now the Natural History Museum) as part of a collection of Solnhofen fossils.

Fifteen years passed before another specimen was found, this time at Blumenberg, near Eichstätt (**254**). The 1876 specimen, complete with skull, was sold to Berlin Museum by Häberlein's son, Ernst. Only nine further specimens have ever been discovered: the Maxberg specimen, found in 1956

and since lost; the Haarlem specimen, found in 1855, but only recognized as *Archaeopteryx* in 1970 (in Teylers Museum, Haarlem, Netherlands, where it had been displayed as a pterosaur); the Eichstätt specimen, a juvenile, found in 1951, but only recognized as *Archaeopteryx* in 1973 (in the Jura Museum, Eichstätt); the Solnhofen specimen, found in the 1960s, but not described until 1988 (in the Bürgermeister Müller Museum, Solnhofen); the Munich specimen, found in 1992, and described as a new species, *Archaeopteryx bavarica* (in the Munich Museum); the Daiting specimen (in a private collection, but on permanent loan to the Munich Museum and from the younger Mörnsheim Beds; **255**); the ninth specimen, a single wing, found in 2004 (also private collection, but currently in the Bürgermeister Müller Museum, Solnhofen); the Thermopolis specimen, arguably the most informative skeleton, which appeared in 2005 in the estate of another private collector (and now in the Wyoming Dinosaur Center, Thermopolis, Wyoming; see p. 155); and the eleventh specimen, revealed in 2011 (currently in a private collection and yet to be described).

STRATIGRAPHIC SETTING AND TAPHONOMY OF THE SOLNHOFEN LIMESTONE

The Solnhofen Limestone (strictly the Solnhofen Formation) is divided into a lower and upper Plattenkalk, both dating from the lower part of the Lower Tithonian (Upper Jurassic, zone of *Hybonoticeras hybonotum*, approximately 150 million years ago; **255**). Its outcrop in the Southern Franconian Alb covers an area of 70 km (c. 45 miles) by 30 km (c. 20 miles) and the 95 m (c. 300 ft) sequence is thought to represent a period of about half a million years (Viohl, 1985).

At the beginning of the Jurassic all of the Southern Franconian Alb lay under a wide shelf sea, which by middle Jurassic times had connections with the Tethys Ocean to the south (Barthel *et al.*, 1990, Figure 2.6). It is generally accepted that the Solnhofen Limestone was deposited in a restricted lagoon within this shelf sea, protected by the Mitteldeutsche Schwelle landmass to the north (see Chapter 6, The Hunsrück Slate, p. 66) and by coral reefs to the south and east, separating the lagoon from the Tethys Ocean.

255 Diagram to show the stratigraphy of the Upper Jurassic succession in the Solnhofen–Eichstätt area (after Barthel *et al.*, 1990).

By the late Jurassic (Oxfordian Stage), sponges and cyanobacteria (blue-green 'algae') began to build reef mounds in this lagoon and micritic lime mud was deposited in basins between them. From the Kimmeridgian onwards the basinal deposits included Plattenkalks, and in the early Tithonian the Solnhofen Plattenkalk was deposited. The regular lamination of the Plattenkalks testifies to their deposition in quiet, protected waters (compare with Burgess Shale, Chapter 2, and Holzmaden Shale, Chapter 11), and there are two distinct facies. Thicker beds of extremely pure and well-lithified micritic limestone (known as 'Flinz') alternate with thinner, less pure, micritic limestones with foliaceous weathering (known as 'Fäule'; **252**). This clearly indicates some sort of cyclicity, but there is disagreement as to the cause. Barthel (1978) believes that the latter represent normal deposition, while the former represent storm deposits (see later text).

Such restricted lagoonal waters would have been stagnant, and high evaporation rates under the semi-arid climate led to salinity density stratification with heavy, hypersaline bottom waters hostile to life (Viohl, 1985, 1996). The toxic bottom waters may also have been anoxic (lacking oxygen) and/or further poisoned by algal blooms. The lateral continuity over large distances of the thin Plattenkalk beds, with no evidence of bioturbation of the sediment by crawling or burrowing organisms, suggests a lack of benthic fauna in the lagoon. Surface waters, on the other hand, were probably better aerated with normal salinities, allowing some planktonic (floating) and nektonic (swimming) organisms (as evidenced by their coprolites), while both nektonic and benthic forms may have survived on top of the sponge–microbial mounds which reached the oxygenated surface waters (Viohl, 1996).

The depositional model proposed by Barthel (1964, 1970, 1978) suggests that violent monsoonal storms periodically caused a mixing of surface and bottom waters with sudden fluctuations in salinity causing the death of any organisms living near the surface. Fossil evidence does point to mass mortality events and sudden death (for example, fish killed while in the process of eating). Marine faunas were also washed into the lagoon from the reef community and from the open sea, airborne pterosaurs and *Archaeopteryx* were caught in high

winds and drowned, while flying insects and plant fragments were blown across the lagoon and sank. The stagnant, hypersaline conditions excluded scavengers and slowed down microbial decay of the corpses. Some animals did live for short periods after being washed into the lagoon (for example, the famous 'death trails' of the horseshoe crab *Mesolimulus* and the crustacean *Mecochirus*, whose bodies are preserved at the ends of their trails).

The storms played a further part by stirring up carbonate ooze that had been deposited around the coral reefs and washing this sediment into the lagoon. The finer particles, resuspended in the turbulent water, were carried north to the Plattenkalk basins where they were finally deposited out of suspension as 'Flinz', rapidly burying any corpses that had fallen to the lagoon floor. (In this model the sediment is regarded as allochthonous. Keupp's depositional model [Keupp, 1977a,b] differs slightly in regarding the sediment as autochthonous, produced by cyanobacteria on the lagoon floor.)

Rapid burial ensured that intricate details of soft tissue, such as the wings of insects, the tentacles of squids, and the feathers of birds, were preserved in the fine mud as simple impressions. Occasionally organic material is preserved unaltered, such as the ink-sacs of cephalopods or the original single feather of *Archaeopteryx*, or may be replaced by calcium phosphate (francolite), most usually in the muscles of fish and cephalopods. A cyanobacterial mat on the lagoon floor (suggested by the presence of hollow spheres of coccoid cyanobacteria in the sediment) may have played a further role in this preservation by encapsulating corpses and by binding together the carbonate ooze, preserving tracks and traces.

DESCRIPTION OF THE SOLNHOFEN LIMESTONE BIOTA

Archaeopteryx. The earliest known bird in the fossil record displays a number of reptilian characteristics (**253, 254, 256**). The hands have three sharply clawed fingers that are not incorporated into the wings, the jaws are edged with sharp teeth, and it has a long, bony, reptilian tail. Avian features include long, slender legs with bird-like feet, a tail fringed by a fan of feathers, a strong furcula (wishbone) near the front of the chest, and wings with asymmetric flight feathers. The latter feature suggests that *Archaeopteryx* was able to fly (Feduccia & Tordoff, 1979). However, Ostrom (1974) suggested that it would not have been an efficient flier as the sternum lacked a keel for the attachment of the pectoralis muscles and the coracoids lacked attachment processes for the supracoracoideus muscles, which lift the wing. Ostrom (1985) thus regarded it as 'a feeble flapper'. As to its lifestyle, Martin (1985) and Yalden (1985) argued for an arboreal habitat, the latter showing that the claws of the manus had a climbing function as with treecreepers and woodpeckers.

Compsognathus. The only dinosaur from the Solnhofen Limestone was a tiny, chicken-sized coelurosaurian theropod (**257, 258**). The long neck and small, swivelling head were balanced by a long tail. Long, powerful hindlimbs contrast with short forelimbs with two-fingered hands. The skeleton of this bird-like dinosaur has many features in common with *Archaeopteryx*. The single Solnhofen specimen contains the skeleton of a lizard in its stomach. A second specimen was discovered in 1972 from the Tithonian of Provence in France, and in 2006 a new compsognathid dinosaur, *Juravenator*, was described from the Kimmeridgian plattenkalks of Eichstätt, just beneath the Solnhofen Limestone.

256 Reconstruction of *Archaeopteryx*.

257 The small theropod dinosaur *Compsognathus longipes* (cast in MM; original in BSPGM). Width of block 310 mm (c. 12 in).

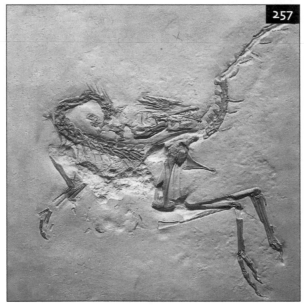

258 Reconstruction of *Compsognathus*.

259 The rhamphorhynchoid pterosaur *Rhamphorhynchus muensteri* (JM). Length 510 mm (c. 20 in).

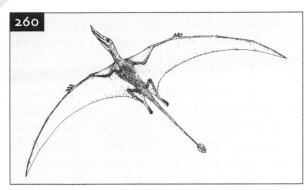

260 Reconstruction of *Rhamphorhynchus*.

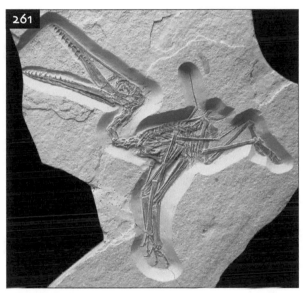

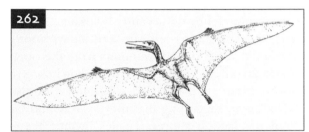

262 Reconstruction of *Pterodactylus*.

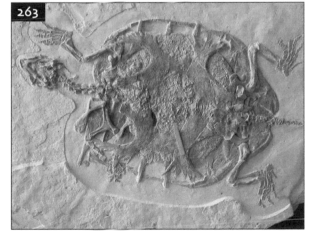

261 The pterodactyloid pterosaur *Pterodactylus kochi* (BMM). Length 220 mm (c. 8.7 in).

263 The cryptodiran turtle *Solnhofia* (BMM). Length 400 mm (c. 16 in).

264 The holostean fish *Lepidotes maximus* (BMM). Length 1.06 m (c. 3 ft 6 in).

265 The predatory fish *Caturus furcatus* with half-swallowed prey (*Tharsis dubius*) (JM). Length (tail to tail) 500 mm (c. 20 in).

Pterosaurs. Solnhofen pterosaurs were generally small compared to the Cretaceous giants of the Santana Formation (Chapter 16). They include 'long-tailed' rhamphorhynchoids, such as *Rhamphorhynchus* (**259, 260**) and *Scaphognathus*, with wingspans up to 1 m (c. 3 ft), and 'short-tailed' pterodactyloids, such as the thrush-sized *Pterodactylus* (**261, 262**). Wing membranes and the rudder-like tail membranes may be preserved as impressions, sometimes showing hair-like coverings, and some specimens show webbing between the toes.

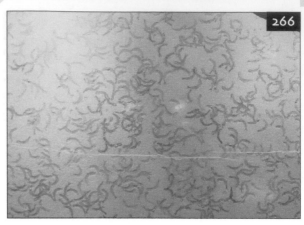

266 The teleostean fish *Leptolepides sprattiformis* (BMM). Length of individual fish about 50 mm (c. 2 in).

Other reptiles. These include ichthyosaurs, plesiosaurs, crocodiles, turtles, lizards, and sphenodons. Ichthyosaurs and plesiosaurs (known from a single tooth) are rare and always poorly preserved. Such strong swimmers from the open sea would only be washed into the lagoon in severe storms. Lizards and lizard-like sphenodons are also uncommon, suggesting that they lived inland far from the lagoon shores. Turtles include fresh-water/coastal types (**263**) and crocodiles include marine and terrestrial forms, but again they are rare in the Plattenkalks.

267 The shark *Phorcynis* (BMM). Length 300 mm (c. 11.8 in).

Fish. The most common Solnhofen vertebrates include actinopterygians (ray-finned fish), sarcopterygians (lobe-finned fish), and chondrichthyans (cartilaginous fish). Actinopterygians are mostly holosteans (the bony ganoid fish such as *Lepidotes* (**221, 264**), up to 2 m (c. 6 ft) long, *Gyrodus*, resembling a parrot fish, and *Caturus*, a giant predator (**265**), but some of the early teleosteans (modern bony fish) are also known, such as the sprat-like *Leptolepides*, commonly found in mass mortality assemblages (**266**). Sarcopterygians are not common, but include the small coelacanth *Coccoderma*, while the chondrichthyans include sharks (**267**), rays (**268**), and chimaeras (ratfish).

268 The rayfish *Aellopos* (BMH). Length approx. 1 m (c. 3 ft).

269 The crustacean *Aeger tipularius* (BMH). Length 65 mm (c. 2.5 in).

270 Reconstruction of *Aeger*.

271 The horseshoe crab *Mesolimulus walchi*, preserved at the end of its death trail (JM). Total length of trackway 3.48 m (c. 11 ft 5 in).

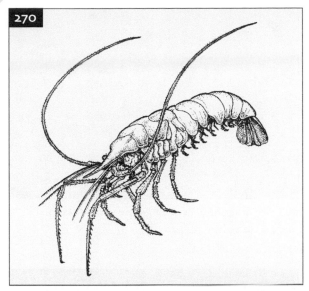

Crustaceans. Decapod crustaceans (shrimps, lobsters, and crabs) are perhaps the best known of Solnhofen's marine invertebrates, and common genera include *Aeger* (**269, 270**), *Mecochirus*, and *Cycleryon*. *Mecochirus* and the horseshoe crab *Mesolimulus* are commonly found fossilized at the end of a spiral or haphazard trail, suggesting that they landed on the toxic lagoon bottom, struggled for a few steps and then died (**271**).

Insects. Terrestrial arthropods are usually preserved as impressions; only winged insects are known, but fine detail of wing venation is often preserved. They include mayflies, dragonflies (**272**), cockroaches and termites, water skaters, locusts and crickets, bugs and water scorpions, cicadas, lacewings, beetles, caddis flies, true flies, and wasps.

Other invertebrates. Most marine invertebrate groups are represented including sponges, jellyfish (**273**), corals (**274**), annelids, bryozoans, brachio-

pods, bivalves, gastropods, cephalopods (squids, belemnoids, nautiloids, and ammonoids), and echinoderms (crinoids, starfish, brittle stars, sea urchins, and sea cucumbers). The squids (such as *Acanthoteuthis*) often have their tentacles and tentacular hooks preserved as impressions (**275, 276**), and their ink sacs preserved as original carbon. The floating crinoid *Saccocoma* is one of the Solnhofen's most common fossils (**277**).

Plants. All the vascular plants from Solnhofen are gymnosperms. They include pteridosperms (seed ferns), bennettitales, ginkgos, and conifers, but there is no evidence of large trees.

Trace fossils. Trace fossils are important in the Solnhofen Limestone and include coprolites, such as the worm-like *Lumbricaria*, which includes disaggregated plates of floating crinoids. Different ichnospecies of *Lumbricaria* probably represent the faeces of fish and squids and indicate that the lagoon was not entirely devoid of autochthonous

272 The dragonfly *Tarsophlebia eximia* (BMH). Wingspan 75 mm (c. 3 in).

273 The jellyfish *Rhizostomites admirandus* (BMM). Diameter 320 mm (c. 12.5 in).

274 A gorgonian coral (JM). Diameter 240 mm (c. 9.5 in).

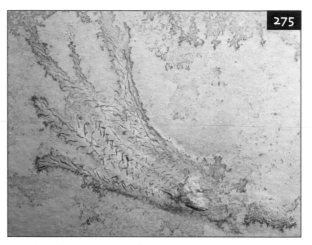

275 The squid *Acanthoteuthis* (JM). Length of longest tentacle 170 mm (c. 6.7 in).

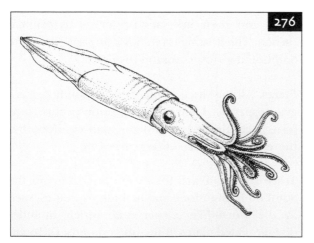

276 Reconstruction of *Acanthoteuthis*.

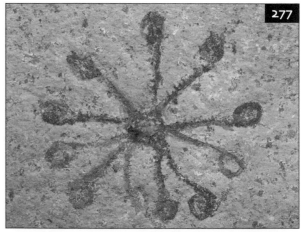

277 The floating crinoid *Saccocoma tenellum* (MM). Width 35 mm (c. 1.4 in).

life. Settling traces, of ammonites for example, show that some bottom waters were stagnant, while drag marks suggest weak currents in some places. The trails of crustaceans have already been described.

The fauna and flora from Solnhofen is well documented and illustrated by Frickhinger (1994).

PALAEOECOLOGY OF THE SOLNHOFEN LIMESTONE BIOTA

The Solnhofen Limestone Plattenkalks represent a shallow, saline lagoon community, situated within the subtropical zone at about 25°–30°N, with a semi-arid monsoonal climate. Stagnant, hyper-saline bottom waters inhibited colonization or scavenging and there was no autochthonous benthic life except for the salt-loving cyano-bacterial mat.

Better-aerated surface waters with normal salinities may have supported limited planktonic and nektonic life, at least for short periods following episodes of water mixing. For example, floating crinoids are found in large numbers at some horizons, and these were clearly prey to both fish and squids within the lagoon whose coprolites containing crinoid remains were deposited on the lagoon floor. Viohl (1996) identified four habitats occupied by fish, and includes the oxic surface waters of the lagoon as well as the tops of the sponge–microbial mounds. The latter projected upwards into the equable surface waters and some benthic fauna, especially crustaceans, also lived on these tops.

The inhospitable lagoon was separated from the open sea by a series of coral reefs, which supported a rich marine fauna of invertebrates, fish, and marine reptiles. The Solnhofen biota is biased towards those forms most likely to be washed into the lagoon by storms (Barthel *et al.*, 1990, p. 89). From the reef community planktonic floaters, such as crinoids, jellyfish, and oysters attached to seaweed, were accompanied by weak swimmers such as ammonites and small fish which swam around the reef. Less common are the stronger swimmers, such as squids, large predatory fish, and marine reptiles (ichthyosaurs, plesiosaurs, and crocodiles), all of which inhabited the open sea. Vagrant benthos such as crustaceans, horseshoe crabs, gastropods, echinoids, and starfish, were more likely to be washed in than the sessile

benthos. Thus, the benthic epifauna, such as sponges, corals, bryozoans, brachiopods, and attached bivalves (e.g. *Pinna*), and the benthic infauna, including burrowing bivalves (e.g. *Solemya*) and polychaete worms (e.g. *Ctenoscolex*), are often represented only by fragments.

An adjacent low-lying landmass to the north of the lagoon supported an abundance of shrubby vegetation of low conifers and other gymnosperms adapted to salty soil. There were no large trees. Among this vegetation lived a variety of aquatic, semi-aquatic, and terrestrial insects, which would have been a rich source of food for small lizards, the beaked rhynchocephalians, and birds (*Archaeopteryx*). Bird-like theropod dinosaurs in turn fed on the lizards.

Coastal waters were home to turtles and crocodiles, while the shores were inhabited by numerous pterosaurs, mostly in search of fish (as evidenced by stomach contents) although some were possibly insectivores (suggested by their teeth). Again the Solnhofen biota is biased, the majority of the terrestrial component being flying forms, either insects, pterosaurs, or birds, which could traverse the lagoon. Land-based animals, such as the lizards and dinosaurs, are under-standably extremely rare.

More than 600 species have been described from the Solnhofen Limestone, representing a number of different environments. Perhaps the most striking feature is that the vast majority of these are allochthonous, having been swept into the lagoon from the reef community, the open sea, or from the adjacent land.

COMPARISON OF THE SOLNHOFEN LIMESTONE WITH OTHER MESOZOIC BIOTAS

Lebanon

Although somewhat younger in age the Cretaceous Plattenkalks of Lebanon are comparable with Solnhofen in their appearance, preservation, and in many aspects of their biota. The four main basins of Hâqel, Hjoûla, en Nammoûra, and Sâhel Aalma are all near the coast to the north of Beirut, and are all Upper Cretaceous in age, the first three being Cenomanian, while Sâhel Aalma is from the slightly younger Santonian Stage.

The large number of fossils preserved in these

Plattenkalks is due to mass mortalities caused by huge seasonal blooms of planktonic micro-organisms. The dead organisms accumulated on the bottom of small marine basins, where, due to low circulation of water, oxygen content was low or absent and there was high acidity and salinity (compare also with the Santana Formation, Chapter 16).

The biota is dominated by fish which include chondrichthyans (cartilaginous fish), actino-pterygians (ray-finned fish), and a single sarcopterygian (lobe-finned fish). Sharks are relatively rare and are usually small, less than half a metre (1.5 ft) in length. Rays, however, are common, their flat bodies lending themselves to exquisite preservation with specimens dorso-ventrally flattened. Actinopterygians are diverse and include species from a number of different habitats, including sandy shoals, coral reefs, deep water, and surface waters of the open sea. Pycnodonts, akin to Solnhofen's *Gyrodus*, are common, as are small teleosteans such as the herring-like clupeomorphs. Other vertebrates include marine turtles, rare pterosaurs, at least one articulated enantiornithine bird (the 'opposite birds'; see p. 176), isolated feathers, and a theropod dinosaur. The invertebrates from Lebanon are dominated by decapod crustaceans, just as at Solnhofen, with the shrimp *Carpopenaeus*, being the iconic fossil of the Lebanese Plattenkalks. Cuttlefish, squids, and octopuses are all found with soft-tissue preservation of their tentacles and ink sacs. Echinoids, crinoids, ophiuroids, and ammonites complete the invertebrate fauna along with occasional polychaete annelids ('bristle worms'), which are extremely rare in the fossil record, but which do also occur in the Solnhofen Limestone.

MUSEUMS AND SITE VISITS

Museums

1. Bayerische Staatssammlung für Paläontologie und historische Geologie, München, Germany.
2. Bürgermeister Müller Museum, Solnhofen, Germany.
3. Carnegie Museum of Natural History, Pittsburgh, USA.
4. Jura-Museum, Eichstätt, Germany.
5. Museum auf dem Maxberg, Solnhofen, Germany.
6. Museum Bergér, Harthof, Germany.
7. Museum für Geologie und Mineralogie, Dresden, Germany.
8. Museum für Naturkunde der Humboldt-Universität zu Berlin, Germany.
9. Naturkunde-Museum Bamberg, Germany.
10. Naturmuseum Senckenberg, Frankfurt, Germany.
11. Staatliches Museum für Naturkunde, Karlsruhe, Germany.
12. Staatliches Museum für Naturkunde, Stuttgart, Germany.
13. Teylers Museum, Haarlem, Netherlands.
14. American Museum of Natural History, New York, USA.
15. Manchester University Museum, Oxford Road, Manchester, UK.
16. The Natural History Museum, London, UK.

Sites

Many of the working quarries exclude fossil hunters. However some quarries are accessible, such as the one adjacent to the small, private Museum Bergér at Harthof. Head north-west from Eichstätt on the B13 across the Altmühl valley towards Weißenburg. After a few kilometres turn left towards Schernfeld and then almost immediately left again to Harthof. The museum displays some excellent Solnhofen fossils and on payment of a small fee you can visit the adjacent working limestone quarry to collect. The floating crinoid *Saccocoma* is common; other fossils are not, but remember that the Berlin *Archaeopteryx* was found here in 1876! Another excellent collecting opportunity is the extensive Haardt Quarry a few kilometres outside the village of Solnhofen (**250**). Visits must be booked at the Bürgermeister Müller Museum in Solnhofen, where hammers and chisels can also be hired. Ask at the museum for directions to the quarry, where fish and ammonites are not uncommon. Hard hats are not required, but the quarries do become very muddy in wet weather.

Chapter Fourteen

THE JEHOL GROUP

BACKGROUND: THE EMERGENCE OF FEATHERED DINOSAURS, BIRDS, AND FLOWERING PLANTS

During the Cretaceous Period the Earth was warmer than at any other time in its history – atmospheric CO_2 was high and the greenhouse effect led to warm, dry climates. Polar ice caps melted and the rise in sea level flooded large areas of the continents with shallow seas. At the end of the early Cretaceous intense volcanic activity, particularly at mid-ocean ridges, lifted ocean floors and sea levels rose even further. The volcanic activity released more CO_2, further exacerbating the greenhouse effect. Atmospheric oxygen levels were also continually rising during the late Jurassic and throughout the whole of the Cretaceous to a peak of around 35%, which is higher than at any other time of the Earth's history apart from the late Carboniferous.

Such a dramatic increase in air density must have been extremely beneficial to any organisms that were evolving powered flight as a means of exploring new niches; it is no coincidence that insects first took to the air during the Carboniferous oxygen high, nor that some of the insects at that time were gigantic, with wingspans of up to 710 mm (c. 28 in; see p. 98). In the late Mesozoic, however, not only did the insects once more undergo a dramatic radiation, it was now the turn of the vertebrates to explore the airways with flight evolving in three separate groups. Although pterosaurs had been around since the Triassic, the onset of the Cretaceous saw a similar gigantism in these flying reptiles with forms such as *Quetzalcoatlus* attaining wingspans of at least 10 m (c. 33 ft). Bats also appeared soon after the end of the Cretaceous

(see Nudds & Selden, 2008, Chapter 11), but of course the most successful group ever to take to the air were the birds. The evolution of birds, the origin of bird flight, and the origin of feathers have puzzled palaeontologists ever since Thomas Huxley first suggested in 1868 that birds may have evolved from dinosaurs, but in the last decade these issues have become hotly debated once more and are still the source of much controversy.

The evidence fuelling this renewed interest has all come from one of the most remarkable Fossil-Lagerstätten ever to have been discovered, that of the Jehol Group of Liaoning Province in north-eastern China (**278**). Here, freshwater lake sediments interbedded with volcanic tuffs have preserved numerous examples of early birds, complete skeletons of primitive mammals, some of the first flowering plants, and a huge array of insects – but most unusual of all are the incredible 'feathered dinosaurs'.

HISTORY OF DISCOVERY OF THE JEHOL GROUP

It was not until the 1990s that the Jehol biota began to hit the world headlines with the discovery of numerous fossil birds and the unexpected 'feathered dinosaurs', but it had actually been known to geologists since the 1920s from the work of the American palaeontologist Professor Amadeus Grabau (1870–1946). In his work on the rather more mundane Chinese Cretaceous molluscs Grabau (1923, 1928) coined the terms 'Jehol Series' for the fossil-bearing strata in western Liaoning, and 'Jehol Fauna' for their fossils. These terms were modified to 'Jehol Group' and 'Jehol

Biota' by Gu Zhiwei (1962), a Chinese malacologist from the Nanjing Institute of Geology and Palaeontology, although the fauna was more usually known as the 'E-E-L Biota' after its most common fossils, the conchostracan *Eosestheria*, the insect larva *Ephemeropsis*, and the fish *Lycoptera*. (The term 'Jehol' literally translates as 'hot river', after the numerous hot springs in this area, and comes from the former Jehol Province, which included what is now western Liaoning, northern Hebei, and south-eastern Inner Mongolia; **278**. This area also falls within the historical region of Manchuria, that part of north-east Asia which fell successively under Russian and Japanese rule between the two World Wars.)

In 1987 the first fossil bird from Liaoning was found by a farmer near Chaoyang City (**278**), and 3 years later another bird from this region was collected during an IVPP (Institute of Vertebrate Paleontology and Paleoanthropolgy) excavation of fossil fishes. These birds, later named *Sinornis* and *Cathayornis*, respectively, were, however, fragmentary, and the first real indication of what was to follow came in 1993 when the earliest beaked bird, *Confuciusornis*, was discovered in profusion near Beipiao City; the first significant mammal, *Zhangheotherium*, and the putative angiosperm *Archaefructus*, were also found at about this same time.

From about 1995 the local farmers of Sihetun Village, Beipiao (**278**), began digging fossils and, coupled with official excavations by the IVPP from 1997 onwards, new discoveries were made at an ever increasing rate. In 1996 the first 'feathered dinosaur' from the Jehol biota was named as *Sinosauropteryx*, while the following year saw two further feathered dinosaurs, *Caudipteryx* and *Protarchaeopteryx*, all from western Liaoning. Two more appeared in 1999 (*Beipiaosaurus* and *Sinornithosaurus*), while the year 2000 saw the discovery of a second species of *Caudipteryx* and the sixth feathered dinosaur genus, *Microraptor*, the smallest adult dinosaur ever found. It was a second species of this genus (*Microraptor gui*) which hit the press in 2003, hailed as 'the four-winged dinosaur'.

In 2002 attention turned to the basal section of the Jehol Group where specimens were discovered with their three-dimensionality preserved, albeit without the soft-tissue preservation of the previous discoveries (see section on Stratigraphic setting and taphonomy, p. 170). Two new dinosaurs, *Sinovenator* and *Incisivosaurus*, were described in that year, with two more, *Mei* and *Dilong*, described in 2004.

By 2005 research was also directed to an adjacent basin in Ningcheng in Inner Mongolia (**278**), where some apparently older deposits began to reveal further surprises (see section on

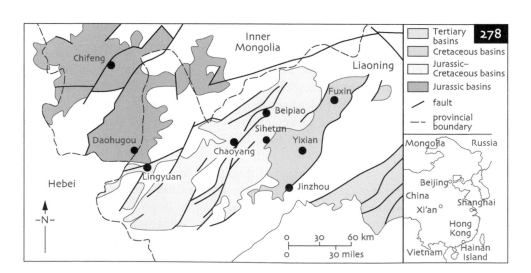

278 Locality map showing the locations of the Jehol biota in Liaoning, Hebei and Inner Mongolia (after Chang, 2003).

Comparison, pp. 181–182) and in the years that followed new discoveries of dinosaurs, birds, mammals, insects, and flowering plants have been made at a rate unprecedented in the history of palaeontology.

The Jehol biota now includes over 60 species of plants, about 100 species of vertebrates, and over 1,000 species of invertebrates, many of which have been beautifully illustrated in the volume edited by Mee-mann Chang in 2003, *The Jehol Biota*. These discoveries have revolutionised our ideas on the origin and early diversification of birds, on the evolution of both feathers and flight, on the co-radiation of angiosperms and their pollinators, the insects, and on the early history of mammals. No doubt much more awaits discovery.

STRATIGRAPHIC SETTING AND TAPHONOMY OF THE JEHOL GROUP

The Jehol Group was traditionally divided into four formations, but recent lithostratigraphic work suggests that the upper two (Shahai and Fuxin formations) are significantly different (coals and clastics) and should be excluded. Thus the Jehol Group now comprises only the lower Yixian Formation and the upper Jiufotang Formation, which outcrop in western Liaoning Province, northern Hebei Province, and south-eastern Inner Mongolia (**278, 279**). These conformable formations both include finely bedded lacustrine sediments (sandstones, shales, mudstones) intercalated with numerous thin volcanic tuffs, but the Yixian Formation succession is further interrupted by four basaltic/andesitic extrusions which subdivide the sediments into four corresponding fossil-bearing beds or 'members'. The overlying Jiufotang Formation is devoid of such intrusions and comprises a single member, making five in total, each with its own distinctive fauna (**279**).

The lowermost Lujiatun Member consists of thick fossil-bearing tuffs lacking obvious bedding planes, suggesting that this deposit represents a single mass-mortality event due to a massive ash fall. Fossils from this horizon are preserved in three dimensions and include numerous examples of the ceratopsian dinosaur *Psittacosaurus*, the troodontids *Sinovenator* and *Mei*, the tyrannosaurids *Dilong* and *Raptorex*, the enigmatic *Incisivosaurus*, and the large mammal *Repenomamus*. It is, however, the succeeding Jian-

shangou Member that has yielded most of the 'feathered dinosaurs' including *Sinosauropteryx, Protarchaeopteryx, Caudipteryx, Beipiaosaurus,* and *Sinornithosaurus,* as well as various species of the bird *Confuciusornis*. The Dawangzhangzi Member has also produced feathered dinosaurs (*Sinosauropteryx* and *Sinornithosaurus*), plus some more advanced birds, and the putative angiosperm *Archaefructus*. The penultimate Jingangshan Member has only a limited vertebrate fauna, but the uppermost Boluochi Member in the succeeding Jiufotang Formation (**279**) has produced two species of the most advanced feathered dinosaur, *Microraptor*, and numerous genera of more advanced birds (see section on biota, p. 172).

There have been a number of attempts to date this important succession using both biostratigraphy and geochronology. Biostratigraphic data have produced conflicting results ranging from late Jurassic to early Cretaceous. The endemic nature of the fish fauna yields no resolution other than late Jurassic/early Cretaceous, and this is also true for the ostracod microfauna. Evidence for a late Jurassic age has been indicated by the presence of non-pterodactyloid (long-tailed) pterosaurs, but conversely the presence of the ceratopsian dinosaur *Psittacosaurus* indicates a Cretaceous age. Vertebrate fossils are, however, generally unreliable for biostratigraphic dating.

Isotopic dating has produced similarly conflicting results, but the most recent work of Swisher *et al.* (1999) and Wang *et al.* (2001) seems to be the most convincing. An argon–argon date of 128.4 Ma was obtained from basalt capping the lowermost Lujiatun Member, while dates of 124.6 Ma and 125.0 Ma have been obtained from sanidine and biotite crystals from tuff layers in the succeeding Jianshangou Member. Finally, at the top of the succession, an argon–argon date of 110 Ma has been obtained from basalt within the Jiufotang Formation (**279**). These dates would suggest that the Jehol biota ranges from late Hauterivian to early Aptian in age and existed for a minimum of 18 Ma (Zhou *et al.*, 2003).

To date there has been little detailed research on the taphonomy of the biota, but clearly there are two distinct scenarios. Fossils from the upper four members (**279**) are preserved as flattened compressions on the upper bedding planes of thin-bedded, fine-grained, grey mudstones, and soft

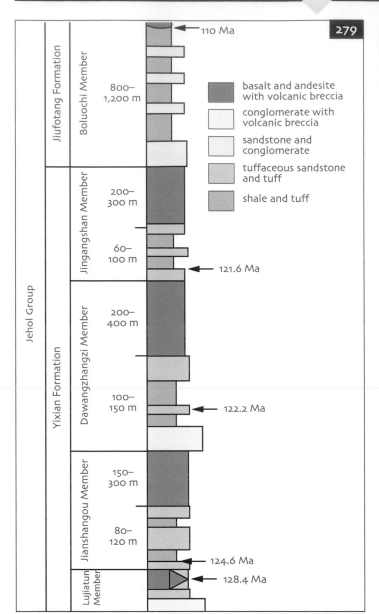

280 The main quarry at Sihetun exposing the lower part of the Yixian Formation (Jianshangou Member).

279 Diagram to show the stratigraphy of the Jehol Group (after Chang, 2003).

281 Detail of bedding (as at 280) showing the alternating layers of grey muds and orange tuffs.

tissue, including integumentary structures (filaments, feathers, and fur) in dinosaurs, pterosaurs, and birds, and wing membranes and colour patterns in insects, is often preserved in great detail. It is significant that the most fossiliferous horizons are overlain by a volcanic tuff, and that the fossils, especially birds, are sometimes found in great numbers on particular bedding planes, suggesting mass-mortality events. The thin bedding of both the mudstones and the tuffs gives a striped appearance to the strata (280, 281) and alternating layers of grey muds and orange tuffs can be traced undisturbed for many tens of metres.

The presence of autochthonous freshwater molluscs and arthropods (see later text) and the absence of the lateral heterogeneity usually associated with river sediments, suggests that the Jehol beds are lacustrine rather than fluvial/deltaic in nature and that the fossils represent carcasses that have fallen to the lake bed. The tuff beds testify to frequent, episodic volcanic eruptions releasing large quantities of poisonous gases into the atmosphere which possibly overcame the birds, dinosaurs, and mammals living on the lake margins or in the trees surrounding the lake. Vast clouds of ash would have been ejected over the water surface and gradually settled out of suspension and buried the carcasses on the lake bed. Rapid entombment in such a fine-grained sediment would have sealed the carcasses in an environment which was both anoxic and toxic, and would not only have slowed down microbial decay of soft tissue, but would also have prevented disturbance of the carcasses by scavengers or burrowing organisms.

Fossils from the lowermost Lujiatun Member, on the other hand, while not preserving soft tissue, do preserve their three-dimensionality. They occur in thick beds of tuff (up to 3 m [c. 10 ft]), which lack any obvious bedding planes, suggesting that this deposit represents a single catastrophic mass-mortality event due to a massive ash fall. It has often been described as the 'Mesozoic Pompeii', and it seems that these animals were buried alive in the fine ash and are thus preserved in life-position, rather than as a death assemblage. These are often referred to as the 'terrestrial beds', and received much press coverage a few years ago from the discovery of the small troodontid dinosaur known as *Mei long* (the 'soundly-sleeping dragon'), which preserves dinosaur behaviour by being fossilized in the classic roosting position of modern birds with its head tucked underneath its 'wing'. The numerous specimens of the small ceratopsian dinosaur *Psittacosaurus*, commonly offered for sale at fossil shows, are also from this horizon.

DESCRIPTION OF THE JEHOL GROUP BIOTA

Dinosaurs. The first feathered dinosaur to be described was *Sinosauropteryx prima*, in 1996. This tiny, chicken-sized theropod, with a large skull bearing sharp teeth and a very long tail, belongs to the compsognathid group of dinosaurs, better known from the Upper Jurassic Solnhofen Limestone of Germany (Chapter 13). Its discovery caused a minor sensation due to the presence of a frill of downy material running down its entire dorsal surface (**282, 283**). Described as 'proto-feathers', and thought to be the forerunner of true feathers, these hair-like structures presumably developed initially for insulation (see Ji & Ji, 1996; Chen *et al.*, 1998). In 2010 Mike Benton and his colleagues recognized different types of melano-somes (colour-bearing organelles) within the protofeathers which suggested that this dinosaur had a ginger and white striped coloration – the first time that actual colour had been identified in a dinosaur (Zhang *et al.*, 2010).

Protarchaeopteryx robusta and *Caudipteryx zoui* both possess long, bird-like legs and small skulls, and are oviraptosaurian dinosaurs. Their tails and other parts of their bodies are adorned, not with 'protofeathers', but with true feathers composed of a shaft and symmetrical vanes, the symmetry suggesting that these animals were incapable of flight (**284–286**). This was the first time that feathers had been discovered on any non-avian animal (see Ji & Ji, 1997; Ji *et al.*, 1998).

While these first three feathered dinosaurs are all rather small theropods, *Beipiaosaurus inexpectus* is the largest feathered dinosaur known, well over 2 m (c. 6 ft) in length. It belongs to the therizino-sauroids, a little-known group of Asian dinosaurs with a peculiar mix of characteristics, which have at times been placed with the ornithischians, the sauropods, and the theropods. Now known to be undoubted theropods (on the basis of beautifully preserved embryos from Henan Province in China; Kundrat *et al.*, 2008), the group is characterized by possessing unusually long, curved, sharp claws on its hands, as in many carnivores, but has un-doubted herbivorous teeth. *Beipiaosaurus* does not possess true feathers, but has large patches of filamentous structures similar to the 'proto-feathers' of *Sinosauropteryx* (see Xu *et al.*, 1999a).

283 Reconstruction of *Sinosauropteryx*.

282 The compsognathid 'feathered' dinosaur *Sinosauropteryx prima* (NIGP). Length of block 670 mm (c. 26.4 in).

285 Reconstruction of *Caudipteryx*.

284 The oviraptosaurian 'feathered' dinosaur *Caudipteryx zoui* (cast in JM). Length 725 mm (c. 28.5 in).

286 Detail of feathers of the oviraptosaurian 'feathered' dinosaur *Protarchaeopteryx robusta* (NGMC). Scale bar = 5 mm (c. 0.2 in).

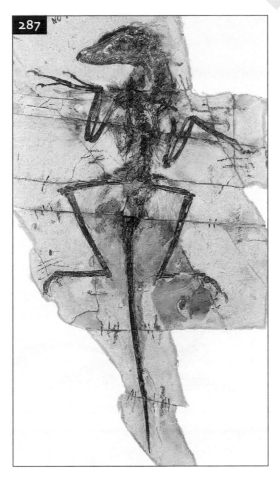

288 Reconstruction of *Sinornithosaurus*.

287 The dromaeosaurid 'feathered' dinosaur *Sinornithosaurus millenii* (NGMC). Length 590 mm (c. 23.2 in).

Sinornithosaurus millenii belongs to yet another group of dinosaurs, the notorious dromaeosaurs. This genus is one of the most bird-like of all the feathered dinosaurs, with long arms that could flap, and a stiffened rod-like tail (**287, 288**). Its entire body was covered by a layer of integumentary filaments, although it is difficult to ascertain if these were true feathers or 'protofeathers' (see Xu *et al.*, 1999b). However, if these early Cretaceous 'raptors' were covered in featherlike structures, what does this say for their later derivatives such as *Velociraptor* and *Deinonychus*? Another interesting feature is their grooved, fanglike teeth which, like modern 'rear-fanged' snakes, were possibly used to inject venom into the wounds of their victims.

Microraptor zhaoianus is another dromaeosaur, and at 390 mm long (240 mm of which is tail) is the smallest dinosaur known. It possessed bird-like teeth and the stiffened tail characteristic of all dromaeosaurs (see Xu *et al.*, 2000). It was, however, a second species of this genus, *Microraptor gui*,

which revealed some most unexpected features. Preserved integument consists of both plumulaceous and pennaceous feathers, the latter occurring not only on the tail and forelimbs, but also on the hindlimbs, earning it the nickname of 'the four-winged dinosaur' (**289, 290**). Not only are these true feathers, but some of them apparently display asymmetry with the leading vane much narrower than the trailing vane. Most probably this was a gliding animal, able to leap from tree to tree in the manner of 'flying' mammals today, before flapping flight had been developed (see Xu *et al.*, 2003).

The most recent dinosaur to be described from the Jehol Group is perhaps also the most surprising. All of the previous examples have been theropods, which belong to the major dinosaur group the Saurischia or 'reptile-hipped dinosaurs' (see p. 144). *Tianyulong confuciusi*, described in 2009, is, however, a heterodontosaurid dinosaur, which is the most basal group of the Ornithischia or 'birdhipped dinosaurs' (**291**). The palaeontological

289 The dromaeosaurid 'four-winged' dinosaur *Microraptor gui* (IVPP). Length 770 mm (c. 30.3 in).

290 Reconstruction of *Microraptor*.

291 The ornithischian dinosaur *Tianyulong* (STMN). Length of skull 54 mm (c. 2.12 in).

world was once more stunned to learn that this animal also bears long filamentous integumentary 'protofeathers', suggesting that the evolution of such structures occurred before the Saurischia–Ornithischia split, and that they were inherited by basal members of each group (see Zheng *et al.*, 2009).

Jinzhousaurus, a second ornithischian, albeit unfeathered, does deserve a mention: it is an ornithopod dinosaur, very similar to its relative *Iguanodon*, even down to the spike-like pollex (thumb), and at 8 m (c. 26 ft) in length is the largest dinosaur from the Jehol Group.

Dinosaurs from the older Lujiatun Member (**279**), which are preserved in three dimensions (see section on Stratigraphic setting and taphonomy, p. 170), do not preserve their soft tissue, and so it is impossible to know if they were feathered. However, two of these, *Sinovenator changii* and *Mei long*, belong to the troodontids, a group of dinosaurs closely related to dromaeosaurs and are extremely bird-like in their morphology with an upright posture, forward pointing eyes, and a large brain. Both are small, less than 1 m (c.

3 ft) long, and both are preserved in the classic 'roosting' position of modern birds (**292, 293**), sitting on their folded hindlimbs with their heads tucked under their elbows against their bodies. This avian strategy reduces exposed surface area and conserves body heat during sleep (see Xu *et al.*, 2002b; Xu & Norell, 2004).

Five further genera have been described from this older horizon. *Incisivosaurus* is a small oviraptosaur with a pair of premaxillary teeth resembling rodent incisors and suggesting a herbivorous diet (see Xu *et al.*, 2002a). *Dilong paradoxus* is a small basal tyrannosauroid dinosaur, only 1.6 m (c. 5 ft) long, and is the oldest known so far. A second specimen of *Dilong*, collected from the younger Jianshangou Member (**279**), shows that this genus possessed 'protofeathers', very similar to those of *Sinornithosaurus* (p. 174), and is the first evidence that tyrannosauroids possessed a hair-like covering (see Xu *et al.*, 2004). A second tyrannosauroid, *Raptorex*, was only slightly larger at 3 m (c. 10 ft) long, but was far more derived, possessing all of the functional specializations of the later giant tyrannosauroids such as the Upper Cretaceous *Tyrannosaurus rex* (Sereno *et al.*, 2009). Finally are two ornithischian dinosaurs. *Psittacosaurus* is a ceratopsian, and hence a distant relative of *Triceratops*, and is not only the most common Jehol dinosaur, but was also the first to be described in the 1970s (**294**). *Jeholosaurus* is an ornithopod, and hence related to *Jinzhousaurus* (p. 175), and also therefore to *Iguanodon*, but is a miniature form, less than 1 m (c. 3 ft) long.

The feathered dinosaurs from Liaoning have provided almost indisputable evidence that birds evolved from small, agile theropods. The vocal minority of ornithologists who reject this theory argue that such a 'ground-up' theory for the origin of flight from fast-running dinosaurs is unlikely, and that flight evolved 'trees-down', from gliding, non-dinosaur reptiles. The recent discovery of the gliding, 'four-winged' dinosaur, *Microraptor*, now suggests that the orthodoxy was only half right: birds were indeed descended from dinosaurs, but flight evolved from the trees down.

Birds. The earliest known bird in the fossil record is *Archaeopteryx* from the Upper Jurassic Solnhofen Limestone of Germany, which retains many primitive reptilian features (Chapter 13). Prior to the discovery of the Jehol birds in the early 1990s there was a large hiatus in the fossil record between *Archaeopteryx* and the essentially modern birds from the Upper Cretaceous of Kansas. The diverse fauna of birds discovered in the Lower Cretaceous Jehol Group of Liaoning Province in the last 20 years has filled this stratigraphic and morphological gap.

Confuciusornis was not the first bird to be discovered from the Jehol biota (see section on History, p. 169), but it is by far the most abundant. Since its discovery in 1993 over one thousand specimens have been collected, often concentrated on particular bedding planes, suggesting gregariousness and/or mass mortality. It first appears in the Jianshangou Member (**279**) and like *Archaeopteryx* is a basal bird, but is considerably more advanced in possessing a toothless horny beak, and an abbreviated tail with a pygostyle (the 'parson's nose') instead of the toothed jaw and long reptilian tail of the Jurassic species. (Indeed, the basal birds are nowadays classified into two groups, the 'long-tailed' variety, with a bony tail, and the 'pygostylians'.) It does, however, retain some primitive features, such as the three long 'fingers' terminating in claws external to the wing (**295**). Many specimens are preserved with beautiful flight feathers showing a shaft and asymmetrical vanes; individual barbs, and sometimes even barbules, may be preserved. The presence of two long tail feathers in some specimens, but not in others, has been interpreted as sexual dimorphism. At least four species have been described, although some of these may be synonymous (see Hou *et al.*, 1995; Chiappe *et al.*, 1999).

So rapid was the diversification of birds in the early Cretaceous that both major groups of derived birds had also appeared by the time of the Jianshangou Member (**279**). The enantiornithine birds ('opposite birds') are the dominant Mesozoic group and are characterized by a unique articulation between the coracoid and scapula that is 'opposite' to modern birds. This group is also characterized by possessing a 'bastard wing' or alula, where three to five feathers on the first digit or 'thumb' provide additional manoeuvrability during slow flight or take off, rather like the slats on an aeroplane wing. *Eoenantiornis* is the most primitive of the known enantiornithine birds; it is slightly smaller than *Confuciusornis* and comes from the same horizon in the Jianshangou Member (**279**).

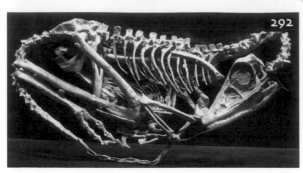

292 The troodontid dinosaur *Sinovenator changii* (PC). Length 400 mm (c. 16 in).

293 The troodontid dinosaur *Mei long* (PC). Length of block 170 mm (c. 6.7 in).

294 The ceratopsian dinosaur *Psittacosaurus* with juveniles (STMN). Length of specimen approx. 1 m (c. 3 ft).

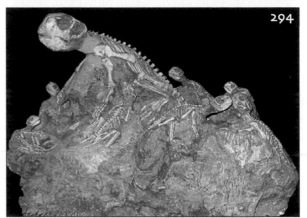

295 The basal bird *Confuciusornis sanctus* (MM). Length (top of skull to tip of wings) 290 mm (c. 11.5 in).

Protopteryx is an enantiornithine bird about the size of a starling from the succeeding Dawangzhangzi Member (**279**), while *Cathayornis* is smaller still, about the size of a sparrow, and occurs in the Jiufotang Formation at the top of the succession (**279**). Conversely it shares many features with *Archaeopteryx*, including a toothed jaw, but it has a much more advanced pectoral girdle and wing than either *Archaeopteryx* or *Confuciusornis*.

The second major group of derived birds is the ornithurines, which includes all extant birds (the 'opposite birds' did not survive the K/T extinction), and are characterized by possessing a well-developed keeled sternum suggestive of more powerful flight. *Liaoningornis* is the earliest known ornithurine bird and is the only one known from the Jianshangou Member (**279**). The Jiufotang Formation, however, has a diverse fauna of ornithurine birds including *Yanornis*, *Yixianornis*, and *Chaoyangia* and many others. At least 15 different genera have been described from this formation, illustrating an amazing diversity in both basal and advanced forms. This assemblage is often referred to as the *Cathayornis–Chaoyangia* avifauna, as opposed to the *Confuciusornis* avifauna from the older Yixian Formation.

297 Pterosaur egg with embryo (CAGS). Length 62 mm (c. 2.4 in).

296 The long-tailed rhamphorhynchoid pterosaur *Pterorhynchus* with soft-tissue head crest (CAGS). Length (top of skull crest to tip of tail) 680 mm (c. 27 in).

Other reptiles. The Jehol pterosaurs also divide neatly into two distinct assemblages coeval with the two avifaunas. The lower assemblage, from the Yixian Formation, is characterized by *Eosipterus* and *Haopterus*, which are pterodactyloids ('short-tailed pterosaurs', see p. 163), *Pterorhynchus*, which is a rhamphorhynchoid ('long-tailed pterosaurs', see p. 163), and *Dendrorhynchoides* and *Jeholopterus*, which were originally referred to the rhamphor-hynchoids, but are now thought to be ptero-dactyloids (Wang *et al.*, 2005). This assemblage is comparable to that from the Upper Jurassic Solnhofen Limestone (Chapter 13), and is associated with the *Confuciusornis* avifauna. As with the birds they are known from the upper three members of the Yixian Formation, but are absent from the lowermost Lujiatun Member (**279**).

The upper pterosaur assemblage occurs in the Jiufotang Formation (**279**), associated with the *Cathayornis*–*Chaoyangia* avifauna, and includes only pterodactyloids, such as *Sinopterus*, *Liaoningopterus*, and *Chaoyangopterus*. This assemblage is more comparable with that from the Lower Cretaceous Santana Formation (Chapter 16), especially as *Sinopterus* is one of the few records of a tapejarid pterosaur outside of Brazil (see **364**). As with the dinosaurs and birds, soft tissue is often preserved and includes wing membranes, body hair, and head crests (**296**). Even pterosaur eggs with contained embryos have been preserved (**297**).

Other reptiles include lizards and turtles, neither of which are common, and the aquatic choristoderes such as the long-necked genus *Hyphalosaurus* (synonym *Sinohydrosaurus*), repre-sented by tens of thousands of specimens and commonly sold cheaply at fossil shows around the world (**298**).

Mammals. Mesozoic mammals are never common and are normally only known from isolated teeth and fragments of jaw (see Chapter 12, The Morrison Formation). True to form the Jehol biota includes not only many articulated skeletons, but also some preserved with a complete covering of body hair. It has also revealed some surprises. Triconodonts, the most primitive mammals, are represented by two genera: the enigmatic *Jeholodens* from the Jianshangou Member (**279**), which seems to have derived characteristics for the pectoral girdle, but primitive characteristics for the pelvic girdle, and by two species of *Repenomamus* from the Lujiatun Member (**279**). *R. giganteus* is by far the largest Mesozoic mammal known, more than 1 m (c. 3 ft) in length, while one specimen of *R. robustus*

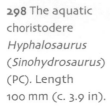

298 The aquatic choristodere *Hyphalosaurus* (*Sinohydrosaurus*) (PC). Length 100 mm (c. 3.9 in).

has the remains of a juvenile psittacosaurid dinosaur preserved in its stomach (Hu *et al.*, 2005). Multituberculates are rodent-like mammals, represented by the arboreal *Sinobaatar*, while symmetrodonts, shrew-like mammals, are represented by *Zhangheotherium* (**299**) from the Jianshangou Member (**279**), which was the first Jehol mammal to be discovered, in 1992, and was the first ever complete symmetrodont skeleton. In 2002 *Eomaia*, a eutherian mammal (the group which includes extant placentals including humans) was collected from the Dawangzhangzi Member and was the oldest known record of this group by some 50 million years (see Ji *et al.*, 2002) until it was superseded by another eutherian, *Acristatherium*, from the underlying Lujiatun Member (see Hu *et al.*, 2009).

299 The symmetrodont mammal *Zhangheotherium* (PC). Width of block 75 mm (c. 3.0 in).

Amphibians and fish. Amphibians include rather rare frogs and very common salamanders, but the most abundant fossils in the Jehol biota are fish. Grabau (1928) described *Lycoptera* and six further genera have been described since. Three of these (*Peipiaosteus*, *Yanosteus*, and *Protopsephurus*) are all sturgeons, the latter genus being the earliest known member of the paddlefish family (Polyodontidae), extant members of which still live in the Yangtze River. *Sinamia* is similar to the living bowfin, *Amia*, a primitive holostean bony fish, wheras *Lycoptera* and *Jinanichthys* are members of the extant 'bonytongues', the former being the earliest teleostean bony fish in China. Finally, is *Longdeichthys*, another teleostean fish.

Insects and spiders. According to Chang (2003) and Ren *et al.* (2010) more than 270 species of insects belonging to over 100 families within at least 16 orders have been described from over

10,000 specimens collected from the Jehol Group. This includes terrestrial, aerial, and aquatic forms, all preserved in exquisite detail, including preservation of colour patterns. Many new forms have yet to be scientifically described and the fauna compares with that of the Crato Formation (Chapter 16) in its vast diversity. At family level the assemblage is essentially modern and includes many familiar groups.

The most basal of the pterygotes (winged insects) include a diverse record of Ephemeroptera (mayflies) and Odonata (damselflies and dragonflies), with both nymphs and adults represented. Nymphs of the hexagenitid mayfly *Ephemeropsis trisetalis* are one of the most abundant fossils of the Jehol biota. They are often preserved in large numbers on a single bedding plane, in association with the large dragonfly *Aeschnidium* (**300**). The latter genus also occurs in the slightly older Solnhofen Limestone of Germany (Chapter 13), although the Chinese species actually exhibit more primitive characteristics. Adult mayflies are rare with only one known genus, whereas dragonflies are represented by at least 14 families.

More derived hemimetabolous pterygotes (those lacking a pupal stage in their ontogeny) include four families of Plecoptera (stoneflies) and the Orthopterida, which includes three families of Orthoptera (grasshoppers, crickets, and locusts), rare Phasmatodea (stick insects) and just two genera belonging to the extinct and obscure Chresmodidae, which is usually placed in the Archaeorthoptera (see p. 191). Hemimetabolous insects also include two families of Blattaria (cockroaches) and the Hemiptera, represented in the Jehol Group by 14 families, including several genera of palaeontinids (**301**), and by some true bugs such as the flower bugs. Dermaptera (earwigs) and the primitive Grylloblattodea are not, however, represented.

Holometabolous insects (those going through a pupal stage in their ontogentic development) are also diverse and abundant. One of the most basal groups, recognized by their large wings and net-like venation, is the Neuropterida, represented in the Jehol biota by two of the three extant orders, Neuroptera (lacewings and antlions), with 11 families, and Raphidioptera (snakeflies; **302**), with three families. Superficially similar are the Mecoptera (scorpionflies), with six families, which are actually closely related to the fleas (Siphonaptera).

More derived holometabolous insects are: the Coleoptera (beetles), the sister group to the Neuropterida, represented by about 20 families including elaterids and cupedids; Hymenoptera (bees, ants, and wasps), represented by 24 families, including ephialtitids, pelecinids, sawflies, and ichneumon wasps; Trichoptera (caddisflies), represented by just two families; and Diptera (true flies, gnats, and mosquitoes), represented by at least 11 families, all distinguished by having only one pair of fully developed wings. These include the brachycerans (see later text) and the small chaoborid mosquitoes, which are related to present-day phantom midges.

Perhaps the most interesting feature of the Jehol insects is that some of them have been shown to possess specialized mouthparts (i.e. a proboscis),

300 The dragonfly *Aeschnidium heishankowense* (PC). Wingspan 130 mm (c. 5.1 in).

301 The palaeontinid *Liaocossus beipiaoensis* (CNU). Wingspan 55 mm (c. 2.2 in).

302 The snakefly *Sibooptera fornicate* (CNU). Wingspan 33 mm (c. 1.3 in).

assumed to have a nectar-gathering function, suggesting their association with the recently evolved angiosperms. Such flower-associated insects are not common, but are vitally important with regard to the co-evolution of these possible early pollinators with the emerging flowering plants. They include the brachyceran flies previously mentioned, and also the flower bugs and tumbling flower-beetles.

Spiders are uncommon and low in diversity, but include orb-web spiders of the superfamily Araneoidea.

Other invertebrates. These include non-marine molluscs (gastropods and bivalves), conchostracans and ostracods (both bivalved crustaceans), and freshwater shrimps (crayfish).

Plants. Most major groups are represented including bryophytes, lycopods, sphenopsids, ferns, ginkgos, conifers, bennettites, gnetales, and angiosperms. The latter are potentially the most important as these only appeared at the start of the Cretaceous and thus any angiosperms from the Jehol biota will be among the oldest on record. However, several that have been described are now considered dubious. The genus which has received most attention is that of *Archaefructus*, a submerged water plant on which the reproductive axes have separate male and female organs, but the presence of undoubted angiosperm features has not been demonstrated and its affinity remains in question. Zhou *et al.* (2003) consider that it does not belong in the angiosperm crown group, but that it may lie on the angiosperm stem lineage. The genus *Sinocarpus* from the Dawangzhangzi Member possesses united carpels, a derived feature in angiosperms, and would seem to be a more likely candidate.

PALAEOECOLOGY OF THE JEHOL GROUP BIOTA

The Jehol Group was deposited in a terrestrial/freshwater lake setting near the eastern margin of Eurasia, which at the time was isolated from western Eurasia by the epicontinental Turgai Strait and from North America by the wide Bering Strait, and situated at approximately 40°–45° N. The climate is interpreted as being seasonally fluctuating between semi-arid and mesic. At this time eastern Eurasia was an emergent landmass and extensive volcanism on its eastern margin

resulted from continuous tectonic activity along the western Pacific Rim, as it does today. Numerous shallow lakes developed in a series of north-east– south-west faulted basins which gradually filled with volcanic ash and lake-bed muds.

The lakes were home to freshwater molluscs, crustaceans (conchostracans and crayfish), abundant fish, amphibians (frogs and salamanders), and aquatic reptiles (turtles and choristoderes), while the lake margins were mostly vegetated with conifers, but also supported various lycopsids, sphenopsids, ferns, benettitaleans, gnetaleans, ginkgos, and possibly early angiosperms. A rich fauna of terrestrial and aquatic insects were the prey of giant mayflies, orb-web spiders, and more particularly of the numerous species of early birds, some of which were gregarious and arboreal. The trees overhanging the lake must have been filled with flocks of these colourful, feathered species as they swooped over the water surface displaying their newly discovered flying skills.

Watching somewhat enviously from the lake shores were their recent ancestors, the small, agile, theropod dinosaurs, also clothed in colourful display plumage, but unable to fly. Some, however, did climb trees, and yet others were able to glide from branch to branch. More proficient gliders were the pterosaurs, scanning the lake surface for insects and fish, while larger, unfeathered ornithischian dinosaurs browsed on the plentiful vegetation. Keeping a low profile were many small, insectivorous mammals, but others, the size of a badger, not only began to compete with dinosaurs for food and space, but actually began to prey on their young: the tables were beginning to turn. All, however, were unaware that the next volcano to erupt would bury their world in a shower of hot, toxic ash.

COMPARISON OF THE JEHOL GROUP WITH OTHER FEATHERED DINOSAUR SITES

Daohugou Formation, Inner Mongolia, China

The only other locality to have yielded anything like the surprising fauna of the Jehol Group is the adjacent Daohugou site in Ningcheng County in Inner Mongolia (**278**). Since 1998 this locality, which is well known for its rich and varied insect fauna, has also produced abundant salamanders, a pterosaur with preserved hair (*Jeholopterus*), an aquatic mammal (*Castorocauda*), a gliding mammal (*Volaticotherium*), and in 2002 an arboreal feathered dinosaur, *Epidendrosaurus*, a member of the scansoriopterygid ('climbing wings') family. This was followed in 2005 by the description of *Pedopenna*, a bird-like dinosaur related to troodontids and dromaeosaurs and characterized by long feathers on the metatarsus, and in 2008 by a second scansoriopterygid dinosaur, *Epidexipteryx*, characterized by four long feathers on the tail, which seem to have been used solely for display.

If the age of the Jehol Group has been controversial, dating of the Daohugou Formation is even more contentious, with suggested ages ranging from middle Jurassic to early Cretaceous. Wang and Zhou (in Chang, 2003) regarded this deposit to be comparable with or slightly lower than the Lujiatun Member of the Jehol Group, but recent uranium–lead dating of zircon crystals from volcanic rocks above and below the salamander horizon seems to support an age of between 164 and 158 Ma, suggesting that this biota is significantly older than the Jehol and most probably dates from the late Middle or early Upper Jurassic (Liu *et al.*, 2006). One would, of course, expect the earliest feathered dinosaurs to pre-date the earliest bird (*Archaeopteryx* from the late Upper Jurassic of Germany – see Chapter 13, The Solnhofen Limestone) and so a late Middle or early Upper Jurassic date for the Daohugou Formation, and the concept of pre-*Archaeopteryx* feathered dinosaurs lend further support to the theropod–bird link.

Further press excitement ensued in 2009 when a feathered troodontid was unearthed from the Tiaojishan Formation, which directly underlies the Daohugou horizon. *Anchiornis huxleyi* (suitably named after Darwin's loyal supporter who first suggested the dinosaur–bird link in 1868) is characterized by aerodynamically pennaceous feathers on its forelimbs, legs, tail, and feet, including the pedal phalanges, and extends the record of feathered dinosaurs even further back into the Jurassic.

MUSEUMS AND SITE VISITS

Museums

A provincial ban on the export of all fossils from Liaoning, passed in 2001, has prevented most European museums from acquiring specimens, even though plentiful birds, psittacosaurid dinosaurs, aquatic reptiles, and insects are frequently offered for sale at fossil shows around the world, and in government-run museum shops within China, and even at Chinese airports! Several enlightened museums in Europe and the USA did manage to acquire specimens of the ubiquitous *Confuciusornis* prior to the ban (for example Manchester University Museum and the University of Kansas Natural History Museum)! Several museums in China have good displays, including:

1. National Geological Museum of China, Beijing, PRC.
2. Shandong Tianyu Museum of Nature, Pingyi, PRC.
3. Yizhou Fossil and Geology Park Museum, PRC (see below).
4. Sihetun Field Museum, PRC.
5. Beipiao Museum, PRC.
6. Jinzhou City Museum, PRC.
7. Changzhou Dinosaur Park Museum, PRC.

Sites

Until recently Liaoning Province in north-eastern China has not been a particularly accessible region for fossil collectors. However in 2008 the owners of the Wyoming Dinosaur Center (WDC) in the USA (see Chapter 12, The Morrison Formation) opened the Yizhou Fossil and Geology Park in He Jia Xin Village, near Tou Tai Town in Yi County, Liaoning, which has opened up the Jehol Lagertätte and its fossils to the public. Not only does it boast an excellent museum (displaying birds, mammals, pterosaurs, insects, and plants), but also operates a dig site for the public in adjacent quarries in the same format as the WDC. The park is a 4-hour drive from Beijing (522 km [c. 325 miles]), but the Beijing to Yixian expressway is one of the newest and easiest roads in China, passing by the ocean, and under the Great Wall. Alternatively one can take the fast train from Beijing, on the Beijing–Shenyang line, alighting after 3 hours at Jinzhou, where a 45-minute taxi ride will take you to the park (http://www.liaoningdinosaurpark.com).

EL MONTSEC AND LAS HOYAS

Chapter Fifteen

BACKGROUND: EUROPE IN THE EARLY CRETACEOUS

The background to the Cretaceous Period was outlined in Chapter 14, The Jehol Group. In this and the following chapter (15 and 16), we look at more early Cretaceous sites around the world, and in both chapters we actually include two Lagerstätten close to each other in space and time. First, we look at the Konservat-Lagerstätten of El Montsec and Las Hoyas from northeastern Spain.

El Montsec and Las Hoyas were situated on the western shores of the Tethys Ocean in early Cretaceous times, where they were under the influence of a subtropical climate with annual alternating wet and less pronounced dry periods. The mean temperature of the warmest month was in the region of 40°C (104°F), and it was at least 20°C (68°F) during the cooler months when precipitation rates were relatively high. A complex of fault-bounded basins was established on the northern boundary of the Iberian Plate during late Jurassic–early Cretaceous times adjacent to the European plate, related to the rifting that led to the opening of the Bay of Biscay. The basin complex was made up of several blocks that subsided at different rates, promoting the development of lakes in which laminated muds (now lithographic limestones) were deposited in areas of greatest subsidence.

HISTORY OF DISCOVERY OF EL MONTSEC AND LAS HOYAS

Towards the end of the 19th century a small quarry was opened at about 900 m (c. 2,950 ft) above sea level on the east end of the Montsec de Rúbies ridge above the scenic Terradets Gorge cut by the Noguera Pallaresa river, in the eastern Pyrennean foothills north of Lleida, Catalonia (**303, 304** *overleaf*). Thinly bedded, fine limestones were extracted from the quarry for use in lithography, like those at Solnhofen, Germany (p. 156). Close to the villages of Rúbies and Santa Maria de Meià, the quarry is known variously as La Pedrera de Meià, La Pedrera de Rúbies, or simply La Pedrera ('the quarry'). The first mention of the El Montsec lithographic limestones in the scientific literature was a report of a field meeting to the Province of Lleida by the Société Geologique de France in 1898, led by the mining engineer Lluís Marià Vidal i Carreras (Vidal, 1899). The excursion took place in October and heavy rain hampered their field work in the area, but the party visited the whole section, from the Cenozoic beds down through the Cretaceous to the Jurassic.

303 View north to the quarry of La Pedrera de Meià on the ridge of Montsec de Rúbies.

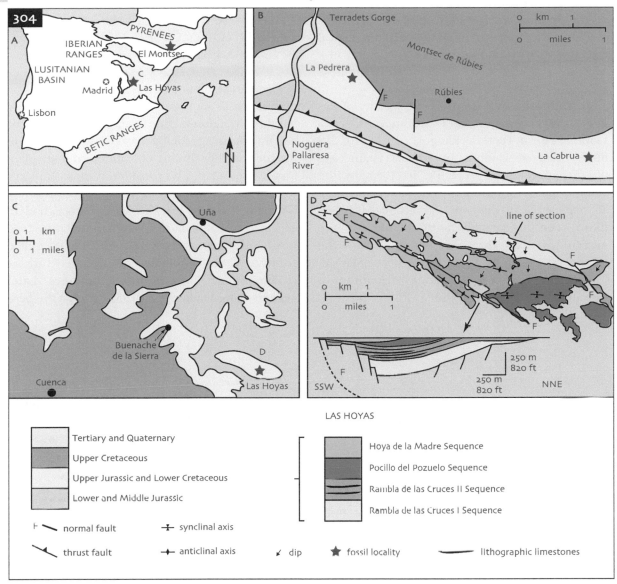

304 A: Location map for the El Montsec and Las Hoyas localities in relation to Mesozoic basins (shown in yellow); **B:** geological map of the Montsec de Rúbies area, showing the locations of the quarries; **C:** geological map of the area north-east of Cuenca, showing the location of the Las Hoyas and Buenache localities; **D:** detailed geological map of the Las Hoyas inlier (after Fregenal-Martínez & Meléndez, 2000; Gomez *et al.*, 2002).

Vidal was aware of the importance of the locality and its magnificently preserved fossils, and began to send specimens to specialists throughout Europe for description. The first scientific descriptions were published by Vidal (1902), by which time plants, insects, fish, and frogs were known. With the closure of the lithographic industry and Vidal's death in 1922 the site was essentially forgotten. Thirty years later, Bataller *et al.* (1953), in the description of the geological map of the area, mentioned another, small outcrop of lithographic limestones along the track from Rúbies to Santa Maria de Meià, just above the Sant Sebastia hermitage (**305**). This outcrop is known as 'La Cabrua'. In the 1980s, an enthusiastic group based around the Institut d'Estudis Ilerdencs called Amics de la Paleontologia (Friends of Palaeontology) made a series of annual expeditions to El Montsec, mainly to the La Cabrua locality, and they also managed to enlist the help of the Spanish army (**306**). These expeditions increased the interest among the palaeontological community with the finding of a great many new specimens, which were distributed to scientists around the world for description.

In contrast to El Montsec, the locality of Las Hoyas (**307**), which lies in the Serranía de Cuenca (south-western Iberian Ranges), 19 km (c. 12 miles) east of Cuenca in central Spain, was discovered relatively recently, in 1985, by the amateur collector Armando Diaz-Romeral. In the 25 years since its discovery, about 30,000 fossils have been unearthed and, thanks to the meticulous research which has gone into the locality by a team from the University of Madrid, Las Hoyas now ranks among the foremost Konservat-Lagerstätten in the world.

STRATIGRAPHIC SETTING AND TAPHONOMY OF EL MONTSEC AND LAS HOYAS

At El Montsec, Vidal was able to correlate the thick Cretaceous sequence in the Montsec ridge using marine fossils such as oysters and rudist bivalves, but the lower, non-marine strata he estimated to be late Jurassic in age, on the basis of the character of the rocks and by finding only indeterminate plant fossils. Presumably, he considered their similarity to the lithographic limestones of Solnhofen, Germany (Chapter 13), and Cerin, France, both of which are late Jurassic in age. When well preserved fossils started to appear from La Pedrera, Vidal confirmed his estimate of a late Jurassic (Kimmeridgian) age for the limestones. Later workers (e.g. Bataller *et al.*, 1953) placed the beds

305 North–south geological cross section through the Montsec de Rúbies ridge just east of the map shown in **304B**, from Vidal (1899). Bed 20 is the lithographic limestones.

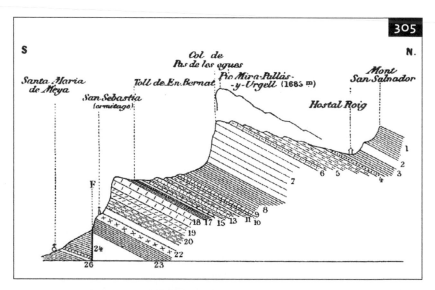

306 Collecting at the La Cabrua quarry in the 1980s by the Amics de la Paleontologia.

307 Las Hoyas Quarry.

close to the Jurassic–Cretaceous boundary, and by the 1970s the age had crept up to Lower Cretaceous (Berriasian– Valanginian) on the basis of the ostracod fauna. More recent studies on regional stratigraphy, charophyte algae, and fossil assemblages have shown that the age of the El Montsec lithographic limestones is latest Berriasian–early Barremian. A Barremian age for the El Montsec beds brings them closer to the age of the La Huérguina Formation of Las Hoyas, which are late Barremian (c. 126 Ma), based on ostracods, charophytes, palynomorphs, and macroplants. Earlier estimates of the age of the La Huérguina Formation had placed it as late Hauterivian?–early Barremian, so the fossils from Las Hoyas have been helpful in dating the formation. Comparative stratigraphic logs are shown in 308.

The lithographic limestones of El Montsec were mapped as a subunit of the Calcaires à Charophytes du Montsech. Marine fossils such as rudist bivalves and foraminiferans occur beneath and above the Calcaires à Charophytes du Montsech (308). Charophytes are freshwater plants

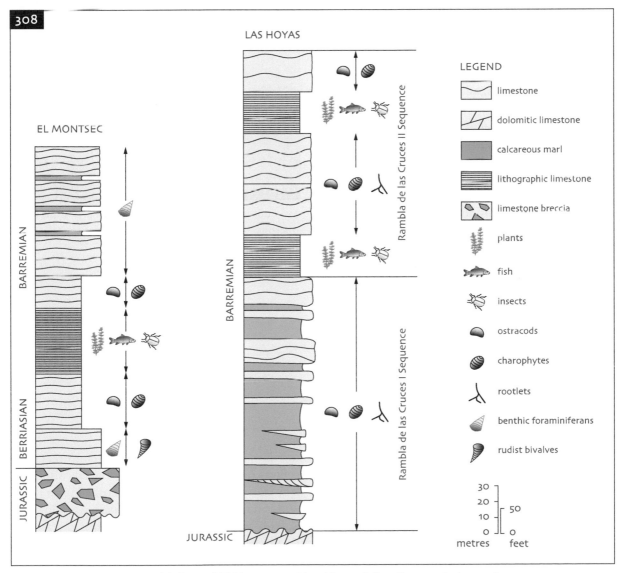

308 Comparative stratigraphic logs of El Montsec and Las Hoyas (after Gomez *et al.,* 2002).

which can be thought of as being phylogenetically between algae and vascular (land) plants. They are alive today and have a good fossil record because they deposit calcium carbonate in their tissues (hence the common name 'stonewort'), and their seed-like oospores have a limy skeleton called a gyrogonite, which can be useful in biostratigraphy. Charophyte remains contribute to the deposition of limestones in fresh water. Because charophytes are green plants, their presence indicates shallow water; ostracods (tiny bivalved crustaceans) are also abundant in this facies. The charophyte beds are interpreted as the development of a lake, and charophytes would have remained close to the edge of the lake, within the photic zone, as the lake deepened. The lithographic limestones are rhythmically laminated carbonate muds without any sign of currents or emersion features such as raindrop marks or mud cracks. The exceptional preservation of soft tissues, skeletons undisturbed by scavengers or decomposers, and random orientation of fossils, all point to a deeper water, probably anoxic lake bottom. The rhythmic laminations are typical of lacustrine conditions and reflect seasonal changes in the lake environment. Eventually the lake basin filled with sediment, became shallow again (more charophytes), and then the sequence reverted to marine sedimentation once more (**308**). Although the lake was close to the sea, there is no evidence of a permanent connection in the form of marine fossils in the lithographic limestones. In this respect, these sediments differ markedly from the lithographic limestones of Solnhofen (Chapter 13).

At Las Hoyas, the La Huérguina Formation reaches its greatest thickness (300 m [985 ft]) and rests unconformably on Jurassic dolomitic limestone karst. Initially, the sequence shows more fine calcareous mud at Las Hoyas (**308**) than at El Montsec, but note a similar sequence of charophyte–ostracod limestones, which indicate shallow freshwater, as do the rootlets which also occur in this facies. As at El Montsec, the lithographic limestones of Las Hoyas show rhythmic laminations characteristic of a seasonal climate (**309**). More work has been done on the Las Hoyas sediments than those of El Montsec, and under the microscope two distinct types of laminated limestone have been distinguished there. The first (Facies 1) is the result of carbonate mud and plant

309 Rhythmically laminated sediments in the Las Hoyas quarry.

310 Characteristic orange-skin surface produced by microbial mats, in the Las Hoyas quarry.

debris washing into the lake. The second (Facies 2) reflects production of carbonate crystals within the lake caused by the growth of microbial mats (**310**), and is thought to be related to periods of low lake water level. Facies 1 was deposited during wet-season flooding, whereas microbial mats grew during dry periods when the water column was drastically reduced to probably just a few centimetres/inches.

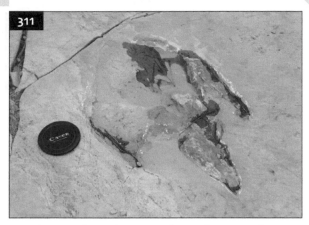

311 Dinosaur footprint, part of a trackway on the floor of the Las Hoyas quarry. Lens cap 77 mm (c. 3 in) in diameter.

312 The salamander *Celtedens* from Las Hoyas, showing skin impression. Length 64 mm (c. 2.5 in).

There is no evidence of complete desiccation of the lake bottom although dinosaur (**311**) and crocodile trackways point to very shallow water at times.

Fossils from both sites are usually articulated and in anatomical connection. For example, plants have stems with leaves and cones in connection, and even the thin, distal parts of the fin rays in fishes are completely articulated. The most exceptional preservation consists of the soft bodies of insect larvae and soft tissues such as the abdominal cavity of the fish *Rubiesichthys* and the skin of amphibians (**312**). The coloration patterns of insect wings and pigmentation in the barbules of bird feathers are also preserved. Most remarkable, at Las Hoyas the soft tissue of the dinosaur *Pelecanimimus* has been preserved by mineralization. Skin and muscle are replicated in iron carbonate and the outline of the dinosaur is

preserved by a phosphatized microbial mat that enshrouded the carcass, confirming the existence of either a throat pouch or dewlap, and soft occipital crest (Briggs *et al.*, 1997). This study draws attention to the importance of microbial mats in the fossilization of soft tissues. Organic geochemical analyses of fish scales from Las Hoyas (Gupta *et al.*, 2008) have shown some degree of chemical alteration during diagenesis.

The two facies at Las Hoyas reflect contrasts in the composition and texture of the sediments, and thus the depositional environment and diagenetic history (Sanz *et al.*, 2001b). Facies 1 is characterized by sediment enriched in organic matter and clays (4–6%). The high concentration of organic matter inhibited decay by creating a reducing environment within which preservation of delicate organic structures became possible. Loss of skeletons was caused by the dissolution of bone in an acid environment; the processes of infilling and cementation occurred at an early diagenetic stage. The content of organic matter and clays favoured strong compaction and lithification of sediments, leading to deformation and flattening of specimens. These wet-period sediments record the maximum biodiversity. In contrast, Facies 2 contains less than 2% organic matter and clays. Decay was more efficient and the preservation of delicate organic structures was mainly the result of sealing by microbial films. The sediments suffered less compression and the resulting rocks are more fissile, which allowed ground-water to circulate preferentially along the lamination planes, which resulted in more dissolution and infilling of moulds, and at a later time, than in Facies 1. These dry-period sediments contain the best preserved and greatest number of fossils. So, at least at Las Hoyas, a combination of bacterial mats, anoxia of the lake bottom, and rapid sedimentation explains most of the taphonomic features of the biota.

DESCRIPTION OF THE EL MONTSEC AND LAS HOYAS BIOTA

Plants. Charophytes (stoneworts) are abundant in both localities in the underlying and overlying marls and limestones. Remains of liverworts have also been found at Las Hoyas. Fern-like foliage can be difficult to assign to pteridophytes (ferns and their allies) or to families within the Filicales without the fertile fronds; nevertheless, numerous

313 Fragment of frond of the fern *Weichselia* from El Montsec (IEI). Length 48 mm (c. 1.9 in).

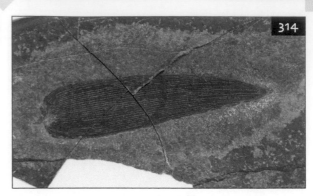

314 The bennettitalean *Zamites* from El Montsec (IEI). Length 50 mm (c. 2 in).

forms are known from El Montsec and Las Hoyas. The commonest fern fragment at both localities is *Weichselia* (**313**). The fronds of this plant are robust and the cuticle is thick, which explains why it is well preserved, and that it is most likely a xerophyte, i.e. living in dry habitats. Only sterile fronds have been found in these Spanish localities, but fertile parts from elsewhere indicate it belongs in the family Matoniaceae. Rare, very large fronds point to it having been a tree-like fern with a massive trunk and roots. Another very common fern at these localities is *Ruffordia*, in the family Schizaceae.

Gymnosperms were more diverse in the early Cretaceous than today. Both the extinct cycad-like Bennettitales and the true cycads, Cycadales, occur at Las Hoyas and El Montsec, but they can only really be told apart by study of the stomata (breathing pores in the cuticle) under a microscope. A common bennettitalean is *Zamites*, whose sharp pinnae are distinctive (**314**). Caytoniales is an extinct group of seed plants with fern-like foliage, represented in the Cretaceous of Spain by the fronds called *Sagenopteris* (**315**). Gnetales are represented today by some bizarre xerophytes, but were probably more diverse in the Mesozoic; the genus *Drewria*, better known from the early Cretaceous of Virginia, is reported from Las Hoyas. A lovely deciduous park and garden tree today, *Ginkgo* is known as a living fossil. Sure enough, Ginkgoales occur at El Montsec. Conifers are better-known gymnosperms today, and numerous genera occur at these early Cretaceous sites. *Cupressinocladus*, *Frenelopsis*, *Pagiophyllum* (**316**), *Brachyphyllum*, *Sphenolepis*, and *Podozamites* are the main ones. These would have formed dense forests away from the lake shores.

315 *Sagenopteris*, a Caytoniales from El Montsec (IEI). Length 60 mm (c. 2.4 in).

316 The conifer *Pagiophyllum* from El Montsec (IEI). Length 10 mm (c. 0.4 in).

The early Cretaceous saw the dawn of the angiosperms (flowering plants), which so dominate modern floras. One of the earliest indicators of the presence of angiosperms in the fossil record, *Tricolpites* pollen, is found at Las Hoyas. Two macroplant fossils from El Montsec and Las Hoyas which have been considered as angiosperms are *Ranunculus* and *Montsechia*. The *Ranunculus* was first described from El Montsec as

Montsechites ferreri and later transferred to the extant *Ranunculus aquatilis*: the water crowfoot, by Blanc-Louvel (1984). To be sure, the fossil plant does look remarkably similar to a crowfoot, with slender stems and feathery foliage, which at least point to an aquatic habit, but the fossil has small spiny reproductive units in the axils of the leaves which are distinctly different from fruits of Ranunculaceae, so botanists are not convinced of its placement in that family, though it could still be an angiosperm. A common but poorly understood plant, *Montsechia vidali* (**317**), was first assigned to the aquatic bryophytes by Blanc-Louvel (1991), but it may have been a true angiosperm (Gomez *et al.*, 2006). It, too, was an aquatic plant.

Molluscs. Molluscs are rare, which might seem strange given their abundance in modern lakes and the availability of calcium carbonate for making shells. The fauna consists of just a few bivalves and gastropods.

Crustaceans. The crustacean fauna is a little more diverse, and comprises ostracods, peracarids, and decapods. Ostracods are tiny bivalved crustaceans which can be abundant in freshwater. The ostracods in the lithographic limestones are dominated by *Cypridea*, which is typical of early Cretaceous freshwater lakes. The decapods are much more showy crustaceans. *Pseudastacus llopisi* (**318**) is a type of crayfish; it is found at both sites, but occasionally occurs in mass-mortality horizons at Las Hoyas. The shrimp-like *Delclosia roselli*, from El Montsec, and *Delclosia martinelli*, from Las Hoyas, differ slightly. The crayfish *Pseudastacus* was a benthic scavenger or predator, while the *Delclosia* shrimps would have swum between the charophytes, feeding on plankton (Rabadà, 1993).

Insects. Insects are by far the most abundant animal fossils in the lithographic limestones. For example, arthropods as a whole provide more than 45% of the genera (132) of the biota collected at

317 *Montsechia vidali* from El Montsec (IEI). WIdth of frond 32 mm (c. 1.3 in).

318 *Pseudastacus llopisi*, a crayfish from El Montsec (IEI). Body length 57 mm (c. 2.2 in).

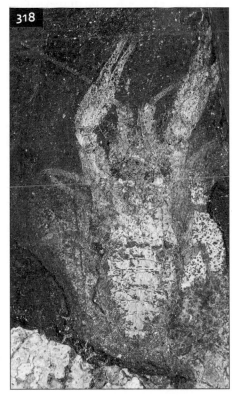

Las Hoyas, of which insects contribute 14 orders and 40 families (Fregenal-Martínez & Buscalioni, 2009). Ephemeroptera (mayflies) are represented by adult *Huergoneta* and *Hispanoneta*, and the burrowing nymph *Mesopalingea*, which was probably a filter-feeder. Modern burrowing mayflies live mainly in rivers where they can maintain a continuous water current through their burrows. The lack of characteristic U-shaped burrows made by burrowing mayflies in the lithographic limestones suggests that the nymph fossils may have been washed into the lake rather than living there. Odonata are represented in the lithographic limestones only by dragonflies (Anisoptera). Aeschniidae (darners) are large, fast-flying dragonflies which are well represented by about half a dozen genera. Gomphids (clubtails) occur as *Ilerdaegomphus*, and there is a single wing called *Condalia woottoni* which might belong to the family Libellulidae (skimmers).

More advanced insects belong to the neopteran orders. Cockroaches are not common at Las Hoyas but are quite frequent at El Montsec. Related to cockroaches are the termites (Isoptera), which are the earliest social insects to appear in the fossil record (Martínez-Delclòs & Martinell, 1995). Moreover, when *Meiatermes* was first described from El Montsec by Lacasa-Ruiz and Martínez-Delclòs (1986) it was the oldest known termite. It has recently been preceded in age by *Baissatermes* from slightly older Cretaceous beds in Russia but, nevertheless, remains part of the primitive stem group which lies basal to all other termites except the Mastotermitidae, a family intermediate between termites and their ancestors, the wood-roaches (Engel *et al.*, 2009). Crickets are represented in the limestones by a couple of poorly preserved genera. A rather striking, large insect resembling a water-strider called *Chresmoda* (**319**) is known from numerous Mesozoic ecosystems, including Solnhofen, Jehol (p. 180) and the Crato Formation of Brazil (p. 209), but its phylogenetic affinity was poorly known until recently. It is now thought to belong to the Archaeorthoptera, a stem lineage of the Orthoptera, so could be thought of as a semi-aquatic grasshopper.

Among the Hemiptera, the Cicadomorpha are rather rare in Las Hoyas, but some rather showy forms are known from El Montsec. Palaeontinidae is an extinct family of primarily Mesozoic

cicadomorphs that bear a superficial resemblance to moths. Three genera have been described from El Montsec: *Pachypsyche*, *Montsecocossus*, and *Ilerdocossus* (**320**). Heteroptera include many aquatic and semi-aquatic forms today. Many occur as fossils in the lithographic limestones, and the aquatic types form a high percentage of the total insect fauna at Las Hoyas. The Belostomatidae are the true water bugs which spend their entire lives submerged. The water boatman *Iberonepa romerali* is the commonest species.

The most primitive Holometabola (insects with complete metamorphosis) are the three orders which comprise the Neuropterida: Neuroptera

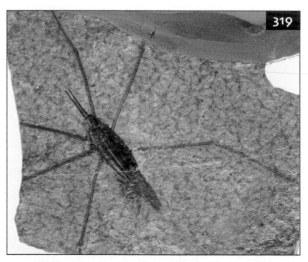

319 *Chresmoda aquatica*, from El Montsec (IEI). Length 25 mm (c. 1 in).

320 Cicadomorph *Ilerdocossus*, from El Montsec (IEI). Wingspan 50 mm (c. 2 in).

(lacewings and antlions), Megaloptera (alderflies and dobsonflies), and Raphidioptera (snakeflies). Snakeflies are rare at both localities. Neuroptera are not known from El Montsec, but antlions (Myrmeleontidae) and six types of Chrysopidae occur at Las Hoyas. Chrysopids are the familiar green lacewings we see commonly today. Kalligammatidae are also well known from Las Hoyas; these are large, showy insects with spots on their wings, like butterflies, which lived in the Jurassic and Cretaceous.

Fossil beetles form the second most abundant insect group at both localities: about 24% at Las Hoyas and 20% at El Montsec (Soriano *et al.*, 2007). They are by far the most diverse insect group at both localities, with more than 35 described species; these localities have yielded one of the most diverse beetle faunas of the Mesozoic. At Las Hoyas the preponderance of isolated elytra (wingcases) makes identification difficult, but at El Montsec they are mainly articulated specimens. Nevertheless, the characters commonly used by neontologists for beetle identification may not be present or may be difficult to see in fossils. Early Mesozoic beetle faunas are dominated by the primitive Archostemata, such as *Cionocoleus* (**321**), which declined in importance through the Jurassic and Cretaceous, and have few Recent representatives. They make up about 10% of Las Hoyas and El Montsec beetle fauna. Most common is the Cupedidae, or reticulated beetles, whose elytra have rows of square punctures between longitudinal ridges (**322**), which are mainly xylophagous (wood-eating). The extinct Schizophoridae were water beetles without swimming adaptations; they are known only from rare, isolated elytra at Las Hoyas.

The Adephaga are the most abundant group of both terrestrial and aquatic beetles at Las Hoyas, where both larvae and adults are preserved, while at El Montsec there is only a scarce record of single specimens of aquatic forms. At Las Hoyas, the terrestrial forms are represented by the family Trachypachidae and the aquatic forms by members of the families Coptoclavidae, Dytiscidae, and Gyrinidae. The extinct family Coptoclavidae was dominant among water beetles during the Upper Jurassic and Lower Cretaceous, and at least seven species are known from Las Hoyas, but only two at El Montsec (Soriano *et al.*, 2007) (**323**). Dytiscidae are

321 Archostemata beetle *Cionocoleus*, from El Montsec (IEI). Length 18 mm (c. 0.7 in).

the characteristic diving beetles, and are represented by one species at each locality. Gyrinids (whirligig beetles) are usually more common than dytiscids in the Mesozoic; one species, represented by isolated elytra and articulated specimens, is known from Las Hoyas, but none from El Montsec.

Most Las Hoyas and El Montsec beetles belong to the Polyphaga, just as today, and all the families found there have living representatives except the extinct Parandrexidae (**324**). Hydrophilidae includes terrestrial and aquatic forms, the latter generally scavengers; one species has been found at Las Hoyas. Staphylinidae (rove beetles) are predators or omnivorous scavengers recognizable by the very short elytra which do not cover the abdomen. They are uncommon, with only three specimens from the El Montsec collection and one from Las Hoyas. The jewel beetles (Buprestidae), so-called because of their beautiful metallic sheen, are relatively abundant today, with herbivorous larvae often boring into wood. Usually they are diverse in Lower Cretaceous localities, but only one species has been described from El Montsec. Pill beetles (Byrrhidae) have an oval, convex shape which conceals the head from above. They are

322 Cupedid beetle *Zygadenia,* from Las Hoyas (MCCM). Length 20 mm (c. 0.8 in).

323 Coptoclavid beetle *Hoyaclava,* from Las Hoyas (MCCM). Wingspan 30 mm (c. 1.2 in).

324 Parandrexid beetle *Martynopsis,* from Las Hoyas (MCCM). Length 15 mm (c. 0.6 in).

often found on lake shores and are a common family in many Mesozoic localities; at Las Hoyas there are at least four different forms and three at El Montsec. The Elateridae (click beetles) are well known for their ability to jump into the air by flexing their bodies when placed on their backs. They have elongate elytra; the larvae live in rotting wood or soil and are relatively abundant today. They are generally the most common family in Mesozoic localities and are represented by at least four forms at Las Hoyas and six at El Montsec. The Scarabaeidae (dung beetles and chafers) is a large family of usually stout-bodied beetles with distinctively clubbed antennae. They are one of the most diverse groups, with up to eight kinds at Las

Hoyas and three at El Montsec. The saprophagous cucujiform beetles are diverse and widespread today, and were similarly so in the Lower Cretaceous of El Montsec, with four species of the family Nitidulidae, and several undescribed forms (Soriano *et al.*, 2007). They were comparatively less diverse at Las Hoyas, where they are represented by the extinct Parandrexidae. The Ptilodactylidae are nowadays mainly a subtropical group living in riparian habitats; they have a very characteristic form of the body and long, flabellate antennae. They exhibit a great diversity at both Las Hoyas and El Montsec. The terrestrial Mordellidae (wasp beetles) live on flowers; a species from El Montsec is only the fourth Mesozoic record of this family.

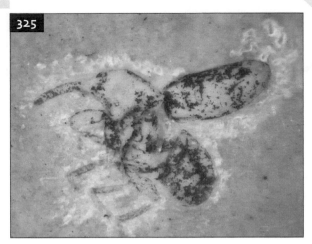

325 The weevil *Montsecanomalus*, from El Montsec (IEI). Length 4.8 mm (c. 0.2 in).

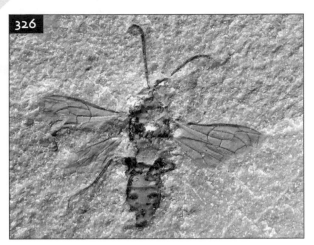

326 The hymenopteran *Cretobestiola*, from El Montsec (IEI). Wingspan 14 mm (c. 0.5 in).

327 Scorpion fly (Mecoptera), from Las Hoyas (MCCM). Length 20 mm (c. 0.8 in).

328 Tangle-veined fly *Hirmoneura*, from El Montsec (IEI). Length 9 mm (c. 0.35 in).

329 The orb-web weaving spider *Cretaraneus*, from El Montsec (IEI). Body length 4.5 mm (c. 0.2 in).

330 Juvenile actinopterygian fish, probably *Notagogus*, from Las Hoyas (MCCM). Length 34 mm (c. 1.3 in).

331 Actinopterygian fish *Caturus,* from El Montsec (IEI). Length 120 mm (c. 4.7 in).

The weevils (Curculionoidea) are distinguished by their elongate snout (rostrum), with mandibles at the end, used for chewing (**325**). This group is comparatively very diverse at El Montsec, with 11 species from the families Nemonychidae, Eccoptarthridae, Belidae, and Anthribidae. In contrast, at Las Hoyas only one specimen, of the family Nemonychidae, has been found.

The two lithographic limestones show markedly different diversity in Hymenoptera. Among the wasps, 28 species are known from El Montsec, but only two from Las Hoyas (Rasnitsyn & Martínez-Delclòs, 2000). One of the more spectacular fossils is *Cretobestiola hispanica* (**326**), which is an angarosphecine, an early Cretaceous offshoot of the Apoidea – a group that includes the digger wasps and the bees. Together with the Hymenoptera, the Panorpida comprises the other half of the holometabolous insects with the beetles (Grimaldi & Engel, 2005). Two mecopterids are known from Las Hoyas, and are interesting because they preserve colour patterns on the wings (**327**). The true flies (Diptera) are mainly represented in the lithographic limestones by their aquatic larvae, such as stratiomyids (soldier flies) and sciomyzids (marsh flies). Adults are represented by Nemestrinidae (tangle-veined flies) (**328**), Mycetophilidae (fungus gnats), and Tipulidae (crane flies).

Spiders. Before the El Montsec spiders came to light in the 1980s, only two specimens of spiders were known from the whole Mesozoic Era. Four genera are now known from El Montsec and Las Hoyas, from five specimens. All belong to modern orb-weaving families: two Tetragnathidae, one Nephilidae (**329**), and one Uloboridae (Selden & Penney, 2003). The first two families weave orb webs with glue strands, while uloborids weave cribellate orbs in which the stickiness of the silk is provided by its 'woolly' nature. This kind of silk traps insects in much the same way that burrs stick to a wool sweater.

Fish. Large fish are rare and are represented by fragmentary remains of hybodontid sharks, and also shark egg-cases. Coelacanths are present, evidenced by the genus '*Holophagus*'. Most of the fishes at El Montsec and Las Hoyas belong to primitive actinopterygian groups; the fauna is dominated by the small (40–50 mm [1.6–2.0 in]) macrosemiid *Notagogus* (**330**). Amiiformes (which includes the living fossil *Amia*, the bowfin) are represented by *Vidalamia* and *Amiopsis*, and the larger actinopterygians include the macrosemiid *Propterus* (140 mm [c. 5.5 in]) and the caturid *Caturus* (250 mm [c. 10 in]) (**331**). Semionoti-formes is represented by at least three forms of *Lepidotes*. Several genera of the deep-bodied

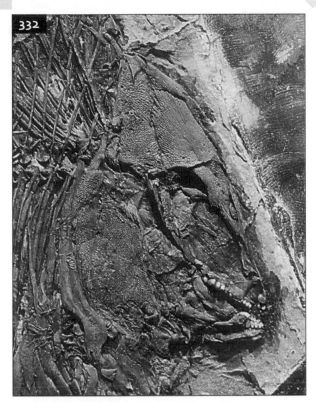

332 The head of the pycnodont fish *Turbomesodon*, from Las Hoyas (MCCM); note the crushing teeth. Height of picture about 120 mm (c. 4.7 in).

pycnodontiform actinopterygians, such as *Turbomesodon* (**332**), are present in the lithographic limestones (see also *Neoproscinetes*, Santana Formation, **375**). Pycnodonts are normally thought of as marine fish, similar (but not related) to the reef-dwelling parrot-fish today. However, evidence from strontium, carbon and oxygen isotope studies at Las Hoyas suggests that these limestones were deposited in non-marine lakes. So, were these pycnodonts (and also the coelacanth) non-marine in the Cretaceous, or are they evidence for a connection to the sea?

Early teleosts are well represented by a stem group of small fishes formerly placed in *Leptolepis*, *Anaethelion* and *Ascalabos* (**333**), but now thought not to belong to these genera but presenting a greater diversity than hitherto recognized and possibly representing a new assemblage of primitive teleosts (Poyato-Ariza, 1997). Two very common small teleosts at these localities are *Rubiesichthys* and *Gordichthys* (**334**), which are the oldest fossil record of the order Gonorynchiformes. Together, these genera form the subfamily

Rubiesichthyinae, which is a sister-group to all other chanids, which form the subfamily Chaninae. Chanids are represented today by the single species *Chanos chanos*, the milkfish; the common fish in the Crato Formation (p. 212), *Dastilbe*, belongs in Chaninae (see **362**).

Amphibians. Three different amphibian lineages have been reported from Las Hoyas and Montsec: albanerpetontids are a group of enigmatic salamander-like fossil amphibians known from deposits of middle Jurassic to Miocene age across Euramerica and Central Asia. More familiar larvae and adults of the true salamanders (Caudata) are represented by *Valdotriton* (**335**), of which several fully metamorphosed adult specimens are known. The wide head of this animal suggests that it had a buccal pump breathing system and lived out of water in damp environments (Evans & Milner, 1996). The other caudatan is *Hylaeobatrachus*, which retained external gills in the adult and was therefore fully aquatic all of its life. The frogs (Anura) are known from three kinds. The commonest is the discoglossid *Eodiscoglossus* (**336**), and there are just single specimens of *Neusibatrachus* and *Montsechobatrachus* known from El Montsec. Tadpoles are also occasionally found (**337**). The evidence suggests that these amphibians bred in the lakes and some spent all of their lives there, feeding on small animals.

Amniotes. The amniote egg is the key to living life entirely on land because it is able to survive and develop out of water, with or without parental care, although some amniotes, such as turtles, have returned to the water. Turtles are anapsids, that is, their skulls have no holes (fenestrae) behind the eye socket (orbit), so the skull is rather thick and heavy, which does not matter in an aquatic animal. A single specimen of a cryptodiran turtle is known from Las Hoyas. All other amniotes found in the lithographic limestones belong to the Diapsida: those with two holes behind the orbit in the skull, which provides strength and room for jaw muscles as well as lightness. Three taxa of lizards are known, of which *Meyasaurus* is by far the best known and occurs at both localities. An odd form, known from just one specimen, is *Scandensia*, whose morphology suggests it was a climbing lizard, like geckos and chameleons. It most probably lived in trees or on rocks and was washed into the lake from afar.

333 Early teleost fish from El Montsec (IEI). Length 97 mm (c. 3.8 in).

334 The common teleost *Gordichthys*, from Las Hoyas (MCCM). Length 36 mm (c. 1.4 in).

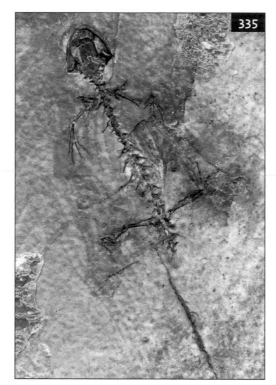

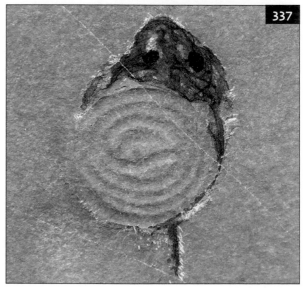

335 The salamander *Valdotriton*, from Las Hoyas; much of the tail is missing in this specimen (MCCM). Length 45 mm (c. 1.8 in).

336 The frog *Eodiscoglossus*, from El Montsec (IEI). Body length 21 mm (c. 0.8 in).

337 A tadpole from El Montsec (IEI). Length 35 mm (c. 1.4 in).

338 The crocodile *Montsecosuchus,* from Las Hoyas (MCCM). Length about 180 mm (c. 7 in).

339 A feather from El Montsec (IEI). Length about 20 mm (c. 0.8 in).

340 The bird *Noguerornis,* from El Montsec (IEI); note the furcula (wishbone) towards the top of the specimen. Furcula about 14 mm (c. 0.5 in) long.

Crocodiles are known from the lakes, but not fully described, apart from *Montsecosuchus* (**338**); other crocodilians are known from isolated teeth, which reflect a fair diversity. Until recently, evidence for dinosaurs at Las Hoyas was sparse but, in 2010, a spectacularly preserved, almost complete skeleton of a theropod dinosaur, named *Concavenator corcovatus,* was unearthed at the site (Ortega *et al.*, 2010). The specimen represents the most completely known individual of the theropod group Carcharodontosauria, a group of large, predatory dinosaurs related to the Jurassic *Allosaurus* (see Chapter 12, The Morrison Formation), which persisted into the Cretaceous. A peculiarity of *Concavenator* is elongated neural spines on the eleventh and twelfth vertebrae, which would have given the animal a strange hump-backed feature of unknown function.

The ornithomimid dinosaur *Pelecanimimus poliodon* from Las Hoyas was the first bird-like theropod found in Europe. The fossil has revealed some unexpected features, including over 200 distinctively shaped teeth and the preservation of skin beneath the lower jaw, resembling the gular pouch of the pelican. However, it is because of their fossil birds that El Montsec and Las Hoyas are among the most famous Lagerstätten in the world of vertebrate palaeontology. Vidal (1902) mentioned the loss of a small bird fossil by the quarrymen at La Pedrera de Meià, and it was not until the 1950s that there was mention of a feather from El Montsec, and formal description of the feathers (**339**), called *Ilerdopteryx,* had to wait until the 1980s (Lacasa-Ruiz, 1985). Unfortunately, the feathers could not be matched with the bird skeletons which started to appear in later years. *Noguerornis* was described by Lacasa-Ruiz (1986, 1989). The modern proportions of *Noguerornis* gives the impression that it was capable of flight, and the acute furcula (wishbone) (**340**), which contrasts with the boomerang shape of that of *Archaeopteryx* and non-avian theropods, suggests that it may have functioned like a spring, which is helpful not only for the flight musculature but also in respiration to supply oxygen during the increased metabolic demands of flight. Sanz *et al.* (1997) described a nestling bird from El Montsec (unnamed because it was considered a juvenile), consisting of skull, neck, and forewings. More recently, however, the bone growth patterns of this specimen have been reinterpreted in comparison with those of modern (Neornithes) birds, and it appears that it was probably nearly adult (Cambra-Moo *et al.*, 2006).

The first report on *Iberomesornis romerali* from Las Hoyas appeared in 1988 (Sanz *et al.*, 1988). Though lacking the skull, parts of the neck, and most of the hands, *Iberomesornis* holds an important position in bird evolution, showing more advanced features than *Archaeopteryx* but more primitive characters than modern birds. A second bird skeleton was discovered at Las Hoyas in the late 1980s and given the name *Concornis lacustris*. Twice the size of *Iberomesornis*, *Concornis* is known from a single, almost complete skeleton with some feathers, but it also lacks the skull and neck (Sanz *et al.*, 1995). One of the most striking features of *Concornis* is the presence of a broad, notched sternum with a caudal carina, similar to that described for *Cathayornis* (p. 177). The third bird skeleton was found at Las Hoyas in 1994. Better preserved than the others, it consists of much of the skeleton and wing feathers but, again, lacks the skull. It was called *Eoalulavis hoyasi*, and its preserved stomach contents show crustacean remains, which suggests that it was one of the earliest freshwater waders. The birds from El Montsec and Las Hoyas are all enantiornithines: an extinct subclass of birds to which most Mesozoic forms belong (see p. 176); *Noguerornis* and *Iberomesornis* are among the most primitive of the Enantiornithes. They are more advanced than *Archaeopteryx* in their greater flight capability and fusion of the tail bones into a pygostyle, but still retained teeth in the jaws and claws on the hands (wings).

Trace fossils. The overall assemblage of trace fossils in the lakes is characterized by low diversity and small burrow size, and is dominated by surface trails and extremely shallow burrows produced mostly by detritus feeders (Buatois *et al.*, 2000; Gibert *et al.*, 2000). This indicates low energy, shallow lacustrine conditions with environmental stress, most likely low oxygen levels. Most trails and burrows probably represent opportunistic coloniz-ation by epifauna and very shallow infauna during brief periods of improved oxygenation related to density flows or dilute turbidity currents. Inverte-brate traces include *Helminthoidichnites* and *Gordia*, probably arthropod grazing trails, *Lockeia*, which is normally interpreted as a resting trace of bivalves but, when small, could be that of bivalved crusta-

ceans like ostracods, *Treptichnus*, feeding traces of worms or arthropods, and *Hamipes*, an arthropod trackway (**341**). A common trail in some horizons is *Undichna*, which is attributed to the marks of fish tails on soft sediment. *U. britannica* from El Montsec consists of two intertwined waves and was probably produced by the primitive teleost *Ichthyemidion* (formerly *Anaethelion*) *vidali*, while *U. unisulca* from Las Hoyas (**342**) is characterized by having only a single sinusoidal wave, and was most probably produced by deep-bodied pycnodont fish.

Not only freshwater animals left their mark in the lime muds of Las Hoyas. Isolated imprints and trackways have been recorded and interpreted as having been made by crocodiles and/or pterosaurs (there is dispute over whether the track called *Pteraichnus* was made by pterosaurs or crocodiles). In fact, both were present: evidence of pterosaur teeth and parts of a skull has recently come to light

341 *Hamipes didactylus*, an arthropod walking trace, El Montsec (IEI). Distance between successive footprints is 20 mm (c. 0.8 in).

342 The trace fossil *Undichna*, made by a fish tail scraping the lake floor, Las Hoyas. Width of view about 300 mm (c. 12 in).

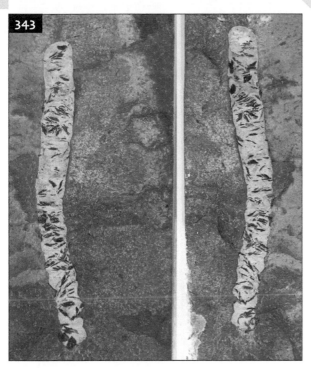

343 A coprolite containing fish debris, from El Montsec (IEI). Length 48 mm (c. 1.9 in).

at Las Hoyas (Vullo *et al.*, 2009). A trackway at Las Hoyas showing characteristic three-toed bipedal dinosaur foot impressions is shown in **311**.

Other kinds of trace fossils include coprolites (fossil dung), which are quite common at El Montsec and Las Hoyas. Figure **343** shows a coprolite containing fish debris, no doubt produced by a predatory fish. A peculiar mass of four juvenile bird skeletons found at Las Hoyas was interpreted by Sanz *et al.* (2001a) as a regurgitated pellet, probably produced by a small, predatory dinosaur.

PALAEOECOLOGY OF EL MONTSEC AND LAS HOYAS BIOTA

The palaeoecological interpretations of Las Hoyas and El Montsec point to a freshwater, wetland-type ecosystem, within which there were many different habitats. While the lake floor, especially in the deeper parts, was anoxic most of the time, occasional turbid flows near the lake shore brought oxygenated water to the lake bottom, which allowed temporary colonization by benthos, as evidenced by trace fossils. At El Montsec, for example, the La Cabrua trace fossil assemblage reflects these occasional oxygenation events with traces of small deposit feeders, while the La Pedrera assemblage lacks grazing or feeding traces, the only trace fossils being trails of benthic crustaceans and fishes, which is consistent with a more distal position.

The Las Hoyas fauna is dominated by obligate aquatic taxa, e.g. fish; about a quarter are amphibious (i.e. needing water for part of their life cycle, like frogs), while the remainder are species found in wetlands but not tied to the water, like dinosaurs and birds. The fish communities show non-teleost neopterygians occupying the higher and more specialized trophic levels, while primitive teleosts occupied the lower, more generalized levels. This represents essentially a relict fish fauna from Jurassic times, comparable to the marine faunas of the late Jurassic of Cerin or Solnhofen.

Among the aquatic insects, benthic forms predominate. For example, burrowing mayfly nymphs are common, as are brachyceran fly larvae. Curiously, no burrows or trails of the latter have been found, so maybe they lived on algal mats, and could have been tolerant of low oxygen levels. Much less common are predatory dragonfly nymphs, three rare mayfly species (probably algal and debris feeders), and rare nematoceran flies, like midges. Caddisflies prefer oxygenated water, and none of their characteristic larval cases have been found. Beetles thrived in the nekton at Las Hoyas, which is characterized by the great diversity of the family Coptoclavidae, and abundance of one species of this family, as well as other aquatic families. Six different feeding strategies are represented: hunters only on the water surface (gyrinids), hunters both on the water surface and in the water column (coptoclavids), hunters only in the water column (dytiscids), benthic hunters (coptoclavid larvae), zooplankton filterers (coptoclavids), and herbivores, possibly among the charophytes or aquatic angiosperms (hydrophillids). The gyrinids (whirligig beetles) would be joined on the water surface by the water-strider-like

Chresmoda, the nekton would include water-boatmen-like *Iberonepa*, and on the lake floor there would be crustaceans, when oxygenation permitted. These arthropods would have been preyed upon by the vertebrates.

The land would have been profusely vegetated. Three groups of halophytic plants grew in the marshes: cheirolepidiaceous conifers, tree ferns, and herbaceous ferns. Among these plants were abundant insects, such as the terrestrial beetle families Cupedidae and Scarabaeidae and the wide variety of flying insects. Preying on these, together with the lizards, birds and other vertebrates, would be the orb-weaving spiders.

COMPARISON OF EL MONTSEC AND LAS HOYAS WITH OTHER LOWER CRETACEOUS LAKE LOCALITIES

It might be expected that lithographic limestones the world over would exhibit some similarities; however, the differences can be marked. The most famous of all, Solnhofen (Chapter 13) represents a coastal lagoon and the preservation of most of the fossils relates to occasional storm events bringing in a marine biota. Indeed, the late Jurassic appears to have been a time particularly prone to the formation of lithographic limestone deposits in lagoons along the northern margin of the Tethys Ocean including, in addition to Solnhofen, Cerin in the French Jura, Nusplingen in the German Swabian Alb, and Wattendorf in the German Franconian Alb. Whereas there are differences between these late Jurassic Lagerstätten, the suggested palaeoenvironment at El Montsec and Las Hoyas was freshwater lacustrine, and so in complete contrast. In palaeobiological terms, there are more similarities between El Montsec/Las Hoyas and other early Cretaceous lacustrine ecosystems such as the Jehol Biota (Chapter 14) and the Crato Formation (Chapter 16). These Lagerstätten have yielded among the most diverse Mesozoic insect faunas as well as exceptionally preserved vertebrates.

It is becoming apparent that there were numerous lakes, at various times, on the Iberian plate during the rifting phase in the early Cretaceous, each giving rise to a potential Lagerstätte. Thus, because of differences in palaeoenvironment, taphonomic processes, and biases in collecting and study techniques, there may be differences in palaeoecological interpretations between these sites that could be real or artefactual. For example, new fossil sites discovered in the Barremian deposits of the La Huérguina Limestone Formation in the Serranía de Cuenca – Uña and Buenache de la Sierra – that compare with those at Las Hoyas (Buscalioni *et al.*, 2008). Together, these sites, though differing in their taphonomy and taxon sampling to some extent, nevertheless paint a picture of a complex low-lying wetland ecosystem at that time, including not only calcareous lakes but also fluvial and marsh habitats. Comparing Las Hoyas and El Montsec, in some cases we see very similar processes; for example, trace fossils at both sites suggest an anoxic lake bottom with occasional influxes of oxygenated water allowing temporary colonization by benthos. However, comparison of the beetle faunas by Soriano *et al.* (2007) has shown major differences. The beetle associations are composed of several different families from the three major suborders of Coleoptera, with very different ecological habits: herbivores (on water plants, among the riparian flora, and feeders on live or dead wood), carnivores (zooplankton filterers, feeding on aquatic and terrestrial fauna, and parasitic), and scavengers. The Las Hoyas beetle fauna is mainly composed of carnivorous aquatic forms, while at El Montsec they are mainly herbivores and terrestrial feeders on wood.

MUSEUM AND SITE VISITS

Museum
Institut d'Estudis Ilerdencs, Plaça Catedral s/n, 25002 Lleida, Spain.

Sites
Fossil collecting is not possible at either of these localities. However, the areas are definitely worth visiting for their geology and scenic attractions. The El Montsec quarries are situated on the track between Rúbies and Santa Maria de Meià. This area is an excellent hiking region, and the quarries lie adjacent to good walking trails. Similarly, the Las Hoyas locality is not open to the public but lies in an extremely attractive limestone region north-east of Cuenca. The nearest small town is Buenache de la Sierra, from where dirt roads will take the visitor exploring this beautiful, forested region.

Chapter Sixteen

THE SANTANA AND CRATO FORMATIONS

BACKGROUND: THE BREAK-UP OF THE PANGAEA SUPERCONTINENT

During the early part of the Cretaceous Period the supercontinent of Pangaea continued to break up, owing to movements deep in the Earth's mantle. Initially North America and Europe pulled apart, opening up the North Atlantic Ocean, and by the end of the early Cretaceous, South America and Africa also began to cleave apart to form the South Atlantic. Meanwhile the Tethys Ocean extended westwards so that the northern continents were divided from those in the south. This loss of land bridges and migration routes led to a diversification of animal and plant life on separated continents.

The dominance of large reptiles on land (dinosaurs), in the sea (ichthyosaurs and plesiosaurs), and in the air (pterosaurs), evident during the Jurassic (Chapters 11–13), continued into the Cretaceous, but this period witnessed significant evolutionary innovations in all these groups before their final extinction at the Cretaceous/Tertiary (K/T) boundary.

As the continents divided, dinosaur groups began to diverge. While at the start of the Cretaceous many dinosaurs (such as *Iguanodon*) were able to migrate between North America and Europe, by the middle of the period, although *Iguanodon* continued to thrive in Europe, the dominant ornithopod (bipedal plant eater) in North America was *Tenontosaurus*. In South America the massive sauropods (giant plant eaters), so successful in the Jurassic (see Chapter 12, The Morrison Formation), continued to dominate, but in northern continents these and

the stegosaurs (plated dinosaurs) quickly declined, their place taken by smaller, ornithischian herbivores, which moved in vast herds across the Cretaceous landscape. They included the ankylosaurs (armoured dinosaurs) and the first ceratopsians (horned dinosaurs, such as *Psittacosaurus*, *Protoceratops*, and *Triceratops*). At the same time the iguanodonts gave rise to the duck-billed hadrosaurs, such as *Maiasaura*.

Among the carnivores the allosaurs (see Chapter 12, The Morrison Formation) began to decrease and were replaced by a group of bird-like theropods, the coelurosaurs, such as *Tyrannosaurus*, *Deinonychus*, *Velociraptor*, and the ostrich-like *Struthiomimus* and *Ornithomimus*.

In the water the dominant predators were massive marine lizards, the mosasaurs, which gradually replaced the ichthyosaurs of the Jurassic. Plesiosaurs continued to thrive, along with giant turtles, but marine crocodiles disappeared early in the Cretaceous. Fish evolved rapidly, but the more primitive holostean bony fish of the Jurassic were gradually replaced by the teleostean bony fish, and by the end of the Cretaceous most holosteans had disappeared. Early teleosts were primitive, including herring-like and salmon-like ancestors. The ancient hybodont sharks made their final appearance in the Cretaceous. Their advanced shark descendants were numerous and included many modern forms; skates and rays had also essentially reached modern conditions. *Mawsonia* is the last known fossil coelacanth and was thought to be the last representative of the group until the discovery in 1938 of a living representative.

In the air the primitive 'long-tailed' pterosaurs

(the rhamphorhynchoids), dominant in the Jurassic, were replaced by the 'short-tailed' pterodactyloids, some with wingspans of at least 10 m (c. 33 ft) – the pinnacle of pterosaur evolution. But from the beginning of the Cretaceous they had to share the airways with their first competitors, the birds, which had developed from feathered dinosaurs during the late Jurassic and earliest Cretaceous (Chapter 14, The Jehol Group). Initially, birds seemed to have thrived around lakes, but due to their greater adaptability than pterosaurs they soon spread to a variety of habitats.

However, arguably the most important evolutionary event of the Mesozoic was the appearance of angiosperms, the flowering plants. The angiosperms had developed a new strategy to defend themselves against grazing animals, by growing and reproducing faster. As the Cretaceous Period progressed, flowers developed a closer relationship with insects. Flowers needed to ensure that the pollen grain reached the protected ovule and so they began to produce nectar to attract insects, which carried excess pollen from one plant to another. As the angiosperms underwent explosive growth at the end of the early Cretaceous, so the insects diversified rapidly in parallel development and the Cretaceous air was full of flying reptiles, birds, and insects.

Much of our knowledge of the fauna and flora during this Cretaceous Period of innovative evolution comes from one of the world's most productive fossil sites, located in the State of Ceará in north-eastern Brazil (344). There, in a succession 700 m (c. 2,300 ft) thick of mainly Lower Cretaceous sediments, can be found not one, but two separate Fossil-Lagerstätten which, by modern stratigraphical nomenclature, are now separately named the Santana Formation and Crato Formation.

Soft-tissue preservation by very different mechanisms has provided an insight into a variety of early Cretaceous biotas. In the Santana Formation these include a variety of fish, spectacular pterosaurs, and other reptiles, including dinosaurs, all preserved inside limestone concretions, while

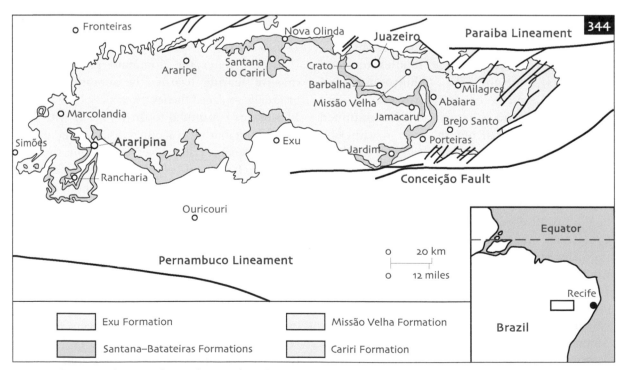

344 Locality map showing the geology and settlements of the Araripe Basin in north-eastern Brazil (after Martill, 1993).

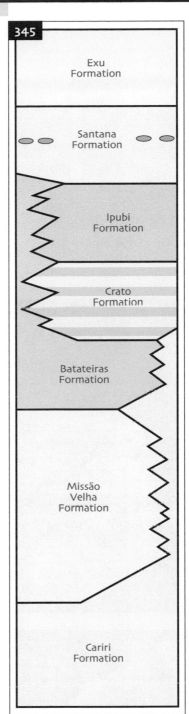

345 Diagram to show the stratigraphy of the Mesozoic succession in the Araripe Basin (after Martill, 1993).

HISTORY OF DISCOVERY OF THE SANTANA AND CRATO FORMATIONS

The discovery of the remarkable fossils of the Santana Formation goes back to Napoleon's destruction of the Portuguese Empire in the early nineteenth century, which led to Portugal's renewed interest in the exploitation of Brazil. Don Pedro, Brazil's first Portuguese emperor, married an Austrian archduchess, which led to an influx of scientists and philosophers coming to Brazil from Austria and Bavaria.

In 1817 and 1820 the German naturalists Johann Baptist von Spix and Carl Friedrich Philipp von Martius, from the Academy of Sciences in Munich, explored much of Brazil, including the north-eastern state of Ceará, where they came across the fish-bearing concretions from what was later described as the Santana Formation. Their report was published between 1823 and 1831 and includes the first illustration of a Santana fish. The fossils are found in sediments exposed around the flanks of an uplifted, flat-topped plateau, known as the Chapada do Araripe, at about 800 m (c. 2,600 ft) elevation. The Chapada represents the remnants of an ancient sedimentary basin, the Araripe Basin, and trends east–west, straddling the states of Ceará to the north, and Pernambuco to the south (**344**).

This region was visited again during the years 1836–41 by Glasgow botanist George Gardner, whose book, *Travels in the Interior of Brazil* (1846), gives a fascinating account of his collection of fossil fishes from 'rounded limestones'. These were sent back to the UK to be exhibited at the British Association in Glasgow, where they were seen by the eminent palaeontologist Louis Agassiz, who described seven species and correlated them to the Cretaceous Period (Agassiz, 1841; Gardner, 1841; Nudds *et al.*, 2005).

Jordan and Branner's 1908 monograph was the first major palaeontological work on the Santana fish, while Small (1913) gave the first stratigraphical account of the Araripe Basin, and introduced the name Sant' Ana Limestones. Beurlen's more recent stratigraphic review (1962) divided the Santana Formation into three members, a lower Crato Member, consisting of shales and laminated limestones (with insects and plants), and an upper Romualdo Member, con-

the Crato Formation displays the world's most remarkable Cretaceous insect fauna, a diverse flora of gymnosperms and early angiosperms, a number of fish, and rare reptiles and amphibians, all preserved in a micritic Plattenkalk limestone, not dissimilar to that of Solnhofen (Chapter 13).

taining the characteristic fish concretions, these being separated by the Ipubi Member of evaporites.

There has been much controversy over the stratigraphic nomenclature of this sequence, but that suggested by Martill (1993), in which the Santana Formation is now restricted to the concretion-bearing sequence previously described as the Romualdo Member, while the Ipubi and Crato members are both elevated to formational status (**345**), seems to be acceptable.

Besides the well-known fish, the Santana Formation (*s.s.*) has recently begun to yield fossil reptiles including crocodiles, turtles, dinosaurs, and spectacular pterosaurs, all found within the concretions. The concretions are 'mined' by peasant farmers (**346–349**), especially around the villages of Cancau (near Santana do Cariri) and Jardim, and are sold cheaply to local commercial fossil dealers, even though trade in fossils is illegal in Brazil.

The spectacular insects and plants of the Crato Formation have only really been known since the 1980s (see Grimaldi, 1990) when commercial exploitation of the limestones began. The fossils have been mostly discovered by the workers, many of them children, who quarry by hand the laminated limestones around the village of Nova

346 Near Jardim, Chapada do Araripe, Ceará, north-east Brazil.

347 Fish 'mine' in Santana Formation, near Jardim (see **346**).

348 Discarded fossil fish debris at fish 'mine' near Jardim (see **347**).

349 'Fisherman' with fossil fish from Santana Formation, Cancau, near Santana do Cariri, Chapada do Araripe, Ceará, north-east Brazil.

350 Manual working of the laminated Crato Formation limestones near Nova Olinda, Chapada do Araripe, Ceará, north-east Brazil.

351 Stone Yard, Nova Olinda, cutting ornamental paving slabs from Crato Formation limestone.

Olinda for use as an ornamental paving stone and for cement manufacture (350, 351). Fossils from the Santana Formation were beautifully illustrated by Maisey in his 1991 book, *Santana Fossils*, while those from the Crato Formation have recently been extensively reviewed by Martill *et al.* in the 2007 publication, *The Crato Fossil Beds of Brazil*.

STRATIGRAPHIC SETTING AND TAPHONOMY OF THE SANTANA AND CRATO FORMATIONS

The Araripe Basin, in which these formations occur, is a fault-bounded interior basin whose evolution is closely related to rifting associated with the opening in Lower Cretaceous times of the South Atlantic Ocean, brought about by South America and Africa pulling apart. The Proterozoic crystalline basement of Brazil is composed of a fractured craton and many of the faults were probably reactivated during the break-up of Pangaea to form coastal and interior sedimentary basins. The Araripe Basin lies between two major east–west trending lineaments (the Paraiba and Pernambuco lineaments; see Martill, 1993, Figures 2.2–2.4), which can be traced across the mid-Atlantic Ridge as transform faults and correlated with their corresponding continental fault zones in west central Africa.

An erosional remnant of the basin has since been uplifted to form the 200 km (c. 125 miles) wide, east–west trending plateau of the Chapada do Araripe, in which horizontal Cretaceous (and possibly also Jurassic) sediments lie unconformably on the Proterozoic/ Palaeozoic basement (344). As noted in the previous section there has been much discussion on the stratigraphy of the Araripe Basin, but the scheme suggested by Martill (1993), which divides the Mesozoic succession into seven formations, seems sensible, as it demonstrates the complex interrelationships between the formations which previous workers have ignored (345).

The four youngest formations are placed by Martill (1993) in the Araripe Group, which includes all the important fossil-bearing beds. At the base of this Group the Crato Formation consists of a series of laminated, micritic limestones (Plattenkalks), interbedded with sandstones, marls, and clays, up to 60 m (c. 200 ft) thick. Martill *et al.* (2007) further divided this formation into four members, and it is the basal Nova Olinda Member of laminated limestones up to 13 m (c. 40 ft) thick, which contains the well-preserved insects and plants. The overlying Ipubi Formation consists of bedded evaporites, mainly gypsum and anhydrite up to 20 m (c. 65 ft) thick, and is devoid of fossils (345).

Above the evaporites, the Santana Formation consists of non-fluvial, deltaic silts and sands which give way to a series of green/grey laminated shales with fossil-bearing concretions. The upper part of the formation, above the level of the concretions, consists of shales with thin limestones containing gastropods and rare echinoids. The overlying Exu Formation, 75 m (c. 250 ft) of coarse, cross-bedded sandstones, forms a resistant, horizontal cap to the Chapada plateau, with the older formations exposed on the plateau flanks (**345**).

The exact age of the Crato and Santana Lagerstätten within the Lower Cretaceous has yet to be determined, partly due to the fact that many of the fossils have been collected by local people who have not recorded their precise provenance. Based on preliminary palynological results the Crato Formation is thought to be of late Aptian/early Albian age (approximately 112 million years), while the slightly younger Santana Formation is probably of Albian age (see Martill, 2007 for extensive discussion).

Although of similar age, the mechanisms of soft-part preservation of these two biotas were quite distinct. In the Crato Formation the preservation of the insects is exceptional, with microstructural detail and even colour patterns preserved (Martill & Frey, 1995; Heads *et al.*, 2005). Scanning electron microscopy has revealed eye facets and fine hairs on cuticle. There are essentially two types of preservation, either as black, carbonaceous replacements with finely disseminated pyrite, or as dark-brown to orange-brown goethite, a hydrated iron oxide. The former is seen when unweathered blue-grey limestones are excavated at depth, while the latter occurs in weathered, surface limestones in which organic matter and pyrite have oxidized, giving the limestones their characteristic yellow, buff or cream coloration.

The environment of deposition of the Crato Formation Plattenkalks was originally considered (on dubious faunal grounds; see p. 209) to have been a freshwater lake developed within the interior basin, but the weight of current evidence suggests that apart from possible brackish conditions at its base, much of the formation was deposited under saline conditions. Overwhelming evidence for marine waters entering the Crato lake/lagoon has been presented by Arai (2000) on geomorphological, stratigraphical, sedimento-logical, geochemical, and palaeontological grounds. However, although the Crato fish are salinity-tolerant, there is a notable absence of open marine nekton (such as cephalopods and plesiosaurs), and Martill (1993) and Martill *et al.* (2007) suggested that the waters were actually hypersaline and inhospitable due to the arid climate. The common occurrence in the Crato sediments of salt pseudomorphs after hopper-faced halite, and the thick sequence of evaporates in the succeeding Ipubi Formation, both seem to support this suggestion, and it seems probable that normal marine salinities did not occur in the basin at any time during the deposition of the Crato Formation.

A salinity-stratified water column with hypersaline, oxygen-deficient bottom waters would have prevented any autochthonous life in the Crato lagoon except in freshwater tongues developed around the mouths of rivers entering the basin. The small gonorhynchiform fish *Dastilbe* probably lived in such an environment (it is related to present-day milkfish, which are usually found in offshore marine waters, but also frequently enter estuaries), but its common occurrence fossilized in large numbers suggests mass mortality occurred when water mixing led to suddenly increased salinities. There is a general lack of benthic organisms and of bioturbation in the finely laminated Plattenkalk layers, while the presence of a benthic cyanobacterial mat, suggested by micro-ripples on bedding planes, also suggests that grazers were absent. All of this suggests that conditions, at least in the deeper parts of the lagoon, were hostile to bottom-dwelling organisms, partly due to excessive concentrations of salt, but also due to anoxia, which is confirmed by organic biomarkers from hydrogen sulphide-loving bacteria (Martill *et al.*, 2007). Such oxygen deficiency would have hindered decomposition of carcasses on the lagoon bed by slowing down microbial decay and allowing the exceptional preservation.

The plants, insects, feathers, and occasional tetrapod carcasses must all therefore have drifted into the lagoon in rivers or may have blown in, so that the vast majority of the fauna and flora of the Crato Lagerstätte should be considered allochthonous. This is not dissimilar to the situa-tion described in Chapter 13 for the Upper Jurassic Solnhofen Plattenkalks, although the Crato Formation lacks any obvious evidence or mechanism for rapid burial.

Soft-tissue preservation within the Santana Lagerstätte concretions is very different and is possibly unique in the fossil record. Martill (1988, 1989) argued that the exquisite preservation of

352 The teleostean fish *Notelops brama*, showing preservation of soft muscle tissue between the ribs (PC). Patch of tissue 47 mm across (1.8 in).

delicate tissues such as gills, muscles, stomachs, and eggs (**352**) is the result not just of rapid burial, but also of rapid fossilization, perhaps instantaneous fossilization.

By the time of deposition of the Santana Formation the increasingly saline lagoons had dried up completely, and a period of evaporite deposition, represented by the Ipubi Formation, was followed by a further activation of the basin either by another marine incursion or by prograding deltas. Huge numbers of fish migrated into the basin before a further mass-mortality event, possibly caused by changes in surface water oxygenation or salinity, or by toxification from hydrogen sulphide released from sediments, toxic dinoflagellates, or cyanobacterial blooms (Martill *et al.*, 2008). The fish are preserved within calcium carbonate concretions, but notably their soft tissue is preserved in calcium phosphate (cryptocrystalline francolite). Martill (1988, 1989) showed that francolite precipitation is enhanced in oxygen deficient, low-pH (acidic) environments, as would be caused by the onset of mass decomposition.

Using scanning electron microscopy of the phosphatized tissue, Martill (1988, 1989) recognized banded muscle fibres with cell nuclei, gill filaments with secondary lamellae, stomach walls, and ovaries with eggs. He observed that the gill tissues of fresh trout became bacterially infested and began breaking down within 5 hours of death and had completely disappeared within 1 week. He therefore concluded that the phosphatization of

the Santana fish must have begun within 1 hour of the fishes' death and he termed this the 'Medusa Effect'. The phosphate was produced by bacteria feeding on proteins in the carcass so that some decay was actually necessary to initiate fossilization.

The next stage in this remarkable preservation is the rapid nucleation after burial of a carbonate concretion around the phosphatized fish to preserve its three-dimensionality. Formation of concretions is known to occur around organic remains in oxygen deficient, low-pH conditions in which lime normally remains in solution, but it requires a local increase in pH in the micro-environment around the decaying carcass (perhaps caused by the release of ammonia?) to allow the lime to precipitate. Both Martill (1988) and Maisey (1991) stressed the contribution to this process made by a putative cyanobacterial mat or scum on the sea floor, as has also been suggested for Ediacara (Chapter 1), Grès à Voltzia (Chapter 10), the Holzmaden Shale (Chapter 11), and the Solnhofen Limestone (Chapter 13).

The same mechanism of fluctuating pH and successive precipitation of calcium phosphate and calcium carbonate has also preserved the segmented limbs in ostracods, the delicate wing membranes of pterosaurs, and comparable structures in other reptiles.

DESCRIPTION OF THE SANTANA AND CRATO FORMATIONS BIOTA

Crato

Insects. Arthropods are the most diverse group in the Crato biota and are dominated by an insect assemblage which, along with that of the Jehol Group (Chapter 14), is the largest and most diverse Cretaceous insect fauna in the world. Approximately 20 orders are represented, all but one of which is still extant, and includes aquatic, semi-aquatic, and terrestrial groups (see Grimaldi, 1990; Martill *et al.*, 2007). The primitive entognath hexapods are represented by a sole japygoid dipluran, which is the oldest definite record of this group (Wilson & Martill, 2001), and the most primitive true insects (wingless apterygotes) by an undescribed genus of Zygentoma (the silverfish and firebrats).

The most basal of the pterygotes (winged insects), as in the Jehol Group (p. 179), includes a diverse

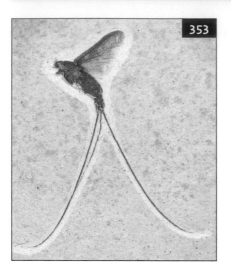

353 A mayfly (PC). Length of body (excluding cerci) 7 mm (c. 0.27 in).

354 A dragonfly (PC). Length of body 54 mm (c. 2.1 in).

355 An earwig (PC). Length of body (excluding cerci and antennae) 19 mm (c. 0.75 in).

356 The long horned grasshopper *Cratoelcana* (PC). Length (excluding antennae) 30 mm (c. 1.2 in).

record of Ephemeroptera (mayflies) and Odonata (damselflies and dragonflies), with both nymphs and adults represented. Mayfly nymphs are restricted to freshwater habitats, and it was their common occurrence that led early workers to consider that the Crato basin was a freshwater lake, not realizing that the insect fauna was allochthonous. Adult mayflies are recognized by the presence of very long cerci, often with a median terminal filament, giving the appearance of a three-pronged 'tail' (353). At rest the wings are folded vertically over the abdomen. The Odonata include four families of Zygoptera (damselflies), which like the mayflies fold their wings vertically at rest; a recently discovered specimen actually preserves the original metallic-green colour seen in modern damselflies (Martill *et al.*, 2007, plate 2c). The Anisoptera (dragonflies) are represented by some 18 genera in 11 families and are known from over 1,000 beautifully preserved specimens with exquisite detail of wing venation preserved. These can be distinguished from damselflies by their habit of holding their wings horizontally over their bodies while at rest (354).

More derived hemimetabolous pterygotes (those lacking a pupal stage in their ontogeny) also compare with the Jehol entofauna (p. 180) and include the Dermaptera (earwigs), which are not common, but are represented by three families (355), and the Orthopterida, which is one of the most diverse groups in the Crato fauna. It includes the Orthoptera (grasshoppers, crickets, and locusts) and the Phasmatodea (stick insects), along with an extinct family of obscure affinities, the Chresmodidae. This is usually placed in the Archaeorthoptera (see p. 191), which is the only extinct insect order in the Crato biota. The phasmatodes are known from a single species, but the Orthoptera are diverse and numerous; the Ensifera (crickets and allies) are known from 16 genera in five families, while the Caelifera (grasshoppers and locusts) are known from eight genera also in five families (356). Hemimetabolous insects also include the Blattaria (cockroaches), Isoptera (termites), and Mantodea (praying mantises). The cockroaches are common (comprising 26% of the insects) and often superbly preserved exhibiting both colour patterns and soft tissues, such as nerves, eye lenses, and digestive tracts (**357**). The largest group of hemimetabolous insects, however, is the Hemiptera (which includes cicadas, leafhoppers, planthoppers, and true bugs), with at least 30 families recorded. The Cicadomorpha includes the palaeontinids, or giant cicadas (**358**), an extinct stem group, while the Heteroptera (true bugs) includes giant water bugs, waterscorpions, creeping water bugs, backswimmers, water boatmen, and some amphibious and terrestrial bugs.

Holometabolous insects (those going through a pupal stage in their ontogenetic development) are also incredibly diverse and not dissimilar to those from Jehol (p. 180). One of the most basal groups, always recognizable by their large wings and net-like venation, is the Neuropterida, represented in the Crato fauna by all three extant orders, Neuroptera (lacewings and antlions), Raphidioptera (snakeflies), and Megaloptera (dobsonflies and alderflies). The neuropterans are represented by over 50 species (**359**), but the megalopterans are extremely rare, and known from a few undescribed specimens. The raphidiopterans are equally intriguing, as the four Crato species constitute the only known records from the southern hemisphere.

Coleoptera (beetles), the sister group to the Neuropterida, are dominated in the Crato fauna by members of the suborder Polyphaga, which are also the largest group of beetles today. They include rove beetles, dung beetles, water scavenger beetles, jewel beetles, click beetles, sap beetles, flat bark beetles, leaf beetles, and weevils. Diving beetles and tiger beetles (suborder Adephaga) are also known from a few specimens. According to Heads *et al.* (2008) beetles are under-represented in the Crato Formation when compared with contemporaneous deposits (for example see Chapter 15, El Montsec and Las Hoyas), but these authors consider that this is a taphonomic artefact.

Hymenoptera (bees, ants and wasps) are another group of holometabolous insects which are diverse in the Crato fauna, as they are today. Many of the major families of wasps are represented along with putative records of an ant (Formicidae) and a bee (Apidae). If confirmed these would constitute the oldest known records, as both families are otherwise first known from the Upper Cretaceous of North America.

Equally diverse among the holometabolous insects are the Diptera (true flies, gnats, and crane flies), which are distinguished by having only one pair of fully developed wings. Although diverse, they are actually very rare in the Crato Formation, comprising only 2% of the insect population, compared with 30–50% recorded from other late Mesozoic insect faunas. Other interesting features of the dipterans include an abundance of Asilidae (robber flies), extant members of which prefer arid environments (which fits well with the palaeoecological interpretation of the Crato Formation), a paucity of Rhagionidae (snipe flies), which are normally one of the most abundant groups in the Mesozoic, and a complete absence of Empidoidea (dance flies) and Nemestrinidae (tangle-veined flies), which were common during the early Cretaceous. Archaic insect groups are also absent, such that the Crato dipterans are more similar to Tertiary faunas than to other early Cretaceous assemblages.

Rare specimens of Trichoptera (caddisflies) and primitive Lepidoptera (butterflies) complete the insect fauna. It is noteworthy, however, that the aquatic caddisfly larvae are totally absent, as this strongly supports the hypothesis that the Crato insects are allochthonous.

Arachnids. The Crato fauna includes one of the most important Mesozoic arachnid assemblages in the world. They are the most diverse of the non-hexapod arthropods and are represented by araneomorph spiders ('true' spiders) and mygalomorph spiders (tarantulas, bird-eating spiders, and trap-door spiders; **360**), hemiscorpionid and chactid scorpions (**361**), and erythraeoid mites. In addition to these main arachnid groups the Crato beds have also produced records of ceromid solifuges (camel spiders), uropygids (whipscorpions), and amblypygids (whipspiders), all of which represent the only known records of these groups from the entire Mesozoic. The most common araneomorph spider, known from hundreds of specimens, is *Cretaraneus*, also known from the Lower Cretaceous of Montsec in Spain (Chapter 15, El Montsec and Las Hoyas; **329**). The mygalomorphs belong to the modern family Dipluridae, characterized by their elongated posterior spinnerets used for weaving funnel-webs to catch jumping insects; the Crato specimens push the appearance of this group back some 90 million years. The scorpions are equally informative; the chactid is the oldest known to date, while the hemiscorpionid represents the first record from the Cretaceous. Similarly, the Crato acariform mite is not only the oldest erythraeoid known, it is also the largest fossil mite ever recorded.

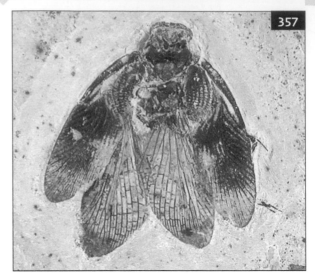

357 A cockroach (PC). Maximum width 9.5 mm (c. 0.4 in).

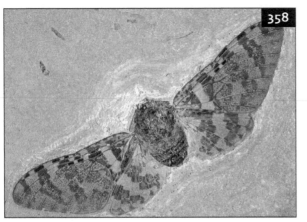

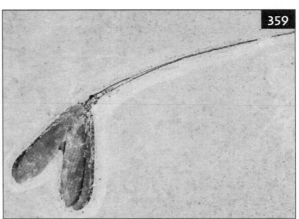

358 The cicada *Baeocossus* cf. *fortunatus* with colour pattern preserved (KMNH). Wingspan 65 mm (c. 2.5 in).

359 The osmylid neuropteran *Nuddsia longiantennata* (SMNS). Length of body (excluding antennae) 30 mm (c. 1.2 in).

360 The diplurid spider *Cretadiplura ceara* (HMB). Length of body (including spinnerets) 15 mm (c. 0.6 in).

361 A scorpion (MM). Total length 21 mm (c. 0.8 in).

Other arthropods. Crustaceans, which are common in other Mesozoic Lagerstätten, are remarkably rare in the Crato Formation and are represented only by ostracods, conchostracans, and a rare caridean shrimp which is the only decapod crustacean. Myriapods include a small number of centipedes including a scutigeromorph (house centipede), the only known example from the Mesozoic.

Fish. The ichthyofauna is dominated by small specimens of the gonorhynchiform fish *Dastilbe crandalli* (**362**), often with several on a single bedding plane (Davis & Martill, 1999). The gonorhynchiforms, which include the present-day milkfish (*Chanos chanos*), are a sister-group to the carps, minnows, and catfish. Other fish taxa include the ichthyodectiform *Cladocyclus*, the semionotiforms *Lepidotes* and *Araripelepidotes*, the coelacanth *Axelrodichthys*, and a single specimen of the aspidorhynchiform *Vinctifer*. These latter genera are all rather rare, but are considerably more abundant in the Santana Formation (p. 214).

Pterosaurs. In recent years the first pterosaurs have been discovered from the Crato Formation, some with soft-tissue preservation including wing membranes and wing fibres, claw sheaths, foot webs, and a heel pad. Only three main lineages, however, seem to be present. Ornithocheirids are represented by *Arthurdactylus* (Frey & Martill, 1994) and *Ludodactylus* (Frey *et al.*, 2003), both with wingspans of approximately 4 m (c. 13 ft), and which may yet prove to be synonymous (**363**). The other two groups were possibly indigenous to the Crato lagoon. Tapejarids are represented by two species of the genus *Ingridia*, erected to include a remarkable specimen complete with its soft tissue head crest described by Campos and Kellner (1997; **364**), while three incomplete specimens have been referred to the tupuxuarids, a group characterized by very long hindlimbs. Both these groups probably fed on the shoals of *Dastilbe* that were common in the lagoon, the latter possibly by wading in the shallow margins of the lagoon. Over 30 specimens of pterosaur are now known, some of which have yet to be described, but the fauna seems to be distinctive in containing few juveniles and no small pterosaurs (Unwin & Martill, 2007).

Other vertebrates. Less common, but nonetheless important, are a few records of frogs, which are often complete and with soft tissue preserved (**365**). Most of these belong to the leptodactylids, but one specimen is an undescribed pipoid. Turtles, also known from just a few specimens, belong to the pleurodires (side-necked turtles), while the two known lizards are both terrestrial. Crocodiles are extremely rare and normally fragmentary (**366**); only two species are known, a neosuchian and a notosuchian, the latter (*Araripesuchus*) also occurring in the Santana Formation and in west Africa (p. 216). Isolated feathers, some with colour patterns, have occasionally been reported (Maisey, 1991; Martill & Filgueira, 1994; Kellner, 2002) including an asymmetrical remex, two or three elongate symmetrical rectrices, and several semiplumes. Naish *et al.* (2007) described two specimens where skeletal remains were associated with feathers, one of which they identified as a possible euenantiornithine bird (see p. 176). To date these are the only known birds from the Crato Formation, which is surprising given their more common occurrence in other Lower Cretaceous Fossil Lagerstätten (see Chapter 14, The Jehol Group and Chapter 15, El Montsec and Las Hoyas).

362 The gonorhynchiform fish *Dastilbe crandalli* (PC). Length 132 mm (c. 5.2 in).

363 The ornithocheirid pterosaur *Ludodactylus sibbicki* (SMNK). Length of skull 450 mm (c. 17.7 in).

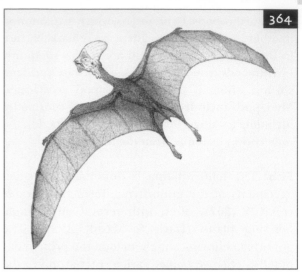

364 Reconstruction of tapejarid pterosaur.

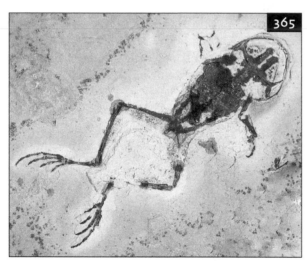

365 A frog with soft tissue preservation (SMNK). Length of body (excluding legs) 30 mm (c. 1.2 in).

366 The neosuchian crocodile *Susisuchus anatoceps* (SMNK). Length 250 mm (c. 9.8 in).

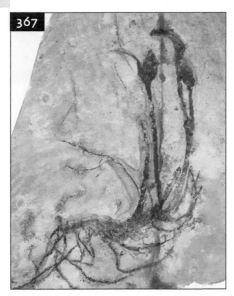

367 A dicotyledenous angiosperm with affinities to Nymphaeales (water lilies) (PC). Scale bar – 50 mm (c. 2 in).

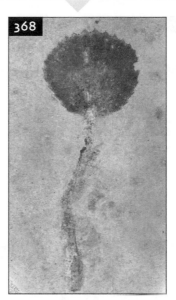

368 The monocotyledenous angiosperm *Klitzschophyllites flabellata* (PC). Scale bar = 10 mm (c. 0.4 in).

369 A dicotyledenous angiosperm belonging to the Magnoliales (magnolias) (HMB). Width of specimen 100 mm (c. 4 in).

Plants. The Crato flora includes some spore-bearing plants (horsetails, lycopods, and ferns), but is dominated by seed-bearing plants including gymnosperms (seed ferns, conifers, cycads, and gnetophytes), and more importantly several early angiosperms, the flowering plants. These include flowers, seeds, fruiting bodies, fruits, leaves, and roots, which in the last few years have begun to be scientifically described (Mohr & Friis, 2000). One of the most common forms is a water plant belonging to the most basal and oldest angiosperm clade, the Nymphaeales, the water lilies (**367**). There is also a record of a single monocotyledon described as *Klitzschophyllites flabellata* (**368**), and a few specimens which have been referred to the Laurales (laurels) and the Magnoliales (magnolias; **369**). *Endressinia brasiliana* is the oldest magnolialean angiosperm so far recorded, often beautifully preserved in three dimensions. Eudicotyledons are also present, but await description, although one specimen seems to have affinities to the Ranunculales (buttercups) and another to the Proteales (proteas, banksias, and macadamias) (Mohr *et al.*, 2007).

Santana

Fish. The well-preserved fossil fish, often in fish-shaped concretions, from the Santana Formation, have been available from commercial dealers around the world for many years. They occur in vast numbers and include over 20 different genera, the most common being actinopterygians (ray-finned bony fish) of the teleostean (modern) group. *Vinctifer* is a long, pike-shaped, aspidor-hynchiform fish, easily recognized by its deep flank scales and its extended rostrum (**370, 371**). It is often found with its back arched, suggesting dehydration of body tissues on death. *Tharrhias*, a gonorhynchiform fish related to *Dastilbe* (Crato Formation), often occurs as several individuals within a single concretion (**372**). Other common genera include the ichthyodectiform *Cladocyclus*, an elongate fish often over 1 m (c. 3 ft) in length, *Rhacolepis*, which has a thin, fusiform body and a very pointed snout (**373**), and the closely related *Notelops* (**352, 374**). *Calamopleurus* is related to the modern bowfin, and belongs to the more primitive holostean ray-finned fish, as does the distinctive pycnodontid fish *Neoproscinetes* (**375**).

Sarcopterygians (lobe-finned bony fish) are represented by two latimeroid coelacanths, *Mawsonia* and *Axelrodichthys* (**376**), among the largest fish known from Santana (over 3 m [10 ft]), while the chondrichthyans (cartilaginous fish) are represented by the hybodont shark *Tribodus*, recognized by its two dorsal fins with spines, and the ray *Rhinobatos*.

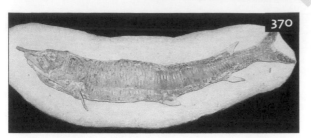

370 The teleostean fish *Vinctifer comptoni,* with elongated flank scales and extended rostrum (PC). Length 600 mm (c. 24 in).

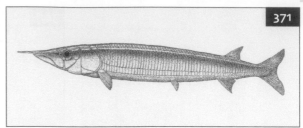

371 Reconstruction of *Vinctifer.*

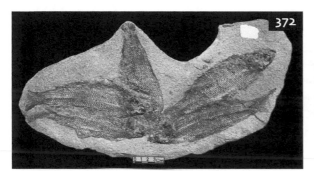

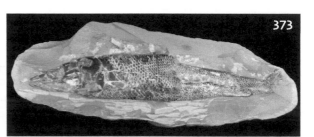

372 The teleostean fish *Tharrhias araripis* (PC). Length of nodule 780 mm (c. 2.5 ft).

373 The teleostean fish *Rhacolepis buccalis* (NMW). Length 190 mm (c. 7.5 in).

374 The teleostean fish *Notelops brama* (PC). Length 637 mm (25 in).

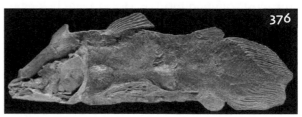

376 The coelacanth fish *Axelrodichthys* (PC). Length 800 mm (31 in).

375 The pycnodontid fish *Neoproscinetes* (PC). Length 305 mm (12 in).

Pterosaurs. In recent years several well-preserved pterosaurs have been described, some with wing membranes (e.g. Martill & Unwin, 1989). They are all pterodactyloids and belong to at least four lineages, the ornithocheirids, tapejarids, tupuxuarids, and ctenochasmatids, mostly large forms with wingspans over 5 m (c. 16 ft) and with bizarre skull crests (**377**). There is some disagreement as to whether these endemic forms (such as *Santanadactylus*, *Araripesaurus*, *Cearadactylus*, *Brasileodactylus*, *Anhanguera*) are related to the pterosaurs from the English Greensand or whether they represent new groups.

Dinosaurs. Recently, a number of theropod dinosaurs have also been described. These include partial skulls of two spinosaurids, *Irritator* (Martill *et al.*, 1996) and its possible junior synonym, *Angaturama* (Kellner, 1996), both large fish-eating dinosaurs with unusual head crests (**378**). Others include the synsacrum of a possible oviraptosaur (Frey & Martill, 1995), and two other small coelurosaurs. One is a possible tyrannosauroid named *Santanaraptor* (Kellner, 1996, 1999), and is known from the ischia, hindlimbs, and caudal vertebrae and includes patches of fossilized skin and muscle fibres. The other, a compsognathid named *Mirischia*, is known from the pelvic girdle, sacrum, and partial hindlimbs, and exhibits soft-tissue preservation of the intestinal tract and a postpubic air sac (Martill *et al.*, 2000; Naish *et al.*, 2004). It is the youngest compsognathid yet recorded.

Other reptiles. Other tetrapods include two genera of pelomedusid turtles (the oldest known examples of pleurodires or side-necked turtles) and two crocodilians (a terrestrial notosuchid and an aquatic trematochampsid). The notosuchid genus *Araripesuchus* is also known in west Africa and testifies to the existence of a continental link with South America after the origin of this lineage.

Invertebrates. Apart from ostracods, invertebrates are not common in the concretions, but do include a few decapods (small shrimps and crabs) and rare occurrences of insects. Conchostracans occur at other levels, particularly near the base of the formation, while corbulid bivalves, cassiopid gastropods (*Paraglauconia*), and two irregular echinoids have been described from the thin limestones near the top of the formation. There are no ammonoids, belemnoids, nautiloids nor corals, crinoids, or brachiopods.

PALAEOECOLOGY OF THE SANTANA AND CRATO FORMATIONS BIOTA

As we have already discussed (p. 207), the Crato Formation Plattenkalks represent a stagnant lagoon community in which increasing salinity inhibited any autochthonous life except for rare crustaceans and the small fish *Dastilbe*, which lived in less saline water where rivers entered the lagoon. The lagoon margins supported an abundance of thick vegetation, among which lived a variety of

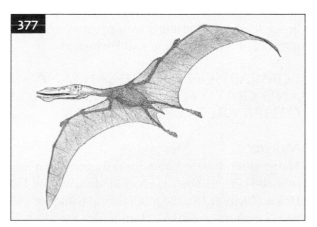

377 Reconstruction of *Anhanguera*.

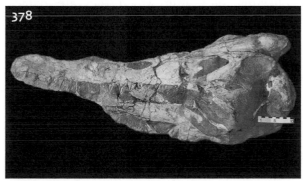

378 The spinosaurid dinosaur *Irritator challengeri*, with faked snout. (SMNS). Estimated true length of skull 800 mm (c. 30 in).

aquatic, semi-aquatic, and terrestrial insects, spiders, and mites, which would have been a rich source of food for lizards, frogs, and scorpions. These must all have drifted or been blown into the lagoon and are allochthonous elements of the biota, as are the rare birds and pterosaurs which flew over the lagoon and the rare crocodiles and turtles which lived in the surrounding hinterland.

Interpretation of the Santana Formation fish fauna is less straightforward; it has been variously described as a fully freshwater, fully marine, and as a marine fauna living in estuaries. Martill (1988) initially cited the occurrence of echinoderms as evidence for marine conditions, but Maisey (1991) showed that these were from a higher horizon than the fish-bearing concretions, suggesting that fully marine conditions did not occur until near the end of the basin's life. The lack of normal marine faunas (such as ammonoids, corals, brachiopods) supports this view.

Most Santana fish species are endemic to the basin, and although most of the recorded families are normally marine, almost all of them do have some freshwater members. Of the two genera that are known outside the basin *Rhinobatus* occurs in marine deposits, while *Mawsonia* is freshwater. These contradictory data have led most authors to describe it as a 'quasi-marine' or perhaps brackish water community and it probably represents either a lagoonal setting or a basin with only restricted connections to waters of normal marine salinity (Martill, 1993).

Maisey (1991) presented evidence for at least three distinct fossil assemblages in the Santana Formation, each characteristic of a particular concretion lithology representing three separate collecting sites after which they are named. 'Santana' concretions are usually oval in shape, of small size, and do not reflect the fossil outline. *Tharrhias* is abundant, *Brannerion*, *Araripelepidotes*, and *Calamopleurus* are common, and *Cladocyclus*, *Axelrodichthys*, *Vinctifer*, and *Rhinobatos* are rare. Crocodiles, turtles, pterosaurs, and plants are also found and the environment is interpreted as clear, oxygenated, near-shore waters.

'Jardim' concretions are large and platy with their shape reflecting the outline of the fossil (**375**). Large specimens of *Rhacolepis* and *Vinctifer* are abundant, *Brannerion*, *Araripelepidotes*, *Cladocyclus*, *Calamopleurus*, *Axelrodichthys*, and *Rhinobatos* are

common, and *Mawsonia* and *Tharrhias* are rare. Turtles and pterosaurs are rare and the environment is interpreted as a muddy/sandy, anoxic bottom, further from the shore.

'Old Mission' concretions (from Missão Velha) are also large, but are thick rather than platy and do not reflect the fossil outline. *Rhacolepis* and *Vinctifer* are abundant, *Brannerion* is common, and *Araripichthys*, *Calamopleurus*, and *Cladocyclus* are rare. Terrestrial and aquatic reptiles are unknown and the environment was deeper, open water with a muddy anoxic bottom.

All three assemblages are dominated by pelagic organisms, mainly fish, but are notable for their absence of pelagic marine invertebrates, such as cephalopods. Moreover, while semi-aquatic reptiles (crocodiles and turtles) do occur, fully marine reptiles such as plesiosaurs or mosasaurs are absent. Epibenthic molluscs (bivalves and gastropods) are known, as are rare echinoids, but other epibenthic marine invertebrates such as corals, crinoids, and brachiopods are not, and Maisey (1991) concluded that the environment was a shallow embayment passing into a coastal region with periodic marine incursions causing mixing of waters.

Whichever scenario is correct, it is clear that immediately after the deposition of the concretion-bearing layer, normal marine salinities did exist briefly, perhaps for the only time in the entire history of the basin, as evidenced by the thin, but widespread echinoid-rich limestones. Martill (2007) has recently suggested that this horizon might represent one of the three mid-Cretaceous high sea stands, which occurred during Albian, Cenomanian, and Turonian times, respectively. Soon after this event the life of the basin came to an end as it finally filled with sediment, either due to a subsequent fall in sea level, or by infilling with prograding fluvial material of the succeeding Exu Formation.

COMPARISON OF THE SANTANA AND CRATO FORMATIONS WITH OTHER CRETACEOUS BIOTAS

Morocco

Many other Lower Cretaceous Lagerstätten are described in this book, such as El Montsec and Las Hoyas (Chapter 15), the Jehol Group (Chapter 14), and Lebanon (p. 166), but none are directly comparable with the Santana and Crato formations with their seemingly endemic fauna. Recent research, however, by Dave Martill at The University of Portsmouth, suggests that the Lower Cretaceous beds of the Kem Kem region of southern Morocco might be comparable with the Brazilian deposits. The age of the Moroccan beds is considered to be early Cenomanian, thus only fractionally younger than the Santana and Crato formations, and it has already been shown (Wellnhofer & Buffetaut, 1999) that the pterosaur faunas are remarkably similar (with tapejarid, tupuxuarid, and anhanguerid [ornithocheirid] pterosaurs all present in Morocco). Although little else has been published, the recent discovery of the ichthyodectiform fish *Cladocyclus* in Morocco would suggest that some of the Santana fish genera are not confined to the Araripe Basin. Further research in this area will hopefully answer some of the remaining questions regarding the palaeogeography of the Crato and Santana Lagerstätten.

Indeed, just prior to the publication of this book, a comprehensive review of the Crato and Santana fish fauna (Brito & Yabumoto 2011) suggested that, contrary to previous ideas that the fauna was endemic to the Araripe Basin and was related to the opening of the South Atlantic Ocean, it was instead related to the fish fauna of the Tethys Ocean, and that a marine transgression in Aptian/Albian times connected several of the Brazilian interior basins with the western part of the Tethys. These authors cited the recent discovery of Araripe genera in Morocco (e.g. *Cladocyclus, Araripichthys*), and also in Columbia, Venezuela and Mexico (e.g. *Vinctifer, Rhacolepis, Notelops, Araripichthys*), as evidence for this brief marine link.

MUSEUMS AND SITE VISITS

Museums
1. Museu Nacional, Rio de Janeiro, Brazil.
2. Museu de Paleontologia da Universidade Regional do Cariri, Santana do Cariri, Ceará, Brazil.
3. Departamento Nacional da Produção Mineral (DNPM), Crato, Ceará, Brazil.
4. Small private museum in Jardim, Ceará, Brazil.
5. Museum für Naturkunde der Humboldt-Universität zu Berlin, Germany.
6. Staaliches Museum für Naturkunde, Karlsruhe, Germany.
7. Staaliches Museum für Naturkunde, Stuttgart, Germany.
8. American Museum of Natural History, New York, USA.

Sites
The book by David Martill published by the Palaeontological Association (1993) gives good advice on the practicalities and logistics of visiting field sites in both the Crato and Santana formations in the State of Ceará in north-east Brazil. There are many flights to Rio de Janeiro, from where it is possible to hire a car or endure a 36-hour coach journey to the north-east. Flying to Fortaleza or Recife cuts out about two days of the three-day drive from Rio. There are hotels in Crato or much cheaper Pousada accommodation in the village of Nova Olinda, which is in walking distance of several small quarries in the Crato Formation. Martill (1993) also gives several field itineraries; the Santana Formation and its fossils are best seen around the villages of Cancau, near Santana do Cariri on the north of the plateau, and Jardim, on the south.

It is important to realize that, while it is not illegal to collect fossils in Brazil, it is illegal to buy them – all commercial trade in fossils is forbidden in Brazil. You will, however, be offered fossils very cheaply by small children who work the Crato Plattenkalks by hand, and by 'fish miners', many of whom live in adobe huts in the poor villages. It is also illegal to export fossils from Brazil unless you have the appropriate authorization from the relevant government department. It is theoretically possible to acquire this from the Departamento Nacional da Produção Mineral (DNPM) in Crato, but such authorization is very tightly controlled and is rarely given, even for *bona fide* research or museum display.

GRUBE MESSEL

Chapter Seventeen

BACKGROUND: THE CENOZOIC ERA

The end of the Cretaceous Period, which was also the end of the Mesozoic Era, was marked by a mass-extinction event which saw the end of the dinosaurs and pterosaurs on land, and ammonites and marine reptiles in the sea. In the succeeding Cenozoic Era, which consists of the Palaeogene, Neogene (together these are commonly referred to as the Tertiary) and Quaternary periods, animals and plants assumed a more modern appearance. Mammals and birds replaced dinosaurs and pterosaurs as the dominant vertebrates on land. After the Cretaceous Period, the great many ecological niches left empty by the extinction of the dinosaurs became filled by mammals and birds in an adaptive radiation. By the middle of the Eocene Epoch (the middle epoch of the Palaeogene Period), nearly all of the orders of mammals and major groups of birds had evolved, and there were also present some mammal groups which have since become extinct. At this time, there were no high Alps, nor North Atlantic, and Europe and North America were still connected by land bridges in the vicinity of the Faeroes. There were marine basins in the North Sea, northern France, the Low Countries and Denmark area, and a complex of islands and basins over much of the rest of Europe. Volcanism was common, some associated with the opening of the Atlantic Ocean, and also along old fracture zones such as the Rhine Rift Valley (graben). Grube Messel (which translates to 'Messel Pit' – it was an oil shale open pit until its closure and subsequent conservation as a World Heritage Site in 1995) was situated where the Rhine Graben cut across the Central European Island, and in this down-warping of the Earth's crust an extensive series of lakes was formed. Grube Messel represents a crater lake within the rift valley.

The mammals are the best-known fossils from Grube Messel. Surprisingly most of the mammals at Grube Messel appear to have originated outside Europe and migrated across the region in Eocene times. Mesozoic mammals are rare and primitive, and few Palaeocene fossils exist. The few finds of pre-Eocene mammals in Europe show that they were insectivore-like remnants of Mesozoic types rather than the more modern forms found at Messel. A few mammal types at Messel seem to represent these primitive European forms: the insectivore-like mammals, early hedgehog relatives, and early ungulates. Invasions from elsewhere brought in modern mammals, for example rodents, ant-eaters, horses, bats, and primates. The Messel birds belong to modern orders and, despite Messel being a lake deposit, there are almost no water-birds, most being forest-dwellers such as owls, swifts, rollers, and woodpeckers. Plant fossils indicate a subtropical climate, with representatives of palms and citrus, for example, but without specialist tropical forest families. Land arthropods do not generally preserve easily, and though rare at Messel, when they do occur they commonly show beautiful colours and patterns. The beetles, in particular, show structural colours (iridescence). Many fish, amphibians (e.g. frogs), and reptiles (e.g. crocodiles and turtles) occur at Messel, which indicates that there was life either in the lake or in tributaries close by.

HISTORY OF DISCOVERY OF THE GRUBE MESSEL

The first pit at Messel was dug in 1859 for iron ore, but in 1875 oil shale was discovered and mining of this deposit began. Later that year the first fossil was found – the remains of a crocodile. Over the next century, a great many fossils were discovered, and after major mining activities ceased in the 1960s, more methodical collecting and preparation techniques began. In the early 1970s, the Hessen government planned to use the Messel Pit as a repository for waste material after oil shale extraction had ceased. Immediately, objections were raised by scientists and amateur palaeontologists. By this time, some Messel fossils were attracting high prices and many fossil dealers were collecting at the site, so the pit was closed to the public for safety reasons. Because of the threat of infilling the pit with refuse, a number of German palaeontological institutions began rescue excavations. As a result of these investigations the site's importance became much better known to the palaeontological world. In the late 1970s, special exhibitions on the Messel biota were put on in the Senckenberg Museum, Frankfurt am Main, and in the Hessen Museum in Darmstadt. In the early 1980s, permission was granted for refuse disposal to begin at the site. In April 1987, with the threat of complete loss of the fossil site imminent, an international symposium on the Messel Pit was organized at the Senckenberg Museum, but it was not until the early 1990s that the plan to fill Grube Messel with refuse was finally abandoned. The site was eventually recognized as being of international importance to science as a World Heritage Site in 1995, thus preserving the pit for future generations of palaeontologists to continue to excavate and study.

The main collecting technique at Messel involves simply splitting the thin shale slabs with large blades. Once a fossil is discovered, however, special techniques are required to protect the find while transferring it to the laboratory for further study. If the shale is allowed to dry out then the fossil disintegrates, so the most urgent treatment is to keep the shale block wet. The fossil is then transferred to the laboratory while still damp and wrapped in watertight plastic. Once in the laboratory, the transfer technique developed by Kühne (1961) is most useful for vertebrate fossils. Still keeping the fossil damp, the shale is carefully removed from around the bones with needles under a dissecting microscope. As much of the fossil is exposed as possible and then the bones are coated with resin to hold them in place. Then the whole slab can be turned over and the other side of the fossil exposed in the same way. Eventually, the fossil is completely free from the rock and encased in resin. Insects and plants have to be left in the shale and kept wet both during preparation and afterwards, when glycerine (which is less volatile than water) is used to maintain dampness. A further technique used in the study of Messel vertebrates is the use of x-rays to visualize details of the skull and fine bones (e.g. bat wing bones), which are otherwise difficult to see or uncover from the matrix without damage.

STRATIGRAPHIC SETTING AND TAPHONOMY OF THE GRUBE MESSEL

The Grube Messel mammal fauna was correlated with the lowermost mammal faunas of the Geiseltal lignite beds, which have themselves been correlated with the marine Lutetian Stage (lower Eocene) of the Paris Basin. The Messel Formation was thus thought to date to the earliest part of the Lutetian Stage, and to be about 49 million years in age. In 2001, a borehole was drilled into the Messel Pit from which underlying igneous rocks were obtained and dated at 47.5 Ma. Therefore, the mammal fauna is within the Lutetian Stage rather than demarcating its base.

The Messel borehole was drilled in the centre of the pit and reached 433 m (c. 1,420 ft). It proved that the origin of the circular-shaped sedimentary basin was a crater lake formed by a maar volcano (**379**). The borehole penetrated sedimentary rocks down to 240 m (c. 790 ft), volcaniclastic rocks such as lapilli tuffs from 240 m to 373 m (c. 790–1,224 ft), and then volcanic breccias to 433 m (c. 1,420 ft). Maar volcanoes are created by an explosion caused by groundwater coming into contact with hot magma. The eruption consists mainly of gas, water, and volcanic debris. A maar volcano produces a crater in which a lake usually forms. The underground part of the maar system is called a diatreme which, when eroded, commonly forms a hill rather than a depression.

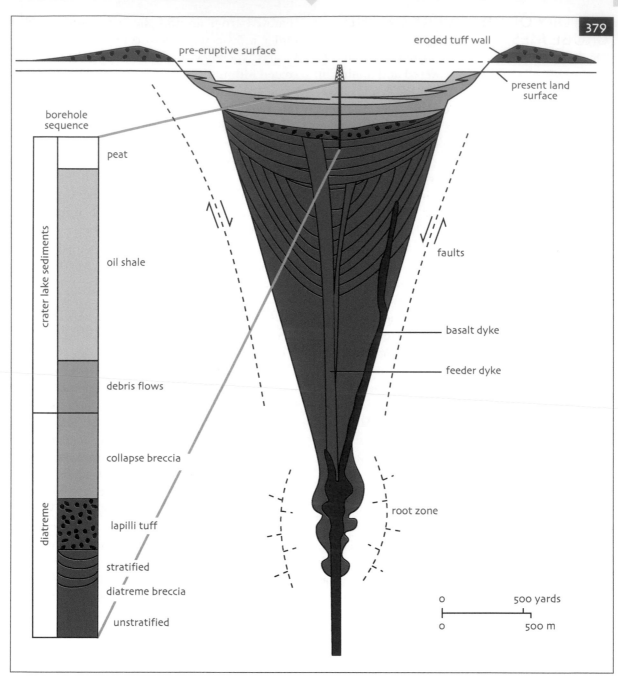

379 Geological cross-section of Grube Messel, as deduced from borehole information (left), reconstructed as a maar volcanic crater lake (after Schulz *et al.*, 2005).

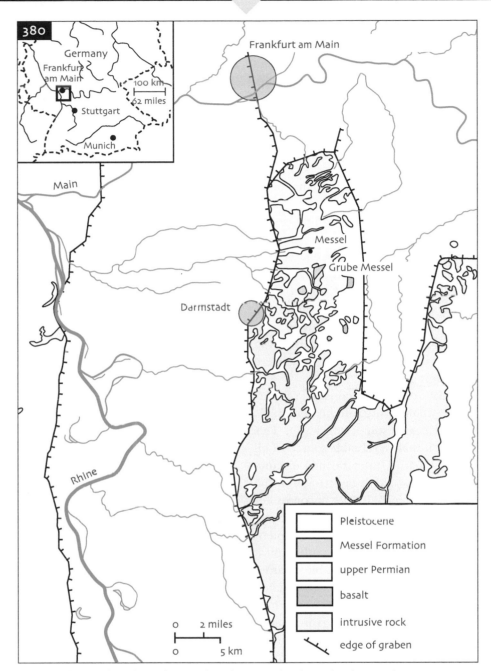

380 Locality and geological map of Grube Messel (after Schaal & Ziegler, 1992).

The Messel Formation can be traced only a few kilometres away from Grube Messel, because it was laid down in restricted lake basins within the Rhine Rift Valley (**380**). The lakes were formed in fault-bounded basins and because the faults were active and there was associated volcanic activity, the preserved sediments reflect this. The Messel Formation consists of a base of gravel and sand, which represents an initial high-energy fill resulting from screes at the edge of the basin and/or river activity. The oil shale overlies the sand and gravel, and represents quieter conditions of sedimentation once the lake had formed. Occasional slumps of coarse material from the lake edge, caused by renewed fault activity, can be seen as lenses of sand and gravel within the oil shale sequence. Slumps of shale within the shale sequence can also be seen. With continued fault movement, the lake basin was constantly being rejuvenated, and its shape and extent were probably also changing.

The present-day oil shale consists of clay minerals (i.e. the 'shale'), organic matter (15%, kerogen – the 'oil'), and some 40% water. This makes the Messel oil shale a valuable source of oil for exploitation by man. Under normal circumstances, organic matter would decay in water by the action of bacteria, through the process of oxidation, but at Messel in the Eocene something prevented this from happening. The environment suggested for Messel at that time was subtropical forest, and a great deal of vegetable matter, especially algal remains, would have accumulated on the lake bed. If the bottom waters of the lake lacked oxygen (anoxia), then the organic matter would decay only partly or not at all, resulting in the accumulation of kerogen and, more importantly for palaeontologists, the preservation of the soft parts of animals and plants. It is possible that the sheer volume of algae blooming in the lake at times resulted in an excess of their decaying remains, which used up all of the available oxygen and rendered the bottom waters anoxic.

DESCRIPTION OF THE GRUBE MESSEL BIOTA

Plants. As today, the angiosperms (flowering plants) dominate the flora at Grube Messel, but there is some evidence of other groups. Planktonic dinoflagellates occur in great abundance at various levels in the borehole core. Fronds found at Messel belong to the modern marsh and mangrove fern families Osmundaceae (Royal ferns), Schizaceae, and Polypodiaceae. Gymnosperms include the families Cephalotaxaceae (Chinese plum-yews), Cupressaceae, Taxodiaceae, and Pinaceae. The Taxodiaceae are the swamp-cypresses, which grow in permanently inundated conditions around the Gulf of Mexico and as far north as Virginia. Pinaceae (pines), on the other hand, are generally found in drier conditions. The relative scarcity of gymnosperms in the Messel lake deposits suggests that they did not live very close to the lake, and the floral composition provides evidence of a subtropical to warm-temperate climate.

Among angiosperms, true grasses (Graminae) had not, by mid-Eocene times, evolved the diversity we see today. However, grass-like plants such as rushes (Juncaceae), sedges (Cyperaceae), and reed-maces (Typhaceae) were present; such plants prefer wetter conditions than grasses. An interesting family, which shows a Gondwanan (mainly southern hemisphere) distribution today in wet habitats, is the Restionaceae. Pollen of this family occurs at Messel, as well as world-wide in late Cretaceous and early Tertiary deposits. It appears that the Restionaceae were displaced by the radiation of the Graminae in mid-Tertiary times. Other monocotyledons present at Grube Messel include the palms (Palmae, Arecaceae), which are typical of tropical and subtropical climates today as well as in Eocene times, and arums and lilies, which typically prefer damp conditions.

An important family of woody, evergreen trees and shrubs in Tertiary times was the Lauraceae, the laurels. Again, their presence indicates subtropical conditions. Another climatic indicator family present at Grube Messel is the Menispermaceae (moon-seed); these plants occur mainly as lianas in tropical and subtropical forests today. Other plant families of mostly tropical–subtropical climates, remains of which have been found commonly at Grube Messel, are the Theaceae (tea), Icacinaceae (a small family of vines), Vitaceae (grape),

381 Leaf of water lily (Nymphaeaceae) (SMFM). Length 160 mm (c. 6 in).

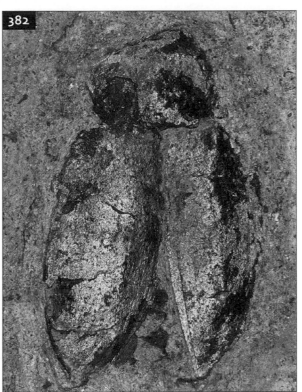

382 Iridescence on the elytra (wing cases) of a leaf beetle (Chrysomelidae) (SMFM). Length 4.5 mm (c. 0.2 in).

Rutaceae (citrus), and Juglandaceae (walnut). One of the commonest leaf fossils found at Grube Messel is that of the water-lily family, Nymphaeaceae (**381**). The presence of this family cannot give us climatic information but it suggests that shallow, open, oxygenated water conditions were present near to the site of deposition of the oil shale. Mention should also be made of a very important plant family which occurs commonly at Messel: the Leguminosae. This family has an almost worldwide distribution today and was clearly common in the Eocene too.

Judging from the abundant leaf, fruit, seed, spore, and pollen remains found at Grube Messel, the forests surrounding the lake must have been diverse and lush in a subtropical climate. However, the lake has clearly sampled a number of different habitats: open water, swamp, bank-side, damp forest, and drier regions. For example, a family of plants restricted today to south-east Asia but common in other Tertiary floras is the Mastixiaceae. However, this family is rare at Messel,

and it has been suggested that they did not occur close to the lake so their large fruits were unable to be carried into the site of deposition. Similarly, Fagaceae (beeches, chestnuts, and oaks) is a widespread family mainly outside the tropics today. Distinctive remains occur commonly at other fossil sites but at Messel only their pollen has been found, suggesting that these trees occupied higher, drier ground further away from the lake.

There is abundant evidence at Messel of plant interactions with other organisms. For example, leaves occur with characteristic fungal (rust) spots, evidence of insect eggs on leaves, and larval chewing marks. Plant remains have been found in the gastrointestinal tracts of some mammals; for example, grape seeds have been found inside the small horse *Propalaeotherium*, together with leaves of laurel, walnut, and other families, indicating a varied diet for this animal. Pollen has been found under the elytra (wing cases) of beetles, which implicates them as pollinators.

Arthropods. Some of the most beautiful of Messel fossils are the beetles, with well-preserved structural coloration (iridescence) on their elytra (**382**). Because of their relatively tough bodies, beetles (Coleoptera) are the most common (63%) of all insect remains found at Messel. Among the Coleoptera in the collection at the Senckenberg Museum, Frankfurt, click beetles (Elateridae) are the most common (15.8%), followed by weevils (Curculionidae, 12.8%), jewel beetles (Buprestidae, 8.4%), dung beetles (Scarabaeidae, 3.9%), stag beetles (Lucanidae, 1.7%), ground beetles (Carabidae, 1.4%), water beetles (Dascillidae, 1.4%), longhorn beetles (Cerambycidae, 0.5%), and rove beetles (Staphylinidae, 0.26%). Other families are present in smaller percentages and include the colourful leaf beetles (Chrysomelidae) and the ground-beetle family Tenebrionidae. Larvae of the water-beetle genus *Eubrianax*, which can only survive in highly oxygenated water, such as at waterfalls, were a surprise find at Grube Messel, and indicate that these animals, as well as many others, perhaps, were washed in from elsewhere.

Hymenoptera (ants, bees, and wasps) are the second-most common (17%) insects from Messel. Of these, ants (Formicidae) are the most frequent, and these are almost entirely known from winged stages. Of special interest are fossils of giant queens with wingspans of up to 160 mm (c. 6 in) across. At this size they exceed all known Hymenoptera, let alone ants. Other Hymenoptera from Messel include the parasitic wasps (Ichneumonidae), Chalcidae, Tiphiidae, Scoliidae, potter wasps (Eumeniidae), Anthophoridae, spider-hunting wasps (Pompilidae), and digger wasps (Sphecidae). Interestingly, no plant wasps (Symphyta) have been found, and this may be explained by their abundance today in temperate, rather than tropical, climates.

Bugs (Heteroptera) form 12.5% of the insect fauna at Messel, and are mostly represented by the family Cydnidae (> 80%); these are generally plant-sucking bugs, but have been observed in the tropics to favour juices from animal carcasses which were, of course, abundant around the Messel lake. Other ground- and plant-dwelling insects occurring at Messel include cockroaches (Blattodea, 1.5%) and crickets (Orthoptera, 0.5%). Diptera (flies, 0.4%) and Lepidoptera (butterflies and moths, 0.25%) are strangely rare at Messel compared to other Tertiary insect localities until it is realized that, with their large wings, these insects would have floated on the water surface rather than sunk to the lake floor, and so it is not surprising that very few dragonflies (Odonata), stoneflies (Plecoptera), and caddisfly adults (Trichoptera) have been found at Grube Messel either.

Arachnids (spiders, harvestmen, mites, ticks, scorpions, and their allies) are rare as fossils outside amber (see Chapter 19, Baltic Amber), even in sediments which preserve good insect fossils. A handful of specimens occur at Grube Messel, and most seem to belong to the orb-weaver spiders. This is not surprising since such species are common in lakeside vegetation today. A single specimen of a harvestman (Opiliones) has been found at Grube Messel.

Fish. All fish preserved at Grube Messel belong to the bony fishes (osteichthyans) but show a wide diversity. One of the most frequently encountered Messel fish is *Atractosteus*, a gar. The gars are distinctive predators with large heads and massive jaws; their scales are of a large, overlapping, ganoid type with shiny enamel, resembling armour plating. The bowfins are represented by *Cyclurus*, which is the most frequently encountered fish at Messel. Both bowfins and gars are about 200–300 mm (c. 8–12 in) in length, although smaller and larger (up to 500 mm [c. 20 in]) specimens do occur. Like gars, bowfins are also formidable predators. A single specimen of an eel, *Anguilla ignota*, about 600 mm (c. 2 ft) long, was a striking find. Classified in the modern genus of catadromous eels (i.e. those which are born in the sea and return to it to spawn but spend much of their lives in freshwater), its presence suggests that the Messel lake had a connection to the sea. Why then, has only one specimen ever been found at Messel, when in modern freshwaters where eels occur, they are abundant?

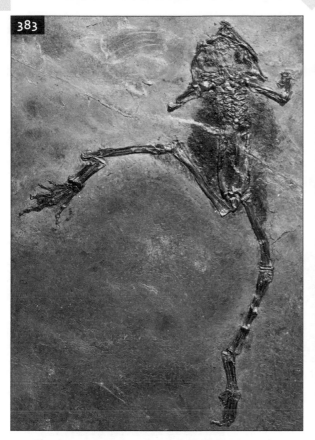

383 The frog *Eopelobates wagneri* (SMFM). Body length (excluding legs) about 60 mm (c. 2.4 in).

384 The turtle *Trionyx* (SMFM). Carapace length 300 mm (c. 1 ft).

Amphibians. Lakes are usually thought of as excellent places in which to find amphibians – tetrapod vertebrates which return to water to breed even if they spend much of their time on land. Yet at Grube Messel only a single specimen of salamander, *Chelotriton*, has been found, and very few anurans (frogs and toads). Nevertheless, the preservation of the frogs is excellent (**383**); some show skin and muscles preserved, and there are also tadpoles.

Reptiles. Freshwater turtles generally make good fossils with their bony shells. Completely removed from the oil shale, the skeleton of *Trionyx* makes a beautiful specimen (**384**). Trionychid turtles today are almost entirely aquatic, leaving the water only to lay eggs. Crocodiles are also associated with water and, not surprisingly, many specimens have been found at Grube Messel. Some complete skeletons have been recovered, from juveniles to 4 m (c. 13

ft) long adults. In all, six genera occur at Messel, which is a high diversity for a single site compared to the present day. It has been suggested that only one genus, *Diplocynodon*, actually lived in the lake, because a full set of juvenile stages of this genus, the commonest, has been found. All the others are presumed to have been washed in from adjacent rivers or other parts of the complex lake system.

Much rarer than the semi-aquatic crocodiles and turtles at Messel are the terrestrial lizards and snakes. A variety of lizards has been found at Messel, from large, predatory monitors, through agile iguanids, to limbless forms. Snakes are relative latecomers among reptiles, the first fossils occurring in Cretaceous rocks. By the Eocene, the familiar constrictors *Boa* and *Python* were present and have been found at Messel as the genus *Palaeopython* (**385**), while the more advanced venomous snakes came later and none occur at Messel.

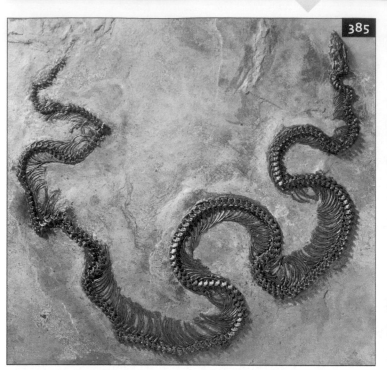

385 The beautifully preserved snake *Palaeopython* (SMFM).
Length about 2 m (c. 6 ft).

386 A roller-like bird displaying fine preservation of feathers
(SMFM). Total length about 200 mm (c. 8 in).

Birds. A great variety of birds can be found at Grube Messel. Palaeognathous birds today include only the ratites (the flightless ostriches, rheas, and allies) and the cursorial tinamous of South America. Grube Messel yields *Palaeotis*, an ostrich-like palaeognathan which could represent an ancestor to modern ratites. All other birds are classified as Neognathae, the sister-group to Palaeognathae. There is some evidence for birds of prey and chicken-like birds. The primitive ibis *Rhynchaeitis messelensis* was the first bird found at Messel, in 1898. Giant, flightless birds dominated the scene in early Tertiary times. A femur imprint of a diatryma (a giant with a powerful beak) has been found at Messel, together with three specimens of the large *Aenigmavis*. Other birds in the same group as these, the Gruiformes (crane-like birds), are found at Messel, including seriemas (South American stork-like birds) and rails. Among Charadriformes (plover-like birds) there is a flamingo preserved at Messel: *Juncitarsus*. The Messel owl, *Palaeoglaux*, preserves a strange plumage unknown in modern owls; possibly it was not yet nocturnal. There are representatives of the nightjars, swifts, and rollers (**386**). The last is a group of birds with colourful plumage: kingfishers, hoopoes, hornbills, and bee-eaters as well as rollers. Some of the Messel rollers are preserved with exceptionally fine plumage (but without colour, unfortunately) and very falcon-like grasping claws. A number of birds at Messel can be assigned to the woodpeckers, which provides further evidence of forest nearby. Indeed, the single specimen of a flamingo is the only evidence of a true water-bird; all of the others are terrestrial or forest-dwelling forms.

Mammals. Grube Messel is best known for its fossil mammals. Many groups are present, including marsupials in the form of opossums. At least two forms are present: a small, tree-climbing form with a prehensile tail, and a larger, shorter-tailed genus which was probably a ground-dweller. Placentals form the bulk of the mammal fauna at Messel. A fascinating creature called *Leptictidium* (**387, 388**) was originally pigeon-holed with the insectivores, but is now included with the Proteutheria: primitive placentals from the Cretaceous and early Tertiary. Three species of *Leptictidium* have been described from Grube Messel, and what makes these animals striking is their locomotion. They had relatively huge hindlegs, small forelimbs, and an exceptionally long tail consisting of more than 40 vertebrae (this is far more than in any known mammal today). The tail was not prehensile, so we can only assume that it acted as a balancing organ and the animal used its large hindlimbs for locomotion. However, unlike kangaroos and jerboas, which use their hindlimbs to jump along, the weak joints of the hindlimbs of *Leptictidium* suggest that it loped along with a running gait unknown in extant mammals. Also in Proteutheria is *Buxolestes*, which has short, strong feet and a thick tail; its resemblance to otters suggests

that it was a good swimmer, and gut contents of fish bones and scales confirm this. Among true insectivores are highly specialized hedgehog relatives, including *Macrocranion* with large hindlegs adapted to jumping, and *Pholidocercus*, which had a scaly tail and a large, strongly innervated organ on its forehead. Another insectivore, *Heterohyus*, was arboreal and had long second and third digits on its forelimbs for extracting insect grubs from tree-holes.

Bats are rarely preserved in sedimentary deposits because of their aerial locomotion and cryptic habits, yet at Grube Messel they are the most common mammal fossils. The suggested reason for this is that they were overcome by noxious gases emanating from the Messel lake and drowned. The hundreds of bats preserved at Messel form a unique resource for the study of bat evolution. All are Microchiroptera (insectivorous, rather than fruit-eating, bats). The excellent preservation in the oil shale includes skin and wing membrane, muscle, fur, and gut contents (**389**). The different wing aspect ratios of the six species described suggest a diversity of habits including high fliers among the trees; lower-level, open-space hunters; and foliage or ground-level skimmers. The gut contents include lepidopteran (moth) scales, trichopteran (caddisfly)

387 The unusual mammal *Leptictidium nasutum* (SMFM). Total length 750 mm (c. 30 in), including tail, which is 450 mm (c. 18 in) long.

388 Reconstruction of *Leptictidium*.

hairs, and coleopteran (beetle) wing-cases. All used echo-location for hunting. Three-quarters of the specimens belong to the group of ground-level fliers, which would be more susceptible to the noxious gases than the high-fliers.

Primates are remarkably rare in the Messel biota; only eight fragmentary specimens were known until recently, when a complete primate skeleton was discovered (Franzen *et al.*, 2009). The specimen was found by private collectors in 1983, who sold it in two parts. The less interesting piece went to the Wyoming Dinosaur Center in Thermopolis, Wyoming, while the other side remained unknown. The much more complete second part was eventually put up for sale in 2007, and bought by the University Museum in Olso, Norway (**390**). While they remain in different institutions, the part and counterpart together form the most complete fossil primate known. Called Ida by its collectors, the specimen was given the scientific name *Darwinius masillae* to com-

memorate the bicentenary of Charles Darwin, 2009. Ida belongs to the Adapiformes; animals half the size of a cat which are possibly relatives of the Madagascan lemurs.

Pangolins (order Pholidota) are nocturnal ant-eaters with body scales instead of hair. When disturbed they roll into a tight ball, protected by their scaly covering. Many of their features are primitive and suggest an early origin, perhaps in the Cretaceous Period. The discovery of the earliest and best-preserved fossil pangolins at Messel, which are almost identical to modern pangolins, is consistent with this hypothesis. However, in spite of what seemed obvious adaptations for ant- and termite-eating (myrmecophagy), the stomach contents of the fossils consisted almost entirely of plant fragments. One intriguing explanation for this paradox is that these early pangolins were actually feeding on pieces of leaf stolen from leaf-cutter ants, common denizens of tropical forests, and that ant-eating evolved from

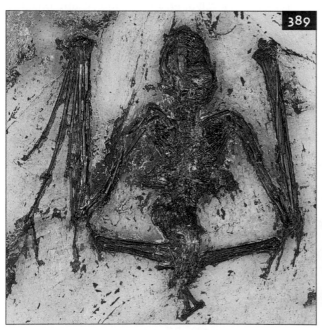

389 The bat *Archaeonycteris trigonodon* (SMFM). Length of forelimb 52.5 mm (c. 2 in).

390 The lemur-like primate *Darwinius masillae* (UOM). Length 580 mm (23 in), including tail, which is 340 mm (c. 13 in) long.

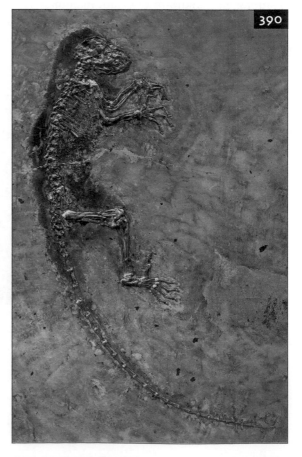

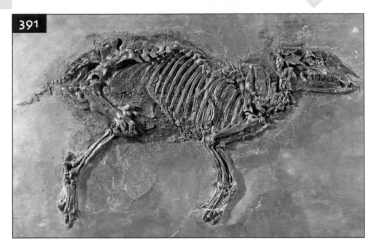

391 The Messel horse *Propalaeotherium parvulum* (SMFM). Height at shoulder 300–350 mm (c. 12–14 in).

this habit. The other order of mammals alive today which specializes in myrmecophagy is the Edentata, of which a single specimen has been found at Messel. Its gut contents included insect cuticle together with sand grains (helpful in grinding up the insects) and woody tissues, suggesting it had been eating wood termites.

Rodents are characterized by their large, single pair of incisors, an adaptation for nibbling vegetation and gnawing seeds. Three genera are known from Messel: a large, squirrel-like rodent with leaves in its gut, and two smaller, mouse-like forms. The rodents would have formed part of the diet of carnivores such as the possibly arboreal *Paroodectes*, a miacid (an extinct family seemingly ancestral to modern Carnivora), and the hyaenodontid creodont *Proviverra*. Hoofed mammals were present at Messel, both early representatives of modern groups and extinct, ancestral forms. An example of the latter is *Kopidodon* which, though many aspects of its morphology indicate that it belongs among the ancestors of hoofed mammals, has strong claws and a long tail consistent with an arboreal lifestyle. The odd-toed ungulates (Perissodactyla) include horses, tapirs, and rhinos, and Messel has provided some of the finest fossil examples of these groups. The Messel horse *Propalaeotherium* (**391**) is known from more than 70 specimens ranging from foals to full-grown adults, and there are two species, one with an adult shoulder-height of 300–350 mm (c. 12–14 in) and one 550–600 mm (c. 22–24 in) – small horses indeed! Other features suggest they were primitive

horses; for example, the forelimbs had four hooves and the hindlimbs three (in modern horses each leg has only a single hoof). The gut contents of these early horses indicate that they ate leaves and seeds (browsers); grasslands and grazing horses had not yet developed in the Eocene. The even-toed ungulates (Artiodactyla) include cattle, deer, pigs, camels, giraffe, and hippopotamus. Two genera of primitive artiodactyls occur at Grube Messel which, like the perissodactyls, were dog-sized browsers or foragers on forest litter.

PALAEOECOLOGY OF THE GRUBE MESSEL BIOTA

The physiographic setting of the Messel lake was hotly debated until 2001, when the borehole proved the maar volcanic origin for the lake. Maar lakes dating from Tertiary times occur throughout southern Germany, e.g. Randecker Maar. In this model, the lake would have been deep and steep-sided – ideal conditions for bottom-water anoxia. In addition, toxic gases emitted from the lake could have brought down over-flying bats and birds, while terrestrial mammals could easily have slipped down the steep banks. Local streams could have been home to caddis larvae and small fish, but lacked larger fish and water-birds, which are rare as fossils. Similarly, while wind-blown leaves and flowers occur, twigs and branches of trees are rare. Previous theories suggested that there was an extensive river system across the area, under which the Rhine Graben began to form, thus isolating parts of the river system as lakes in depressions in the hilly terrain. Evidence supporting this hypothesis came from the distribution and orientation of fish and caddisfly fossils, which are both inhabitants of flowing water. They indicate that water flowed into, and presumably out of, the Messel lake. Another proposal, based on studies which suggested that not only was the lake in existence for some 100,000 years but the sediments across the exposed area are remarkably uniform, is that Messel lake was a large, persistent basin and that the fossil-bearing regions represent deeper, anoxic depressions within this much larger structure. In this model, many of the fossils were inhabitants of the larger lake, which had oxygenated waters. It is more likely that the long age and extensive, but patchy, deposits represent a long period of maar volcanic activity in the region.

Analysis of samples from different horizons of the drilling core showed that small fish coprolites were quite abundant in the oil shales down to a drilling depth of 132 m (c. 433 ft). It was only in the bottom-most layers that no coprolites were found. Richter and Wedmann (2005) explained this by the hypothesis that when the lower layers were deposited no fish were living in the young Lake Messel. The lower horizons also contain cladoceran ephippia (haploid cysts produced by water fleas). The absence of these crustaceans in the upper horizons might be due to the increased predation they suffer where fish are present. Analysis of the fish coprolites showed that the larvae of phantom midges (Chaoboridae) were by far the most important element of the ecosystem over the whole life span of the lake.

Throughout this chapter, reference has been made to what the fossil finds tell palaeontologists about the ecology of the Messel area. The evidence comes from comparison with the habits of their modern relatives, sedimentological setting, and consideration of the taphonomy (preservational history) of the fossils. The picture painted by the evidence is a complex but fascinating one. Messel appears not to preserve a typical lake biota, but to have sampled widely from the surrounding forests as well as different parts of the lake. It is undoubtedly the forest-dwelling forms that have attracted most interest among palaeontologists, because such a habitat is rarely preserved in the fossil record. In this respect, comparison could be made with amber biotas (see Chapter 19, Baltic Amber).

COMPARISON OF GRUBE MESSEL WITH OTHER TERTIARY BIOTAS

The Lutetian–Bartonian brown coal seams of Geiseltal, some 265 km (c. 165 miles) to the north-east of Messel, have been excavated for three centuries, and over that time have yielded a vast quantity of fossils which can be compared to those found at Grube Messel. The vegetation, for example, is similar in many ways, but there are important differences that can be explained because Messel was a lake that sampled vegetation from a wide variety of habitats; the Geiseltal lignite represents a bog habitat, an ecosystem which has a characteristically impoverished biodiversity. There are interesting differences in the fauna between the sites too. More than 300 specimens of an axolotl-like amphibian have been found at Geiseltal, but none at Messel, though the reptilian faunas of the two sites are comparable. Because of its unique setting, bat fossils are relatively common at Messel, but not at Geiseltal; yet other mammals are more comparable – for example, primates, anteaters and ungulates.

MUSEUMS AND SITE VISITS

Museums

1. Naturmuseum und Forschungsinstitut Senckenberg, Frankfurt am Main, Germany.
2. Fossilien- und Heimatmuseum, Messel, Germany.
3. Hessische Landesmuseum, Darmstadt, Germany.
4. Staatliche Museum für Naturkunde, Karlsruhe, Germany.

Sites

A new Visitor and Information Centre opened at Messel quarry recently. Address: UNESCO World Heritage Site Messel, Roßdörfer Straße 108, D-64409 Messel, Germany. Phone: 0 61 59-71 75 9-0; email: service@welterbe-grube-messel.de. The site is situated a few kilometres east of Darmstadt on Dieburger Straße: follow the signs with the fossil horse *Propalaeotherium*. For a small fee you may join a guided tour of the pit conducted in both German and English. Fossil collecting is not, however, permitted.

Chapter Eighteen

THE WHITE RIVER GROUP

BACKGROUND: TERTIARY MAMMALS

In the previous chapter (Grube Messel) we witnessed the major diversification of mammals following the mass-extinction event at the end of the Mesozoic. Early Eocene times, as illustrated by Grube Messel, were characterized by a mixture of primitive types (insectivore-like mammals, early ungulates, and early carnivores) and some more modern mammals such as rodents, horses, bats, and even some lemur-like primates. The even-toed ungulates (artiodactyls) were still rare, but among the odd-toed ungulates (perissodactyls) the titanotheres, a sister group to the horses, had appeared and quickly increased in size to become the largest land mammals on the planet, perhaps filling the niche left by some of the larger dinosaurs.

In this chapter we will see that the later Eocene was characterized by the rise of modern groups of carnivores, including the dog and cat families, the diversification of horses and rhinos, especially the 'running rhinos', and, at the same time, a major expansion of artiodactyls which included pig-like types and early camels. Moreover, we will observe how the change from forests to grasslands at the end of the Eocene, induced by the onset of global cooling, triggered a rapid change in the land mammal faunas. Many of the early groups which had dominated the initial part of the Tertiary were now in decline; the huge titanotheres, for example, made their last appearance in the early Oligocene, and from this point onwards the mammals took on a much more modern appearance.

The world's richest deposit of Oligocene mammals is undoubtedly the White River Group,

which has its type section in the Badlands National Park of South Dakota, USA. This vast deposit, which extends into North Dakota, Wyoming, and Nebraska (**392**), actually straddles the boundary between the latest Eocene and the early Oligocene at this most vital time for the evolution of mammals.

HISTORY OF DISCOVERY OF THE WHITE RIVER GROUP

The first two records of fossils from the White River Group both date from 1847 when Dr Hiram Prout of St Louis described the lower jaw of a huge titanothere as *Paleotherium*, and Dr Joseph Leidy of Pennsylvania described the skull of a camel as *Probrotherium*, both specimens having been found by members of the American Fur Company. Joseph Leidy later also had the honour of describing the first fossil fish from the Green River Formation after being sent a specimen by Dr John Evans (see Nudds & Selden, 2008, p. 188) and it was this same John Evans who in 1849 became the first geologist to visit the White River to collect additional specimens in order to determine the age of the strata.

The following year, 1850, saw Thaddeus Culbertson collect samples for the Smithsonian Institution, while Ferdinand Hayden, Director of the United States Geological Survey of the Territories, undertook field work in 1853, 1855, 1857, and 1866. Joseph Leidy was then called upon to describe these various collections and in 1869 published his monumental work, *The Extinct Mammalian Fauna of Dakota and Nebraska*. This inspired others to explore the region, including

232

392 Map showing the outcrop of the White River Group and the Arikaree Group in western Nebraska (after Harmon *et al.*, 2003).

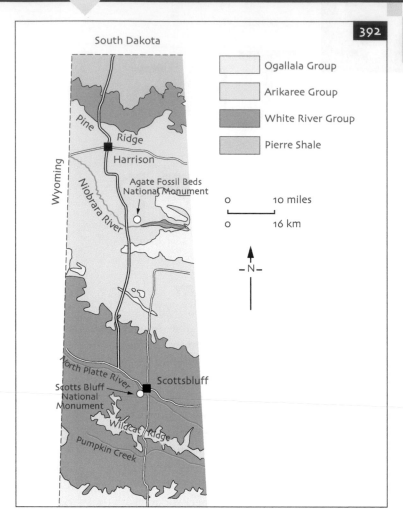

Yale University, the American Museum of Natural History (AMNH), and Carnegie Museum, among several others.

Field parties from Yale visited the area in 1870–1874 under the direction of Professor Othniel Charles Marsh, the well-known dinosaur hunter (see Chapter 12, The Morrison Formation), and continued to contract collectors up to 1898. Marsh refined collecting and preparatory methods and mounted several articulated skeletons.

Further field parties included the University of Princeton from 1882 to 1894, under the direction of Professor W.B. Scott; the AMNH from 1892 to 1916, under the direction of Professor H.F. Osborn; the University of Nebraska from 1892 to 1908, under the direction of Professor E.H. Barbour; and the Carnegie Museum from 1902 into the 1920s, under the direction of O.A. Peterson, who in 1904 discovered the important deposit at Agate Springs (see section on Comparison, pp. 241–242).

In addition, the South Dakota State School of Mines sent student parties to this area from 1899 on an annual basis and published *The Badland Formations of the Black Hills Region* in 1910, revised as *The White River Badlands* in 1920, both written by their President and Professor of Geology, Cleophas C. O'Harra. These volumes have now served as definitive works on the White River fauna for 100 years (O'Harra, 1910, 1920).

STRATIGRAPHIC SETTING AND TAPHONOMY OF THE WHITE RIVER GROUP

The White River Group dates from the late Eocene to the early Oligocene (from approximately 37 to 30 million years ago; **393**) and outcrops in North and South Dakota, Nebraska, and into Wyoming (**394**).

Ironically it was Professor Edward Drinker Cope, the great rival of Othniel Marsh (see Chapter 12, The Morrison Formation), who in 1883 introduced White River terminology to the outcrops in North Dakota, these being assigned to the 'White River Formation' by Quirke in 1918. Powers (1945) recognized a three-fold division of this formation, which he named, in ascending order, the *Titanotherium* beds, the *Oreodon* beds and the *Protoceras* beds. Denson *et al.* (1959) elevated the White River Formation to group status and Moore *et al.* (1959) formally identified two formations within the White River Group, the late Eocene Chadron Formation and the early Oligocene Brule Formation (applying names that had previously been used informally). Sandstone beds capping the Brule Formation were also described by Moore *et al.* (1959) as the Arikaree Formation, now referred to as the Arikaree Group (**393**). This has become the accepted stratigraphy of the White River Group in North Dakota, with the Chadron Formation further divided into a lower Chalky Buttes Member and an upper South Heart Member, and the Brule Formation further divided into a lower Scenic Member and upper Poleslide Member (Murphy *et al.*, 1993). Away from North Dakota, however, the lithostratigraphy varies considerably and in Nebraska, for example, the Brule Formation is divided into a lower Orella Member and an upper Whitney Member (**393**).

In Wyoming, on the other hand, it is still common practice to refer to the White River 'Formation', which is simply divided with reference to the traditional North American Land Mammal Ages (NALMAs). The Chadronian NALMA roughly equates to the Chadron Formation, the Orellan NALMA roughly equates to the Orella Member, and the Whitneyan NALMA corresponds to the Whitney Member (**393**). The boundaries between these three 'ages' can be easily recognized by reference to 13 conspicuous beds of white volcanic tuff, the base of the Whitneyan being about 5 m (c. 16 ft) below 'Ash 7', dated at 31.2 million years, and the base of the Orellan (which is also the base of the Oligocene) being 1 m (c. 3 ft) above 'Ash 5', dated at 33.9 million years (**393, 395**).

Taphonomically the White River Group cannot be considered as a Concentration Lagerstätte since fossils are not that common, nor is there any evidence of mass mortality. Neither is it a Conservation Lagerstätte in the strictest sense, since soft tissue is not preserved. However, Conservation Lagerstätten may also include sites which yield articulated vertebrate skeletons, and the White River Group unquestionably merits inclusion on these grounds.

393 Diagram to show the stratigraphy of the White River Group and Arikaree Group in Dakota, Nebraska, and Wyoming.

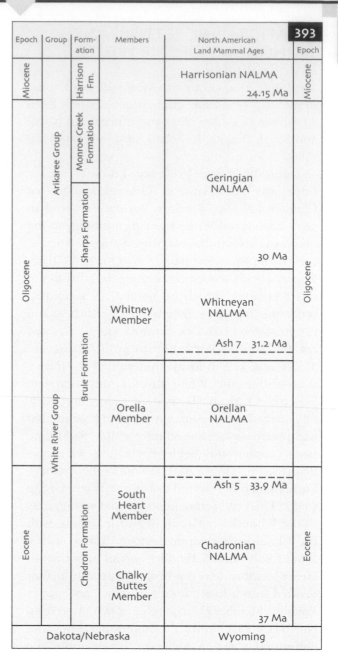

393

Epoch	Group	Formation	Members	North American Land Mammal Ages	Epoch
Miocene	Arikaree Group	Harrison Fm.		Harrisonian NALMA 24.15 Ma	Miocene
Oligocene	Arikaree Group	Monroe Creek Formation		Geringian NALMA	Oligocene
Oligocene	Arikaree Group	Sharps Formation		30 Ma	Oligocene
Oligocene	White River Group	Brule Formation	Whitney Member	Whitneyan NALMA Ash 7 31.2 Ma	Oligocene
Oligocene	White River Group	Brule Formation	Orella Member	Orellan NALMA	Oligocene
Eocene	White River Group	Chadron Formation	South Heart Member	Ash 5 33.9 Ma Chadronian NALMA	Eocene
Eocene	White River Group	Chadron Formation	Chalky Buttes Member	37 Ma	Eocene
	Dakota/Nebraska			Wyoming	

394 The White River Group outcropping in the badlands near Douglas, Wyoming, with Mount Laramie and the Precambrian foothills of the Rockies in the background.

395 White River Group (as in **394**) with the Eocene/Oligocene boundary marked by a conspicuous bed of white volcanic tuff (Ash bed 5).

DESCRIPTION OF THE WHITE RIVER GROUP BIOTA

More than 90 different mammals have been described from the White River Group; they include several ancestral modern lineages, including artiodactyl and perissodactyl ungulates, carnivores, insectivores and rodents, plus a number of extinct mammal lineages.

Oreodonts. This extinct group of artiodactyls (the even-toed ungulates) are the most common mammals of the White River Group, but have no living relatives. From their sheer numbers it is assumed that they lived in large herds. They are often known as the 'ruminating hogs', having previously been regarded as sheep-like or pig-like ruminants, but they were probably more closely related to camels. Kent Sundell from the Tate Museum in Caspar has looked at countless skeletons of these animals and considers them to be burrowing mammals, like modern-day prairie dogs, due to the enlarged canine teeth in both jaws, and the discovery of several specimens apparently preserved in their burrows (Sundell, 1997). They vary in size, but tend to be somewhat smaller than a sheep; the most common genus is *Merycoidoden* (**396–398**).

Entelodonts. Often known as the 'giant killer pigs', and distantly related to modern pigs, these extinct artiodactyls were grotesque, swine-like animals with huge skulls well over 1 m (c. 3 ft) in length, bearing two pairs of knob-like processes on the underside of the lower jaws and tusk-like canines on both upper and lower jaws. The 'tusks' were probably used for digging up roots, but their diet would have been omnivorous, including carrion. The body was strong and compact, but with surprisingly long legs. The most well-known example is *Archaeotherium* (**399, 400**).

Camels. Another very abundant artiodactyl from White River is the primitive camel *Poebrotherium* (**401**). Lightly built, and resembling a llama, it was, however, only about 700 mm (c. 28 in) tall, and was the first animal to be described from the biota. Its discovery caused something of a surprise since primitive camels were not expected in North America.

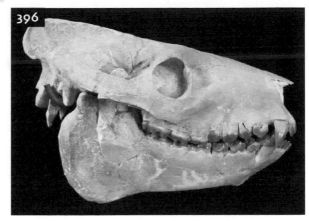

396 The oreodont *Merycoidoden culbertsoni* (BHIGR). Length of skull 200 mm (c. 7.9 in).

Rhinoceroses. Rhinos belong to the perissodactyls – the odd-toed ungulates. They are represented by three families, the rhinocerotids (true rhinos), the hyracodontids ('running' rhinos), and the amynodontids (aquatic rhinos), all of which occur in the White River Group, although only the first are still extant. Hyracodonts were small, fast and hornless, resembling the horses of the day, and possibly competition by horses led to their extinction. Amynodonts were heavy, short-bodied, and hornless, with elevated nostrils in the manner of a present-day hippopotamus, suggesting an aquatic lifestyle, and like the hyracodonts became extinct in the Oligocene. Common examples include *Hyracodon* (a 'running rhino'; **402**) and *Subhyracodon* (a 'true' rhino, belonging to the present-day family).

397 Oreodonts (*Merycoidoden*) preserved in burrow (TMC). Length of specimen approx. 1 m (c. 39 in).

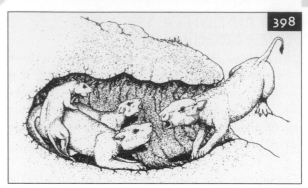

398 Reconstruction of *Merycoidoden*.

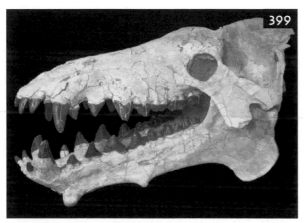

399 The entelodont *Archaeotherium* (BHIGR). Length of skull 575 mm (c. 22.5 in).

400 Reconstruction of *Archaeotherium*.

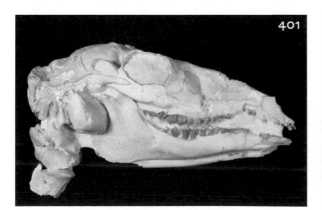

401 The camel *Poebrotherium* (BHIGR). Length of skull 160 mm (c. 6.3 in).

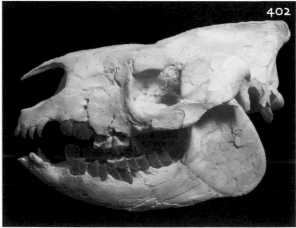

402 The 'running rhino' *Hyracodon* (BHIGR). Length of skull 270 mm (c. 10.6 in).

Horses. The evolution of the horse from the Eocene to the present day has been well documented and generally followed a pattern of increasing size, gradual development of grazing teeth, and a reduction in the number of toes. The most common horse found in the Oligocene beds of the White River Group is *Mesohippus* (**403, 404**), which is about 500 mm (c. 20 in) in height, has short crowned teeth (unlike the long teeth of modern horses) and stands on three toes. This represents an early stage in horse evolution, but is more advanced than those seen in the previous chapter from Grube Messel. Like the rhinos, the horses are perissodactyls.

Titanotheres. Also known as brontotheres and related to horses, these perissodactyl herbivores are by far the largest animals found in the White River Group, approaching 3 m (c. 10 ft) in height,

comparable to a present day elephant. In general appearance they resembled a rhino with a short body, stout limbs, and a huge, grotesque skull, up to 1 m (c. 3 ft) long, usually bearing bizarre bony protuberances in a variety of different shapes and sizes. They are the most abundant White River mammals after the oreodonts and the best known example is *Paleotherium* (**405, 406**).

Hyaenodonts. Among the top carnivores of the White River biota were the hyaenodonts, which belong to the creodonts, an extinct order of carnivorous mammals. They were never abundant, but some species were large, approaching the size of a black bear. Like modern hyaenas they most probably fed on carrion, since they were not adapted for speed. A good example is *Hyaenodon* (**407**).

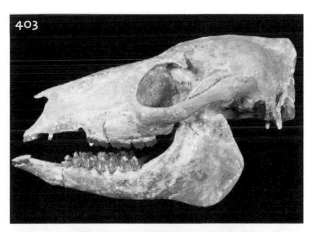

403 The early horse *Mesohippus bairdi* (BHIGR). Length of skull 150 mm (c. 5.9 in).

404 Reconstruction of *Mesohippus*.

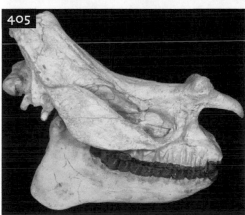

405 The titanothere *Paleotherium* (BHIGR). Length of skull 575 mm (c. 22.5 in).

406 Reconstruction of *Paleotherium*.

Cat-like carnivores. Perhaps the most specialized of the White River carnivores was another extinct group belonging to the family Nimravidae, better known as the 'false cats' or the 'false sabre-tooths'. These superficially resemble the true sabre-toothed cats in possessing two sword-like canine teeth in the upper jaw, but the two groups were not closely related – they merely developed a similar form through parallel evolution. Some of the nimravids, such as *Hoplophoneus*, were about the size of a leopard, while others, such as *Nimravus* and *Dinictis* (**408**) were smaller and faster.

Dogs and dog-like carnivores. Canids make their first appearance in North America in the White River Group with the genus *Hesperocyon* (**409**). This small animal, only about 800 mm (c. 31.5 in) long, had an elongated weasel-like body with a fox-like skull. It was probably more omnivorous than carnivorous, and may have lived alone or in small family groups rather than in packs. Dogs are well represented in the White River biota, with over 20 species known. Also appearing at this time were the amphicyonids, a group of dog-like carnivores, often known as the 'bear-dogs' (which were neither bears nor dogs, but were related to both). These were excavated by the Carnegie crews in the early twentieth century and then reinvestigated by Bob Hunt from the University of Nebraska State Museum beginning in the 1980s. The disposition of the fossils indicates they were buried in a den. These denning carnivores raised their young in the valley walls overlooking the ancient river, undoubtedly an opportune location for hunting prey drawn to the water. The largest species, over 2 m (c. 6 ft) in length, looked more like a grizzly bear with a thick neck and robust limbs (**410**).

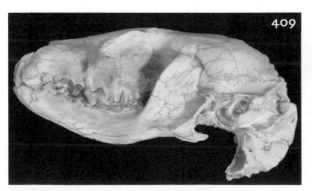

407 The hyaenodont *Hyaenodon crucians* (BHIGR). Length of skull 240 mm (c. 9.4 in).

408 The 'false sabre-tooth' *Dinictis felina* (BHIGR). Length of skull 100 mm (c. 3.9 in).

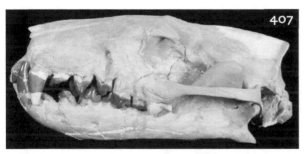

409 The canid *Hesperocyon gregarious* (BHIGR). Length of skull 90 mm (c. 3.9 in).

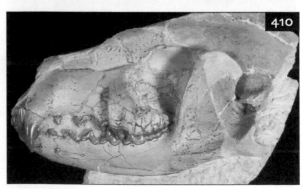

410 The amphicyonid *Daphoenus* (BHIGR). Length of skull 90 mm (c. 3.9 in).

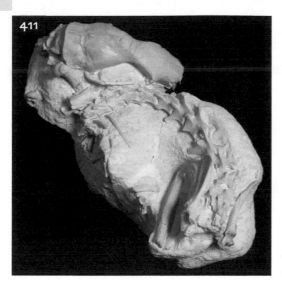

412 The rabbit *Paleolagus haydeni* (BHIGR). Length of skull 55 mm (c. 2.2 in).

411 A mustelid (weasel) (BHIGR). Length of skull 55 mm (c. 2.2 in).

Other mammals. Smaller carnivores comprise the mustelids (badgers, minks, skunks, otters, and weasels; **411**), while other mammals include insectivores (hedgehogs, shrews, and moles) and rodents (squirrels, beavers, rats, and rabbits; **412**). Other ungulates worthy of mention include the tapirs (perissodactyls) and deer (artiodactyls).

Other vertebrates. A few amphibians and about a dozen fish and reptiles have been recorded, including land turtles (**413**), crocodiles, snakes, lizards (amphibacnids, glyptosaurs, peltosaurs; **414**), and one bird (**415**), but apart from the turtles these are never common. Turtle eggs, and occasionally bird eggs, may be found (**416**).

Invertebrates. These include freshwater and terrestrial gastropods, ostracods, and insect traces (such as dung-beetle pellets and insect cocoons).

PALAEOECOLOGY OF THE WHITE RIVER GROUP BIOTA

Thirty-five million years ago the area of outcrop of the White River Group in the Dakotas, Nebraska, and Wyoming was a broad flood plain draining from the Rockies and the Black Hills, and situated somewhat south of its present latitude. During the late Eocene and early Oligocene, volcanic ash from volcanoes associated with the uplift of the Rockies was carried by westerly winds across the area and deposited with river muds and sands to form a horizontal blanket of white volcaniclastic claystones, siltstones,

mudstones, and tuff beds over the high plains.

The basal member of the Chadron Formation consists of yellow/green, cross-bedded sandstones, representing the fill of meandering channels, and bearing Precambrian clasts eroded from the foothills of the Rockies (**394**). The upper member of the Chadron Formation consists instead of light-coloured, tuffaceous claystones which, according to Evanoff (2006), were deposited on a surface of low relief locally cut by deep palaeovalleys. These claystones indicate a time of high weathering associated with warm and humid conditions, but from the end of the Eocene global climates became cooler and drier and the subsequent palaeoecological history of the White River Group is one of decreasing humidity. The Scenic Member of the Brule Formation is composed of thick sequences of light-coloured mudstone, indicating sub-humid conditions in the early Oligocene, while the subsequent Poleslide Member is composed primarily of massive siltstone beds representing dry-land wind-blown loess deposited in semiarid conditions. The sandy siltstones of the Arikaree Group, which form the capping to the Brule Formation, also accumulated in relatively dry aeolian conditions before deposition ceased in the middle Oligocene (Evanoff, 2006).

The warm, humid conditions of the Eocene encouraged the growth of lush forests, but as the climate cooled and became drier, these forests quickly reduced and were replaced during the Oligocene by vast open grasslands. This led to a dramatic change in the mammalian fauna. The

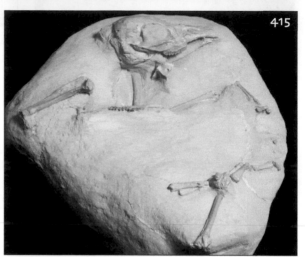

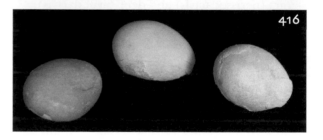

413 The turtle *Stylemis* (BHIGR). Length 140 mm (c. 5.5 in).

414 The lizard *Peltosaurus* (BHIGR). Length 65 mm (c. 2.5 in).

415 A rare bird (BHIGR). Length of skull 55 mm (c. 2.2 in).

416 Birds' eggs (BHIGR). Length of egg 55 mm (c. 2.2 in).

titanotheres, which had evolved rapidly in size from their Eocene origins and which were the most numerous mammals in Chadronian times, suddenly became extinct in the Oligocene when the forests turned to grasslands. Both the horses and the rhinos, on the other hand, quickly adapted from being small forest browsers feeding on leaves and seeds, into efficient grazing animals, leading to a continual increase in size. They had to compete with the ruminants (cud-chewing herbivores), such as the camels and deer, which also diversified quickly once the Oligocene grasslands had become established.

The most abundant mammals during the Oligocene, however, were the oreodonts, hiding from their predators in a system of burrows beneath the flood plain. All of these herbivores were prey to packs of nimravid 'cats' which used their sabre-like teeth to slice up their prey, if not to kill it. Large and small dogs hunted alone or in packs, while the hyaenodonts and entelodonts, the so-called 'killer-pigs', most probably relied more on carrion. The smaller mammals, such as mustelids, insectivores, and rodents, along with snakes, lizards, and turtles, occupied a similar position in

the food chain as they do today. Rivers and freshwater ponds on the flood plain were home to amphibians, fish, snails, and ostracods, but were dominated by the amynodontid (aquatic) rhinos and occasional crocodiles.

COMPARISON OF THE WHITE RIVER GROUP WITH OTHER NORTH AMERICAN TERTIARY MAMMAL SITES

Agate Fossil Beds, Nebraska

The next stage in the evolution of mammals can be seen at the adjacent site of Agate Fossil Beds National Monument to the south of Harrison in Nebraska (**392**). Discovered by the Carnegie Museum in 1904, this site exposes similar light-coloured, volcaniclastic sediments to those of the Eocene/Oligocene White River Group, but which are slightly younger in age, belonging to the Harrison Formation of the Arikaree Group (**393**) of Miocene age (Harrisonian NALMA, 24.15–19.04 Ma). A second outcrop was excavated in 1905 by the University of Nebraska and these two exposures are still known as Carnegie Hill and University Hill.

As would be expected the mammal fauna is similar at family level. Perissodactyls are represented by the horse *Miohippus* (which is larger than the White River *Mesohippus*, has a more specialized ankle joint, and more efficient chewing teeth with an additional crest on the upper molars), and also by *Menoceras*, a pony-sized true rhino (for the running rhinos and aquatic rhinos were by now extinct), which had two horns side by side at the end of its nose. Artiodactyls are represented by *Stenomylus*, a gazelle-like camel, and even the 'giant killer-pigs' are present with the genus *Daeodon* (formerly known as *Dinohyus* – the 'terrible pig'), which was as big as a bison and had a huge head with a large mouth full of fearsome, bone-crunching teeth.

Another interesting group are the land beavers (*Palaeocastor*), which dug large spiral-shaped burrows called *Daemonelix*, the 'Devil's corkscrew'. Why the corkscrew shape? Well, it's a lot easier to descend or, especially, ascend an incline than it is to climb a vertical shaft, and a helical path is the easiest way to go nearly straight down to a significant depth. It is unlikely that these rodents would have dug such deep burrows if flooding were likely, so the water table must have been fairly deep and the climate fairly dry.

Perhaps the top carnivore was *Amphicyon*, one of the bear-dogs (amphicyonids), making a last appearance before their extinction in the late Miocene, but the most bizarre animal of all was *Moropus*, a huge perissodactyl, well over 2 m (c. 6 ft) tall, which belonged to an extinct group known as the chalicotheres. These peculiar browsing animals were as large as a modern horse, but with different proportions – taller, but shorter from head to tail. They had long, gorilla-like front limbs which bore claws rather than hooves, a horse-like head, and a long giraffe-like neck! They may have used the claws to grasp small tree branches and pull them down to eat, or they may have used the claws to defend themselves from predators or to compete for mates. Most likely the claws had multiple uses.

It is thought that the site is an ancient watering hole which attracted drought-stricken animals from the increasingly dry Serengeti-like grasslands.

MUSEUMS AND SITE VISITS

Museums

1. American Museum of Natural History, New York, USA.
2. Tate Museum, Casper, Wyoming, USA.
3. Black Hills Institute of Geological Research, Hill City, South Dakota, USA.
4. South Dakota School of Mines Museum, Rapid City, South Dakota, USA.
5. University of Kansas Natural History Museum, Kansas, USA.

Sites

Scotts Bluff National Monument and the nearby Agate Fossil Beds National Monument in Nebraska (**392**) are the primary visitor sites for the White River Group. Scotts Bluff preserves 150 m (c. 500 ft) of Tertiary sedimentary rocks in a spectacular butte overlooking the North Platte River. The caprock belongs to the Arikaree Group (see section on Comparison, above), while the Brule Formation of the White River Group is well exposed along the trail which leads up (or down – there is a convenient shuttle bus to the top!) to the summit of the north butte.

Agate Fossil Beds National Monument, to the north of Scotts Bluff (**392**), preserves the Miocene part of the Arikaree Group with the original field sites of E. H. Barbour (via a 3 km [c. 2 mile] hike). It has a splendid visitor centre with a museum of fossil displays, and the short, 1-mile *Daemonelix* Trail, near the entrance to the monument, takes visitors through the succession with interpretive signs showing fossil burrows, palaeosols, and the spectacular, spiral burrows *Daemonelix*, made by the large rodent *Palaeocastor*.

BALTIC AMBER

Chapter Nineteen

BACKGROUND: FOREST LIFE IN THE CENOZOIC ERA

In Chapter 17 (Grube Messel) we saw how the Messel lake sampled the plants and animals from a wide variety of habitats both near and far from the lake site. While the larger mammals and birds are the most important fossils from Messel, more delicate flying insects, such as flies (Diptera) and butterflies (Lepidoptera), are rare because they tended to float on the lake surface rather than sink to the bottom. However, the tendency of these insects to adhere to the water surface is what makes them far more likely to be preserved in amber – fossilized tree resin – to which delicate insects and other animals are attracted and by which they become engulfed. Tree resin is a very localized deposit and, while some trees produce copious amounts of exudate, it is most likely to preserve animals and plants that are associated with trees. Rapid removal of a carcass from a decaying environment is the best way of preserving it, and what could be quicker than trapping and engulfing an insect in seemingly impermeable resin? There is no initial transport of the carcass, apart from some flow down the tree trunk and struggling by the animal itself, although later transport of the amber is usually necessary for its concentration into a sedimentary deposit.

Amber samples forest life from different sources than a lake deposit; its method of preservation is far better than that of lacustrine sediments for delicate invertebrates such as insects, while vertebrates are rarely preserved in amber. Insects living on tree bark are the most likely to be found in amber. Many insects are attracted to tree resin, possibly sensing the volatile oils given off by the exudate, but whether this attraction benefits the tree or the insects is not known. Once attracted to, or overcome by, the resin, insects become trapped in it owing to its adhesiveness. Predators such as spiders are attracted to the struggling insects and then they, too, become trapped, in a similar manner to the predators preserved in the tar pits of Rancho La Brea (Chapter 20). Animals living in bark crevices, in moss on or at the foot of a tree, flying insects in the amber forest, and their predators are the most common animals preserved in amber, and a wide range of plant material, such as spores, pollen, seeds, leaves, and hairs, is also commonly found embedded in the amber. Small drops of resin are unlikely to collect many organisms, but some trees exude vast quantities of resin from wounds; the large masses of resin produced from these cracks, termed 'Schlauben' (Schlüter, 1990), flow down the tree trunk and form ideal traps.

417 Young Kauri pine, *Agathis australis*, a prolific producer of resin since Tertiary times, North Island, New Zealand.

Resin is produced by a variety of trees today, as it was in the past. A prolific producer of resin is the araucarian (monkey-puzzle family) Kauri pine, *Agathis australis* (**417**), which grows in northern New Zealand. Amber deposits from this tree are known in New Zealand from some 40 Ma in age. Younger deposits of copal (resin which is not as well fossilized as amber), 30,000–40,000 years old, also occur in New Zealand and formed the basis of a copal mining industry in the last century. Copal will melt with low heat and it was used in the past to make varnish and moulded into objects such as trinkets and even false teeth. During the amberiz-ation process, fresh resin first loses its volatile oils, then polymerization begins. Once the resin has hardened (to 1–2 on the Mohs Scale) and is no longer pliable it is called copal. However, in this state it will dissolve in some organic solvents and melts at a low temperature (below 150°C [302°F]). Copal (especially African) does contain insect and other inclusions but, being much younger than amber, it is generally of less interest to the palaeontologist. To form true amber, polymeriz-ation and oxidation must continue for a longer period of time, until the material has reached a hardness of 2–3 on the Mohs Scale, will not melt

below about 200°C (392°F), and is not soluble in organic solvents. These physical properties of amber and copal are useful to remember because genuine amber with inclusions can command a good price in the gem trade, so fakes, often simply copal or made with melted copal, can be recognized. One excellent example of a forgery was discovered by Andrew Ross at the Natural History Museum in London (Grimaldi *et al.*, 1994). He was interested in a fly in apparently genuine amber which belonged to an advanced modern family otherwise unknown in the fossil record. On examination under the microscope with a rather warm lamp, a crack appeared. Further inspection revealed that a piece of amber had been cut in half, one side hollowed out and the (modern) fly inserted, then glued carefully back together again!

HISTORY OF DISCOVERY OF BALTIC AMBER

Amber was familiar and of special significance to ancient civilizations; it has been found in jewellery dating to before 10,000 BC. Amber was called *succinum* (sap-stone) by the Romans and *elektron* by the ancient Greeks; the English word *electricity* is derived from the static electric effect produced when amber is rubbed by a soft cloth. Pliny, in the first century AD, was the first person to describe the properties of amber and to determine correctly its origin as the fossilized resin from trees; other ideas of its origin concerned the tears of deities or dried excretions of beasts. Pliny recognized that the traded amber originated from the north of Europe, and so Baltic amber is both the oldest recorded and best known of all amber deposits.

Owing to its beauty (**418**), and hence value, trade in amber has been an integral part of Baltic cultures, and from ancient times to the present day, amber has been collected from the shores of the Baltic Sea. The Teutonic Knights, who occupied the Baltic Sea coast in the thirteenth to fifteenth centuries, appropriated control of the amber trade from the Prussians. After the Teutonic Knights were defeated at the battle of Grunwald (Tannenberg) in 1410 by the combined armies of Poland and Lithuania, their monopoly over the amber market crumbled, and other powers struggled for its control. In the middle of the nineteenth century, an enterprise emerged to dredge amber from the sea and mine it from the

418 Baltic amber with inclusions (PC). Length about 70 mm (c. 2.8 in).

419 Map showing the occurrence of Baltic amber: the presumed location of the amber forest, outcrop of the Samland Blue Earth deposits, and deposits derived from the Blue Earth outcrop (after Schlüter, 1990). (Arrows denote movement of amber from original forests by geological action [e.g. glaciation].)

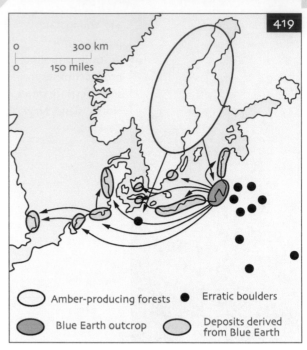

Amber-producing forests   Erratic boulders

Blue Earth outcrop   Deposits derived from Blue Earth

ground. A factory was established near the village of Palmnicken (now Jantarny) on the west coast of Samland region of Russia. (Samland is the old name for a peninsula in the Russian enclave called Kaliningrad Oblast – capital Kaliningrad – squeezed between Poland and Lithuania.) Between 1875 and 1914 the factory produced between 225,000 and 500,000 tons of raw amber per year. The best quality was used for jewellery and sculptures but the inferior grades were melted down for varnish. The finest amber carving ever produced was the Amber Room, commissioned in 1701 by King Frederick I of Prussia. This consisted of wall panels in variously coloured amber pieces depicting scenes. It took 10 years to complete, and was then transported to Peter the Great's summer palace in St Petersburg. However, during the Second World War the room was removed to Königsberg Castle, on the Baltic coast of (at that time) Germany. Later, when the Russians were advancing on Königsberg (Kaliningrad), the room panels were again dismantled and apparently packed into crates in the castle dungeons. What happened to the Amber Room after that remains a mystery; at least one person among those searching for it since the war is known to have lost his life in mysterious circumstances. In 2003 a reconstructed Amber Room was inaugurated at the Catherine Palace in Saint Petersburg.

Awareness of amber inclusions for their biological interest is also very old. The finest collection ever amassed was that of the Königsberg University Geological Institute Museum in Samland. This collection was started when dredging and mining began in 1860. Originally thought to have been destroyed by bombing during the Second World War, the collection was actually dispersed to other museums. Huge collections of Baltic amber can now be found in, for example, the Natural History Museum in London, the Museum of the Earth in Warsaw, the Zoological Institute in St Petersburg, and the Humboldt University Museum in Berlin.

STRATIGRAPHIC SETTING AND TAPHONOMY OF BALTIC AMBER

Figure **419** shows the presumed location of the Baltic amber forest, the coastal outcrop in Samland, and the places where reworked amber can be collected along the shoreline. Amber lumps washed out of the forest in streams were incorporated into a glauconitic clay called the Blue Earth. Glauconite is an iron mineral with a typically green colour but which in a dark clay could appear bluish. A number of layers of Blue Earth occur in a sequence of sandstones and mudrocks (**420**) which, by their marine fossil content, have been dated to mainly between the mid-Eocene and early Oligocene Epochs (Paleogene Period). It is the Blue Earth which is quarried for amber near Palmnicken. During the Pleistocene Epoch, ice sheets scoured the area now occupied by the Baltic Sea, and Blue Earth became incorporated into the glacial boulder clay (till) which now covers much of northern Poland, Germany, and Denmark. Because of its low density, amber from outcrops of Blue Earth or boulder clay derived from the deposit is easily washed out by currents and carried to the beach. Thus, Baltic amber can be found on the coastline of not only the Baltic Sea but also the North Sea.

There has been much discussion about the type of tree which produced the resin which became the Baltic amber. The first studies in the 1830s placed it as a species of the extant pine genus *Pinus*, but later anatomical investigations placed it in an extinct genus and species: *Pinites succinifera*. Other anatomical studies in the last century suggested that the wood was closer to the spruce genus, *Picea*. The more modern technique of infra-red spectroscopy, which is useful for distinguishing different types of amber, reveals a characteristic flattish region to the resulting spectroscopic curve for Baltic amber (called the 'Baltic shoulder'). Care must be taken when comparing spectroscopic curves for ambers with those for resins because the spectrum varies with degree of fossilization. Nevertheless, comparisons have been made which reveal that the infra-red spectrum of Baltic amber is closer to that of the New Zealand araucarian *Agathis australis*. More recent analyses using pyrolysis gas spectrometry have produced similar results. So, the spectrum of Baltic amber points to Araucariaceae, morphology to Pinaceae. Furthermore, while *Agathis* is a copious producer of resin today and *Pinus* produces far less, there is fossil evidence of *Pinus* in Baltic amber but none of araucarians. A possible compromise, suggested by Larsson (1978), is that the tree belonged to an extinct group of gymnosperms with characteristics of both Pinaceae and Araucariaceae (it could not be ancestral to these families because they were both present in the Cretaceous Period). The most recent work used solid-state Fourier-transform infra-red microspectroscopy coupled with multivariate clustering and palaeobotanical observations (Wolfe *et al.*, 2009). These authors proposed that conifers of the family Sciadopityaceae, which has a single living species, *Sciadopitys verticillata*, were responsible for the production of Baltic amber.

At first sight, insects in amber appear to be preserved in three dimensions, with their original cuticle and coloration, but as empty husks lacking any internal organs. This appearance was first shown to be misleading as long ago as 1903, when

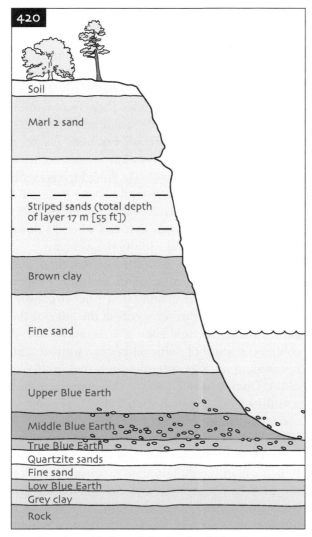

420 Stratigraphic log of the Samland Blue Earth showing the location of amber-rich layers (after Poinar, 1992).

421 Wedge-shaped beetle (Coleoptera: Rhipiphoridae) in Baltic amber (GPMH). Note the covering of white emulsion. Length about 5 mm (c. 0.2 in).

Kornilovich described striated muscles in Baltic amber insects. Later, Petrunkevitch (1950) recognized internal organs in Baltic amber spiders, and the scanning electron microscope studies of Mierzejewski (1976a,b) showed that the preservation of the internal structure of spiders in Baltic amber (including spinning glands, book-lungs, liver, muscles, and haemolymph cells) was better than that shown by the insects. Using the transmission electron microscope, Poinar and Hess (1982) revealed muscle fibres, cell nuclei, ribosomes, endoplasmic reticulum, and mitochondria in a Baltic amber gnat. Organisms are preserved in amber by a process known as mummification. In this process, dehydration of the tissues results in their shrinkage to some 30% of their original volume (thus giving the fossil the appearance of an empty husk). The organic material is not removed from oxidation because amber does allow slow gaseous diffusion (so cannot be used in the study of ancient atmospheres as had been hoped) but amber has fixative and anti-bacterial properties, like many resins. The ancient Egyptians used resins in the preparation of their mummies, and the anti-bacterial properties of many resins are well known; the distinctive flavour of Greek *retsina* wine is due to the use of resin to prevent it going off.

Because of the excellent preservation of structures at the subcellular level there has been considerable interest in the possibility of recovering pieces of the macromolecule deoxyribonucleic acid (DNA) from amber-preserved organisms. In the movie *Jurassic Park* it was suggested that if fossilized dinosaur blood could be extracted from the gut of a gnat preserved in Mesozoic amber, the DNA sequence of the dinosaur thus revealed could then be used to generate living dinosaurs. Like all the best science fiction, this idea is based on possibility. However, while attempts have been made to extract DNA from ambers, none has been successful, and since it is now known that amber is not as impermeable as was once thought, it is highly unlikely that the molecule could have survived for millions of years (or even for more than a few hours after death – it is known that DNA breaks down rapidly after cell death) without degradation.

Even if an inclusion is well preserved, there may be cracks, cloudiness, or other inclusions which obscure important features of the organism. Sometimes the inclusion occurs at the edge of the amber piece, so only part of it is preserved. A common feature of Baltic amber is a whitish fuzz, resembling mould, surrounding the inclusion (**421**). Under the microscope, this material, called 'emulsion' by Petrunkevitch (1942), is seen to consist of tiny air bubbles (Mierzejewski, 1978). It is believed that the emulsion is caused by moisture escaping from the carcass and reacting with the resin early in the taphonomic process. Though it is especially characteristic of Baltic amber, it has been noted in other ambers such as the early Cretaceous amber of the Isle of Wight (Selden, 2002), and its presence or absence may be related to differences in water solubility of the resins from different amber-producing trees (Poinar, 1992).

DESCRIPTION OF THE BALTIC AMBER BIOTA

Since the first scientific studies of Baltic amber some 250 years ago, a huge variety of organisms have been reported as inclusions. The bulk of Poinar's (1992) book is devoted to descriptions of the variety of plants and animals reported from amber inclusions, and Baltic amber figures in nearly every section. For more detailed reviews the reader is referred to this work, to the books by Larsson (1978) and Weitschat and Wichard (2002), which are restricted to Baltic amber, and to Keilbach (1982) which covers arthropods.

Some 750 species of plants have been described from Baltic amber over the past two-and-a-half centuries, principally between 1830 and 1937, and were summarized by Conwentz (1886), Göppert and Berendt (1845), Göppert and Menge (1883), and Caspary and Klebs (1907). The main studies on the Baltic amber tree were by Conwentz (1890) and Schubert (1961). However, according to a study by Czeczott (1961), only 216 of these 750 are valid species, including five species of bacteria, one myxomycete (slime mould), 18 fungi, two lichens, 18 liverworts, 17 mosses, two ferns, 52 gymnosperms and 101 angiosperms.

Fungi. Most of the fungi known from Baltic amber are saprophytic growths on other organisms. Lichens are a symbiotic intergrowth of algae and fungi, and are commonly found on tree trunks.

Bryophytes (mosses and liverworts) are generally rare in the fossil record, and some of the best preserved bryophytes occur in Baltic amber, described by Czeczott (1961). Pteridophytes (ferns) are rarely preserved in amber, but Czeczott (1961) reported two species: *Pecopteris humboldtiana* and *Alethopteris serrata*.

Gymnosperms. The gymnosperms in Baltic amber mainly belong to living genera, and it is interesting that the most closely related living species to the Baltic amber species occur in North America, east Asia, and Africa. The genera represented in Baltic amber include the cycad *Zamiphyllum*; the podocarp *Podocarpites*; Pinaceae *Pinus*, *Piceites*, *Larix*, and *Abies*; Taxodiaceae *Glyptostrobus*, *Sciadopitys*, and *Sequoia*; and the Cupressaceae *Widdringtonites*, *Thujites*, *Librocedrus*, *Chamaecyparis*, *Cupressites*, *Cupressinanthus*, and *Juniperus*.

Angiosperms. About two-thirds of the angiosperms (flowering plants) in Baltic amber have been identified from flowers, fruits, or seeds (**422, 423**); the remainder were described from leaves and twigs. Most belong to extant genera, and a great diversity of families is represented, belonging to temperate, Mediterranean, subtropical, and even tropical groups. As with the gymnosperms, the distribution of closely related living species of angiosperms is interesting. For example, the fossil genus *Drimysophyllum* (Magnoliaceae) most closely resem-

422 Rose in Baltic amber (Angiospermae: Rosaceae) (GPMH). About 5 mm (c. 0.2 in) across.

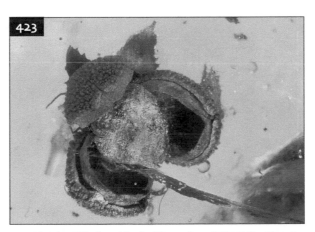

423 Unidentified nut in Baltic amber (GPMH). About 5 mm (c. 0.2 in) across.

bles the extant *Drimys* of the Winteraceae, which now occurs in the Malay Archipelago, New Caledonia, New Zealand, and Central and South America. The nearest present-day locality for *Clethra*, found in Baltic amber, is Madeira. Other tropical or subtropical families represented in Baltic amber are: Palmae, Lauraceae, Dilleniaceae, Myrsinaceae, Ternstroemiaceae, Commeinaceae, Araceae, and Connaraceae. According to Czeczott (1961) about 23% of the families are tropical, 12% are restricted to temperate regions, and the remainder are cosmopolitan or have a discontinuous distribution. One very common inclusion in Baltic amber, which can be helpful in diagnosing amber from this source, are stellate hairs (trichomes) from oak leaves. Look at the underside of an oak leaf with a hand-lens and these hairs can easily be seen. Oak bud scales are particularly thickly covered by them and, in the spring, oak woods become carpeted with masses of these little star-shaped objects.

Nematodes and molluscs. Microscopic round-worms (Nematoda) are common just about everywhere, so it is not surprising that they have been reported in Baltic amber. Figure **424** shows a parasitic nematode emerging from its host, a midge. Land snails (Mollusca: Gastropoda: Pulmonata and Prosobranchia) are, perhaps surprisingly, rare in Baltic amber, though a few genera have been reported. Klebs (1886) gave a table showing where related living species

originated, and the areas include central and southern Europe, North America, central Asia, south China, and India.

Crustaceans. Arthropods are, of course, the most common inclusions found in amber. Perhaps surprisingly, Crustacea are known from Baltic amber in the form of a few specimens of the amphipod *Palaeogammarus* and the isopods *Ligidium*, *Trichoniscoides*, *Oniscus*, and *Porcellio*. Amphipods (sandhoppers and their allies) occur today in moist litter as well as freshwater habitats, and isopods (woodlice) are common inhabitants of litter and damp places.

Myriapods. Myriapods include the predatory centipedes (Chilopoda) and mainly detritivorous millipedes (Diplopoda), both of which are common in terrestrial habitats, especially where it is damp, such as beneath decaying logs and in leaf litter. A wide variety of chilopods have been described from Baltic amber, including the extant genera *Scutigera*, *Cryptops*, *Geophilus*, *Scolopendra*, and *Lithobius*. Among millipedes, bristly, soft-cuticled members of the primitive subclass Penicillata, which are common on tree trunks, are represented by three genera. Other millipedes in Baltic amber include the pill millipede *Glomeris*, and other recent genera such as *Craspedosoma*, *Julus*, and *Polyzonium*. The specimen shown in Figure **425** appears to be a chordeumatid.

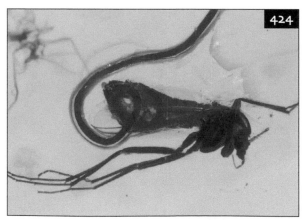

424 Mermithid nematode emerging from the body of its host, a female chironomid midge (GPMH). Length about 6 mm (c. 0.2 in).

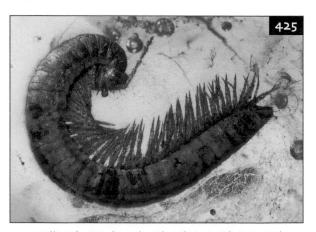

425 Millipede (Diplopoda: Chordeumatida?) in Baltic amber (GPMH). Length about 8 mm (c. 0.3 in).

Hexapods (insects). Primitive hexapods (insects and related groups) are represented in Baltic amber by two species of dipluran and many Collembola (springtails). The latter group is quite likely to get preserved in amber because of their saltatory locomotion. Silverfish (Thysanura), common today under bark and stones, are represented in the Baltic amber by numerous genera in three families.

More advanced, winged insects are the most common inclusions in amber. A great many mayflies (Ephemeroptera) have been described from Baltic amber. The larvae of these insects are aquatic – indicating that there must have been fresh water in the amber forest – but the adults are short-lived and often swarm in great abundance. Like mayflies, stoneflies (Plecoptera) have aquatic larvae, and representatives of four families occur in Baltic amber. In contrast to mayflies and stoneflies, dragonflies and damselflies (Odonata) are strong fliers, and much less likely to be blown onto sticky resin. It is not surprising, therefore, that they are rare in Baltic amber. Common detritivores, the cockroaches (Blattaria) are relatively common and diverse in Baltic amber. The genera represented are mainly subtropical and tropical forms today. Crickets and grasshoppers, too, occur in Baltic amber, perhaps partly because they are likely to jump onto the resin. Praying mantids and stick insects also have representatives in Baltic amber. Turn over any piece of bark and one insect which is almost certain to be encountered is an earwig (Dermaptera), and therefore it is not surprising to find that they are known from Baltic amber too.

Isoptera (termites) are well-known inhabitants of rotting wood, and they are represented quite commonly in Baltic amber by three families and numerous genera, the commonest species being *Reticulotermes antiquus*. Embioptera are a small group of insects with a thin cuticle and poor powers of flight. They live communally in silken tubes beneath bark and under stones, and one species, *Electrombia antiqua*, is known from Baltic amber. Among other small orders of insects with Baltic amber representatives are the Psocoptera (book-lice) (**426**). These little creatures are detritivores on dead plant and animal matter; they are commonly found feeding on the binding paste of old books – hence their common name. In the wild they occur on tree trunks and under bark, so are common in amber. Representatives of eight families and numerous genera are known from Baltic amber. The majority of psocids from Baltic amber are closest to species which today live in tropical and subtropical Asia, America, and Africa. Thrips (Thysanura) are also tiny insects, with sucking mouthparts for extracting sap from living vegetation, as well as fluids from other animal and plant tissue; many are abundant pests. Six families have been reported from Baltic amber, and most belong to the family Thripidae.

The order Hemiptera (bugs) is characterized by sucking mouthparts, and can be conveniently divided into two suborders: Homoptera and Heteroptera. Homopterans include cicadas, aphids, leafhoppers, and scale insects. Homopterans in Baltic amber include many species of aphids, some scale insects, and leafhoppers. In contrast to the abundance of these small homopterans, the large, powerful flying cicadas are rare in Baltic amber.

426

426 Parasitic mite (Arachnida: Acari: Erythraeidae: *Leptus* sp.) on a book-louse (Psocoptera: Caecilidae) in Baltic amber (GPMH). Body length about 0.5 mm (c. 0.02 in).

Heteropterans are all the other kinds of bugs, such as shield bugs, water boatmen, water striders, plant bugs, bed-bugs, and assassin bugs. They all have piercing mouthparts for feeding on plant sap or animal fluids. Baltic amber includes representatives of nearly all of the groups of Heteroptera. It is understandable that plant-sucking bugs of the large family Miridae, for example, are common in amber, but the records of such creatures as *Nepa* (the large, predatory water-scorpion), water boatmen, and water striders are curious. However, these animals do fly regularly, so could be attracted to insect prey trapped in resin.

The Neuropterida covers a range of related groups: Megaloptera (alderflies and dobsonflies), Raphidioptera (snakeflies), and Neuroptera (lacewings, antlions, and mantispids). These are large insects, predatory, but generally slow fliers. A few species of most of these groups have been recorded from Baltic amber. Mecoptera (scorpionflies) have characteristic curved ends to their abdomens, hence their name, but otherwise resemble neuropterans. Four genera have been reported from Baltic amber.

Beetles (Coleoptera) are the most diverse order of insects, and hence the most diverse group of terrestrial animals. Not surprisingly, there are many records of beetles from Baltic amber. Ground beetles (Carabidae) are common on tree trunks and so in amber too. Like the water bugs, water beetles (Dytiscidae, Gyrinidae) have been found in Baltic amber, indicating water bodies in the forested areas. Anobiidae (death-watch beetles) are common wood-boring beetles, and so occur quite commonly in amber. Other wood-boring beetles also occur in amber: the beautiful Buprestidae, the longhorn beetles of the family Cerambycidae, for example. The large family of often brightly coloured leaf beetles (Chrysomelidae) have more than two dozen genera know from Baltic amber. A few genera of ladybirds (Coccinellidae) are known from the amber. A number of families of beetles are associated with bark, especially through their habit of eating bark fungi, and these families are therefore common in amber. Mention should also be made of the superfamily Curculionoidea, the weevils, of which about 50 genera are known from the Baltic amber; and the Elateridae (click beetles), which are very common in the amber (about 40 genera), perhaps because they were attracted to

the resin. Scarabaeidae (scarab beetles) are generally associated with animal dung, but a few genera occur in Baltic amber. One of the largest groups of beetles is the family Staphylinidae (rove beetles), and so not surprisingly about 50 genera have been reported from Baltic amber. Figure **421** shows an adult wedge-shaped beetle (Rhiphoridae), whose modern relatives live on flower heads and whose larvae are parasitic on hymenopterans which visit flowers.

Caddisflies (Trichoptera) have aquatic larvae but their imagines (adults) fly. Ulmer (1912) produced an extensive monograph on the Baltic amber caddises in which 152 species in 56 genera and 12 families were described. This huge diversity (and great number, since they are extremely common in Baltic amber) must reflect a generous supply of rivers and ponds in the amber forest to support the larval stages. Interestingly, while a quarter of all living caddisflies belong to the family Limnephilidae, none from this family occurs in the Baltic amber. The reason for this may be that limnephilids inhabit temperate rather than tropical or subtropical climates. Caddisflies have hairy wings while Lepidoptera (butterflies and moths) have scaly ones. The larger Lepidoptera (macrolepidopterans) are strong fliers and, while many are attracted to resin seeps, are unlikely to become entrapped. Indeed, most representatives of the Lepidoptera found in Baltic amber are microlepidopterans – small moths. Members of the families Eriocraniidae, Nepticulidae, Incurvariidae, Psychidae (bagworms), Tineidae (clothes moths), Lyonetiidae, Gracillariidae, Plutellidae, Yponomeutidae (ermine moths), Elachistidae, Oecophoridae, Scythrididae, Torticidae (tortrix moths), Pyralidae (plume moths), and Adelidae have all been found in Baltic amber. Of these, tineids and oecophorids are the most common, while the large family of tortrix moths is under-represented, apparently because they were not easily trapped in the amber. There have been reports of macrolepidopterans captured in Baltic amber, including Sphingidae (hawkmoths), Arctiidae (tiger moths), Noctuidae, Papilionidae (swallowtail butterflies), and Lycaenidae (blues).

Diptera (flies) are the second largest order of insects after the Coleoptera, and are easily recognized by having only a single pair of wings (forewings); the hindwings are reduced to balancing organs called halteres. Three suborders are

427 Adult wood-gnat (Diptera: Anisopodidae) just hatched from its pupal case, Baltic amber (GPMH). Pupa about 4.6 mm (c. 0.2 in) long.

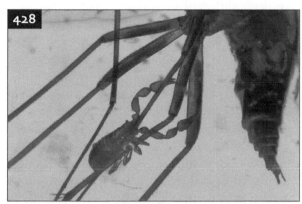

428 Phoresy: pseudoscorpion hitching a ride on the leg of a rhagionid fly, Baltic amber (GPMH). Pseudoscorpion body about 2.5 mm (c. 0.1 in) long.

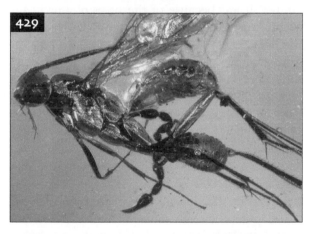

429 Phoresy: pseudoscorpion (*Oligochernes bachofeni*) on a braconid wasp (Hymenoptera: Braconidae) in Baltic amber (GPMH). Body of pseudoscorpion (excluding chelae) about 2.5 mm long (c. 0.1 in).

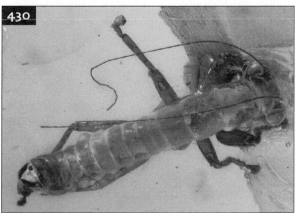

430 *Raptophasma kerneggeri*: type specimen of the recently described order Mantophasmatodea, Baltic amber (GPMH). Body length 11.7 mm (c. 0.5 in).

recognized: Nematocera, Brachycera, and Cyclorrhapha. Nematocera are the midges and fungus gnats and among the commonest insects found in amber. Among the many families found in Baltic amber, it is worth mentioning the Anisopodidae (wood-gnats) (**427**), whose larvae live on decaying organic matter; Bibionidae (March flies), the adults of which appear in large numbers in the spring; Cecidomyiidae (gall midges), whose larvae cause leaf galls; Ceratopogonidae (biting midges), whose adults inflict painful bites on animals (including humans) and whose larvae are detritivores in aquatic habitats; Chironomidae (true midges), whose adults swarm but do not bite and whose larvae are aquatic (see **424**); Culicidae (mosquitoes), infamous blood-suckers with aquatic larvae; Mycetophilidae and Sciaridae (fungus gnats), tiny flies ubiquitous in woodlands with larvae feeding on fungi and decaying wood (though some mycetophilids prey on insects by attracting them to their glowing tails); Psychodidae (moth flies and sand flies), whose larvae feed on decaying vegetation while the adults are blood-suckers; Scatopsidae (scavenger flies), with larvae that scavenge on rotting organic matter and excrement; Simuliidae (blackflies), more blood-suckers; Tipulidae (crane flies), the familiar 'daddy-long-legs' flies with about 40 genera in Baltic amber; and the Trichoceridae (winter crane flies), which inhabit dark places and are seen mainly in the winter.

Baltic amber Brachycera include representatives of the following families: Acroceridae (small-headed flies), internal parasites of spiders; Asilidae (robber flies), large, predatory flies; Bombylidae (bee flies), which suck nectar and generally resemble bees; Dolichopodidae (long-legged flies), whose larvae and imagines are both predatory, the former frequently occurring under bark; Empididae (dance flies), whose larvae live under bark (hence their abundance in amber) and whose adults characteristically form little dancing swarms; Rhagionidae (snipe flies), with predatory larvae and imagines (see **428**); Stratiomyidae (soldier flies), large, colourful predatory flies with larvae that live under bark; Tabanidae (horse flies), which have familiar, painfully blood-sucking, female adults and aquatic larvae; Therevidae; Xylomyidae; and Xylophagidae. Eighteen families of Cyclorrhapha have been reported from Baltic amber, the best known of which are Drosophilidae (fruit flies)

and Muscidae (house flies); each family is represented in Baltic amber by only a single species.

Siphonaptera (fleas) are rare in the fossil record, yet two species are known from the Baltic amber, both belonging to the genus *Palaeopsylla*, which are ectoparasites of shrews and moles.

Hymenoptera can be subdivided into two suborders: Symphyta (sawflies and horntails) and Apocrita (ants, bees, and wasps). Seven families of Symphyta have been recorded from Baltic amber, mainly from single examples of their larvae. There are many different groups of Apocrita: parasitic, solitary, colonial, social, winged, and wingless. The parasitic forms (Parasitica) are important controllers of insect numbers, and many are known from Baltic amber, particularly braconids (**429**) and ichneumons. The presence of gall wasps is usually indicated by their galls, such as oak apples, on vegetation; a number of these occur in Baltic amber. Aculeate Apocrita includes the ants, bees, and wasps. Ants are abundant in ambers, probably because of their habit of crawling up and down tree trunks. More than 50 genera of ants are known from Baltic amber. It was because of the types of ants in the Baltic amber – one group associated with a subtropical climate and the other with a temperate one – that Wheeler (1915) suggested the deposition of amber over a long period of time when the climate cooled and the forest type changed. Two species of wasp (Vespidae) have been reported from Baltic amber. Sphecids and pompilids are solitary wasps which dig burrows into which they deposit their eggs alongside narcotized prey (usually spiders) to provide nourishment for the developing larvae when they hatch. A number of sphecids and fewer pompilids have been reported from Baltic amber. Bees (Apoidea) include both the solitary and social forms. A number of species of both types are known from Baltic amber.

Before leaving the insects, mention must be made of an order of orthopteroids discovered as recently as 2002: the Mantophasmatodea (Klass *et al.*, 2002; Zompro *et al.*, 2002) (**430**). First described as a member of an unknown order of insects from Baltic amber in 2001 (Zompro, 2001), extant specimens were later identifed from Tanzania and Namibia. Nocturnal, pedatory insects, their modern habitat (dry, stony mountains) differs considerably from their habitat in the Baltic amber forest.

Arachnids. The fossil record of arachnids (spiders, scorpions, mites, and their allies) would be much poorer were it not for Baltic and other amber inclusions. However, most Baltic amber specimens belong to modern families and genera, so tell us little about the evolution of the groups. Scorpions are represented in Baltic amber by six specimens of buthoids, each described in its own genus, and which are most closely related to extant genera from Africa and Asia (Lourenço & Weitschat, 2000). Scorpions are much rarer in Baltic amber than in other ambers. Resembling miniature scorpions but without a tail are the pseudo-scorpions. These little animals live in moss, under bark, and in leaf litter, and quite a number of genera in nine families have been found in Baltic amber (Schawaller, 1978). Pseudoscorpions have a clever way of dispersing from one habitat to another. They wait for a flying insect to land on the moss, bark, or litter, then clamp onto the leg of the insect with their clawed pedipalps (**428, 429**). When the insect flies off, it takes the pseudo-scorpion with it, and when it lands in a suitable habitat, the pseudoscorpion unhitches. This type of 'hitch-hiking' is called phoresy, and examples of pseudoscorpions phoretic on braconid wasps have been found in Baltic amber (Bachofen-Echt, 1949). Harvestmen (Opiliones) (**431**) are familiar long-legged arachnids which commonly aggregate in huge numbers under loose bark. Nine genera have been described from Baltic amber, but the actual number of specimens is far greater because of the harvestman's habit of shedding a leg if it gets caught, so there are many isolated, unidentifiable harvestman legs in the amber. Mites and ticks are the most diverse animals on land, after the four largest insect orders. Many are parasitic on plants and animals and have sucking mouthparts for this. Many of the parasitic forms are host-specific, so it is sometimes possible to infer the presence of the host from the fossil evidence of the parasite (**426**). One tick species, *Ixodes succineus*, has been described from Baltic amber, and more than 60 mite genera. Most of the mites belong to the free-living forms which are common detritivores and fungivores today under bark and in moss and litter.

More than 90% of all fossil spiders (Araneae) known are from amber. Representatives of some 33 families have been described from Baltic amber, originally mainly by Koch and Berendt (1854), revised by Petrunkevitch (1942, 1950, 1958), with additions by Wunderlich (1986, 1988). One family, Archaeidae (**432, 433**), was first described from Baltic amber and only later found living in the Afrotropical and Australian regions. The families are those which would be expected in a forest setting, and many of the genera suggest a subtropical climate. Apart from some small specimens, those living under bark, and moulted skins, it is quite likely that many of the spiders preserved in amber were attempting to prey upon insects already trapped in resin. Amazingly, spider silk and webs (**434, 435**) have been preserved in Baltic amber, and a pair of spiders were even caught in the act of mating (Wunderlich, 1982).

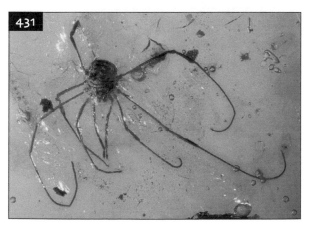

431 Harvestman (Arachnida: Opiliones: *Dicranopalpus* sp.) in Baltic amber (GPMH). Body (excluding legs) about 3.25 mm long (c. 0.13 in).

432 Male archaeid spider in Baltic amber (GPMH). Length (including chelicerae) 5.6 mm (c. 0.22 in).

433 Front view of female archaeid spider in Baltic amber (GPMH). Length from back of head to end of chelicerae about 3 mm (c. 0.1 in).

434 Close-up view of spider spinnerets showing silk strands emerging from the spigots, Baltic amber (GPMH). Spinnerets about 0.2 mm (c. 0.008 in) long.

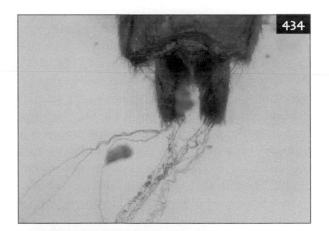

435 Spider web in Baltic amber (GPMH). Picture about 30 mm (c. 1.2 in) long.

Vertebrates. These animals are generally too large to be trapped in resin, but some remains have been found. Unfortunately, most reports of frogs and lizards in Baltic amber have turned out, on closer inspection, to be forgeries; or else the specimens have been lost. Bird feathers occur in Baltic amber (e.g. Weitschat, 1980), and they have been referred to sparrows, woodpeckers, nuthatches, and tits. Mammals are represented in Baltic amber only by a footprint and some hairs. The hairs have been identified as those of dormice, squirrels, and bats.

PALAEOECOLOGY OF THE BALTIC AMBER BIOTA

A number of suggestions have been put forward attempting to explain the assortment of Baltic amber plants that occur today in different climatic zones and widespread parts of the world. Heer (1859) explained the mixture of floral types by proposing that the forest was vast, extending over an area from present-day Scandinavia to Germany and Poland. In this way, the trees could occupy temperate to subtropical climates, possibly extending up the sides of mountains in the northern part of the region. Wheeler (1915) suggested that the Baltic amber forest occupied the same, restricted, geographical area but that the climate changed over the period of amber production. The production of Baltic amber probably extended over several million years, so the forest could have subtly changed from subtropical to temperate (or *vice versa*). Ander (1941) suggested that the forest occupied a humid, mountainous region in which tropical species could live on the southern slopes, while those preferring lower temperatures would be restricted to cooler valleys with a northerly aspect. Abel (1935) and other workers compared the palaeoecology of the Baltic amber forest with present-day southern Florida, where isolated islands (called 'hammocks') of tropical plants occur among subtropical and temperate vegetation. In this region one can find three groups of plants, palms, pines, and oaks, occurring together, as in the Baltic amber vegetation. Baltic amber sampled the plants and animals of a forest. Climatic evidence indicates that it was temperate to tropical, and the forest was certainly wet. One-quarter of the animals in Baltic amber are aquatic insects which inhabit freshwater ecosystems at least in their larval stages. These include the insect orders Odonata, Ephemeroptera, Plecoptera, Hemiptera (Heteroptera), Neuroptera, Megaloptera, Coleoptera, Trichoptera, and Diptera, as well as amphipod and isopod crustaceans.

COMPARISON OF BALTIC AMBER WITH OTHER AMBERS

Schlüter (1990) compared the Baltic amber fauna with that of the Dominican Republic, the only other amber to have produced sufficient inclusions to make quantitative comparisons with Baltic amber possible, with informative results. Diptera account for some 50% of inclusions in the Baltic amber and less than 40% in the Dominican. In Dominican amber, Hymenoptera (mainly ants) are second only to Diptera (also nearly 40%) in abundance, but these insects account for only about 5% of the Baltic amber fauna. The reason for this is that ants make up a more disproportionate percentage of the fauna in the tropics than they do in temperate zones, and Dominican Republic amber shows every sign of representing a tropical forest (Poinar & Poinar, 1999). The Dominican Republic amber was produced by a leguminous tree, *Hymenaea protera*, which has as its closest relative *H. verrucosa* of East Africa. It is generally rather paler than Baltic amber and clearer, lacking the abundant oak hairs and the annoying emulsion which often obscures inclusions in Baltic amber. Amber and copal come from a variety of localities in the Dominican Republic, ranging in age from 15 to 45 Ma. Today, Dominican Republic amber is turning up inclusions which equal or, in some cases (e.g. vertebrates), exceed the significance of those of Baltic amber for enlightening us about life in the forests of the Cenozoic Era.

MUSEUMS AND SITE VISITS

Museums

1. The Natural History Museum, London, England.
2. Museum of the Earth, Warsaw, Poland.
3. The Zoological Institute, St Petersbourg, Russia.
4. Museum für Naturkunde der Humboldt-Universität zu Berlin, Germany.
5. Staaliches Museum für Naturkunde, Stuttgart, Germany.
6. Swedish Amber Museum, Höllviken, Sweden.
7. The Amber Museum, Palanga, Lithuania.
8. Amber Museum Gallery, Nida and Vilnius, Lithuania.
9. Ravmuseet, Oksbøl, Denmark.
10. Skagen Ravmuseum, Skagen, Denmark.
11. Geological Museum, Copenhagen, Denmark.
12. Kaliningrad Amber Museum, Kaliningrad, Russia.
13. Yantarnyi Amber Mine Museum, Kaliningrad, Russia.
14. American Museum of Natural History, New York, USA.

Sites

The Yantarny opencast amber mine and museum figures on most tourist excursions to the Kaliningrad region. If you go under your own steam, there are six trains a day from Kaliningrad to Yantarny; the journey takes about an hour. You can buy pieces of the locally produced gold and silver amber jewellery at the factory outlet. Otherwise, amber can be collected from many parts of the Baltic and North Sea coasts, from Russia to East Anglia. Being light, it tends to collect along the tide-line. It is considered best to look after storms.

Chapter Twenty

RANCHO LA BREA

BACKGROUND: THE PLEISTOCENE IN NORTH AMERICA

By the onset of the Quaternary Period the continents were close to their present positions, although the mid-Atlantic ridge was continuously spreading. At this time, 2.5 million years ago (Bowen, 1999), there was a dramatic deterioration in the climate which over much of North America and northern Europe and Asia remained cold to glacial during the whole of the Quaternary, with temperate to warm intervals of short duration. This is the period of Earth's history known as the 'Great Ice Age', when the climate was the dominant geological force. Icebergs began to appear in northern oceans and vast continental ice-sheets covered much of the northern continents. The North American ice cap covered 13 million sq. km (c. 5 million sq. miles) of the continent; it carved the landscape of northern Canada, with meltwaters carrying the debris south as far as the Great Lakes.

The Quaternary Period, lasting for only 2.5 Ma, is much shorter than any other geological period, and is too short to be subdivided on the traditional basis of faunal and floral evolutionary changes. Instead it is subdivided on the basis of climatic changes, which were dramatic at this time. The term 'Ice Age' often gives the wrong impression; the Quaternary was not one continuous glaciation, but was a period of oscillating climate with advances of ice and growth of glaciers punctuated by times when the climate was not very different from that of today.

In North America there were four major cold periods during the Quaternary (compared to six in the British Isles). The Nebraskan, Kansan,

Illinoian, and Wisconsinan glaciations alternated with intervening warmer periods of the pre-Nebraskan, Aftonian, Yarmouthian, and Sangamonian interglacials. The Nebraskan glaciation began around 1 million years ago and lasted about 100,000 years, but it was the final Wisconsinan glaciation, which began about 100,000 years ago, that included the coldest time during the whole of the Ice Age.

Sea level was much lower during the glaciations because millions of cubic kilometres of water from the oceans were turned into ice, lowering sea level eustatically by as much as 120 m (c. 400 ft). The Bering Strait, which today is a shallow sea separating Alaska and Siberia, emerged periodically as a land bridge connecting north-eastern Asia and north-western North America. Although this prevented exchange of marine organisms between the Pacific and the Arctic oceans, it permitted terrestrial species to migrate between North America and Eurasia. North American species, such as the camel and horse, migrated to Eurasia, while Eurasian mammals, such as mammoths, bison, and Man, entered North America. Gradual changes in the North American mammalian fauna resulting from such migration are used to define a succession of North American Land Mammal Ages (NALMAs) already discussed in Chapter 18, The White River Group.

At the close of the Wisconsinan glaciation, some time between 12,000 and 10,000 years ago, the mammal fauna across the globe underwent severe changes. In North America 73% of the large mammals (33 genera) became extinct, including mammoths, mastodons, horses, tapirs, camels, and ground sloths, together with their predators such as sabre-toothed cats. Whether this was caused simply by climatic changes at the end of the Ice Age, or whether it was the effect of excessive hunting by Man, is still the focus of much debate.

Within the City of Los Angeles, at a locality known as Rancho La Brea (**436, 437**), can be found one of the world's richest deposits of Ice Age fossils. Preserved in asphalt-rich sediments, so numerous are the remains that they can truly be considered as a Concentration Lagerstätte. Their diversity provides a virtually complete record of life in the

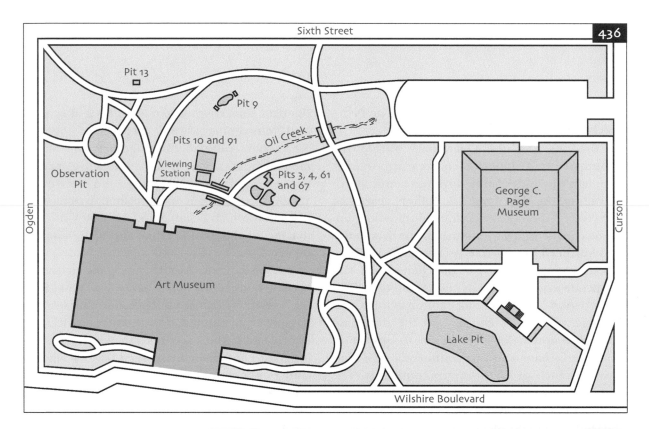

436 Locality map showing the position of Rancho La Brea (Hancock Park) within the city of Los Angeles (after Stock & Harris, 1992).

437 Rancho La Brea, Hancock Park, Los Angeles, showing methane gas bubbling through the oily water of a flooded asphalt working, surrounded by life-size models of Ice-Age mammoths.

438 Naturally occurring asphalt pool at Hancock Park, Los Angeles.

Los Angeles Basin between 10,000 and 40,000 years ago, during this vital period at the close of the Ice Age in North America. This exceptional biota defines the Rancholabrean NALMA (Savage, 1951), and includes approximately 60 different mammal species ranging in size from huge mammoths to the Californian pocket mouse. This virtually entirely preserved ecosystem also includes reptiles such as snakes and turtles, amphibians such as frogs and toads, birds, fish, molluscs, insects, spiders, and numerous plants including microscopic pollen and seeds.

HISTORY OF DISCOVERY AND EXPLOITATION OF RANCHO LA BREA

The naturally occurring asphalt in this area (**438**) has been used by man since prehistoric times. The local Chumash and Gabrielino Indians used the sticky 'tar' both as a glue for making weapons, vessels, and jewellery, and as waterproofing for canoes and roofing (Harris & Jefferson, 1985), but the first record of these deposits was that of the Spanish explorer Gaspar de Portolá, who noted 'muchos pantamos de brea' (extensive bogs of tar) in 1769. In 1792 José Longinos Martínez recorded '…twenty springs of liquid petroleum' and '…a large lake of pitch…in which bubbles or blisters are constantly forming and exploding' (Stock & Harris, 1992).

In 1828 the area became part of a Mexican land grant known as Rancho La Brea (which literally means 'the tar ranch', although the term 'tar' is not strictly accurate – the naturally occurring bituminous substance derived from petroleum is asphalt). During the nineteenth century, and especially during the 1860s and 1870s, the asphalt began to be mined commercially for road construction and during these operations workers began to find bones, but disregarded these as the remains of recent animals which had become trapped in the sticky bogs.

It was not until 1875, when the owner of the ranch, Major Henry Hancock, presented the tooth of a sabre-toothed cat to William Denton of the Boston Society of Natural History that the true age of the fossils was appreciated. Denton visited the area and collected further specimens of horses and birds. No further interest was shown, however, until 1901, when the Los Angeles geologist W. W. Orcutt visited the area with a view to oil production. His scientific excavations between 1901 and 1905 produced specimens of sabre-toothed cat, wolf, and ground sloth, which were passed to Dr John C. Merriam of the University of California. Merriam realized the importance of the deposit and excavated between 1906 and 1913, but in that year Captain G. Allan Hancock, the son of Henry Hancock, gave the County of Los Angeles exclusive rights to excavation (Harris & Jefferson, 1985). More than 750,000 bones were removed in the first 2 years and in 1915 Captain Hancock donated the fossils to the Los Angeles County Museum. At the same time the ranch, later renamed Hancock Park, was also donated to the museum for preservation, research, and exhibition. In 1963 Hancock Park was declared a National Natural Landmark (**437**) and in 1969 excavation resumed at Pit 91 in order to recover some of the smaller elements of the fauna and flora, such as insects, molluscs, seeds, and pollen, which had hitherto been ignored. Excavation continues to the present day.

The name most associated with the research of these deposits is that of Chester Stock (1892–1950), who was a student of John C. Merriam and who joined some of the early excavations at La Brea from 1913 onwards. He was associated with the Los Angeles County Museum from 1918 until his death and published the first comprehensive monograph of the La Brea fossils (Stock, 1930), which had reached its seventh edition by 1992 (Stock & Harris, 1992).

STRATIGRAPHIC SETTING AND TAPHONOMY OF THE RANCHO LA BREA BIOTA

Most of the fossils excavated at Rancho La Brea have been estimated by carbon-14 dating to be between 11,000 and 38,000 years old, which means that the sediments in which they are buried were deposited during the final stages of the Wisconsinan glaciation, at the very end of the Pleistocene Epoch (**439**). In terms of the NALMAs, the fauna belongs to the latter part of the Rancholabrean Land Mammal Age, which began 500,000 years ago, defined by the first occurrence of bison in North America.

Prior to the start of the Wisconsinan glaciation, 100,000 years ago, this part of California was submerged by an extended Pacific Ocean. The fall in sea level at the onset of this glaciation exposed a flat plain between the reduced Pacific and the Santa Monica Mountains, on which were numerous interconnected freshwater lakes. Erosion of the mountains by rivers led to the accumulation of fluvial sands, clays, and gravels between 12 m (c. 40 ft) and 58 m (c. 190 ft) thick, which gradually raised the level of the plain.

Beneath this alluvial plain Tertiary marine sediments of the Fernando Group, consisting of shales and sandstones interbedded with oil sands, acted as a reservoir for the Salt Lake oilfield. These sediments had been faulted, folded, and eroded during the early Pleistocene to form a north-east–south-west trending anticline, and from around 40,000 years ago crude oil began to seep upwards towards the crest of the anticline and into the overlying horizontally bedded Pleistocene fluvial deposits. The lighter petroleum evaporated, leaving sticky pools of natural asphalt at the surface. Many of the asphalt pools within Hancock Park (**438**) are aligned along a north-west–south-east axis, suggesting that the oil seepages may have originated from a subsurface fault.

These shallow asphalt pools formed natural traps for animals and plants, especially during the warm summers when the asphalt would have been viscous. Cooler winters may have solidified the asphalt and covered it with river sediments before the trap was reset the following summer. Repetition of this annual cycle produced conical bodies of asphalt (Shaw & Quinn, 1986). Carcasses accumulated in large numbers and it is likely that

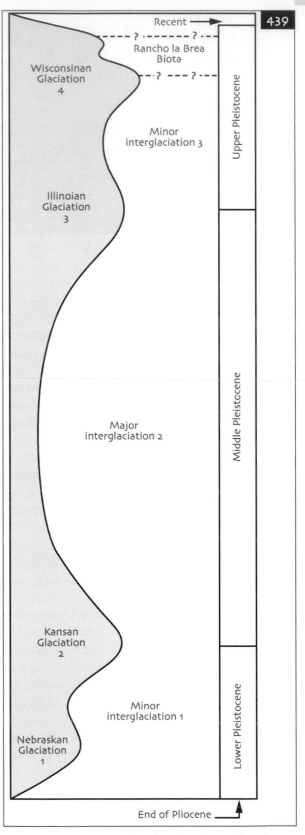

439 Diagram showing the glacial and interglacial stages of the Pleistocene, with the position of the Rancho La Brea biota indicated (after Stock, 1930).

many scavengers were lured to the pools by an initial victim.

The excellent preservation of the biota seems to be the result not only of rapid burial, but also, and unusually, of impregnation of the bones by asphalt. Soft tissues are generally not present so the deposit cannot strictly be considered as a Conservation Lagerstätte, but the sheer numbers of bones preserved (**440**) define it as a Concentration Lagerstätte; more than 50 wolf skulls and 30 sabretoothed cat skulls have been collected within just 4 cubic metres (c. 140 cubic ft).

Bones and teeth are preserved almost in their original state, apart from their penetration by oil, which gives a brown or black coloration. Up to 80% of the original collagen is retained (Ho, 1965) and microstructure is well preserved (Doberenz & Wyckoff, 1967). Surface markings on bones still show the positions of nerves and blood vessels and the attachment points of tendons and ligaments. Often oil has accumulated in skull cavities preserving, for example, the tiny bones of the middle ear or even the remains of small mammals, birds, and insects. Curiously, epidermal structures are rarely preserved; occasional hairs and feathers are known, but nails and claws of mammals or talons and beaks of birds are not. Chitinous bodies of insects retaining the iridescent colours of wing cases, and fleshy leaves and pine cones, thoroughly impregnated with oil, are not uncommon.

DESCRIPTION OF THE RANCHO LA BREA BIOTA

Human remains. The skull and partial skeleton of a human female have been recovered and carbon-14 dated at 9,000 years BP, thus postdating the bulk of the biota. 'La Brea Woman' was between 20 and 25 years of age (Kennedy, 1989) and stood approximately 1.5 m (c. 5 ft) tall. The fractured skull suggests that she may have been murdered and her body dumped in a shallow tar pool (Bromage & Shermis, 1981), although an alternative hypothesis suggests a ritual burial (Reynolds, 1985). Many human artefacts have also been found, mostly less than 10,000 years BP, and include shell jewellery, bone artefacts, wooden hairpins, and spear tips.

440 Observation Pit at Rancho La Brea, Hancock Park, Los Angeles, showing concentration of bones in tar deposits.

Dire wolf and other dogs. The dire wolf (*Canis dirus*) is the most common mammal from La Brea, known from over 1,600 individuals (**441, 442**). They probably fed in packs on animals stuck in the tar and became trapped themselves. The large head with strong jaws and massive teeth made it the major predator of La Brea. Other dogs include the grey wolf (*Canis lupus*) and coyote (*Canis latrans*), the third most common La Brea mammal, which is slightly larger than the modern representatives of this species (**443**). Domestic dogs are also present, one associated with the skeleton of La Brea Woman (Reynolds, 1985).

Sabre-toothed cat and other cats. The state fossil of California, *Smilodon fatalis*, is the best known of all the La Brea mammals and is the second most common (**444, 445**). About the same size as those of the African lion, the exact function of the large upper canine teeth is disputed. Conventionally believed to have been used to stab and kill their prey, recent studies suggest that their fragility would have been more suited to slicing open the soft underbelly of the prey after the kill (Akersten, 1985). Other large cats include the American lion, puma, bobcat, and jaguar.

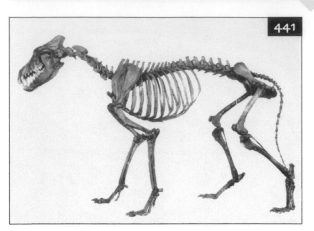

441 Skeleton of the dire wolf, *Canis dirus* (GCPM). Length of body 1–1.4 m (c. 3–4.5 ft).

442 Reconstruction of dire wolf.

443 Reconstruction of coyote.

444 Skeleton of the sabre-tooth cat *Smilodon fatalis* (GCPM). Length of body 1.4–2 m (c. 4.5–6.5 ft).

445 Reconstruction of sabre-toothed cat.

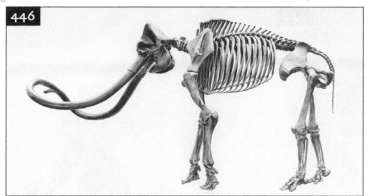

446 Skeleton of imperial mammoth, *Mammuthus imperator* (GCPM). Height up to 4 m (c. 13 ft).

447 Reconstruction of mammoth.

448 Reconstruction of mastodon.

Elephants. The imperial mammoth, *Mammuthus imperator*, was the largest of the La Brea mammals, standing almost 4 m (c. 13 ft) tall and weighing almost 5,000 kg (5 tons) (**446, 447**). The American mastodon, *Mammut americanum*, was smaller, at 1.8 m (c. 6 ft) tall (**448**) and fed on leaves and twigs, unlike the mammoth, which fed on grass.

Ground sloth. This large, primitive mammal, *Glossotherium harlani*, is related to the modern South American tree sloth, but its flat grinding teeth suggest that it fed on grass. At 1.8 m (c. 6 ft) tall these animals had dermal ossicles embedded into the skin of their necks and backs as a protection against predators, similar to the ankylosaur dinosaurs (**449, 450**).

Other large mammals. Additional carnivores include three species of bear: the short-faced bear, black bear, and grizzly bear. Herbivores are diverse and include bison (**451**), horses, tapirs, peccaries, camels, llamas, deer, and pronghorns (**452**).

Small mammals. These include the carnivorous skunks, weasels, racoons, and badgers; the insectivorous shrews and moles; numerous rodents including mice, rats, gophers, and ground squirrels; the lagomorphs (jackrabbits and hares); and bats.

Birds. The protective asphalt coating has preserved more fossil birds than at any other location in the world. Many were predators or scavengers, such as condors, vultures, and teratorns, which became trapped while feeding on carcasses. The extinct, raptor-like teratorn, *Teratornis merriami*, was 0.75 m (c. 2.5 ft) tall with a wing-span of 3.5 m (c. 11.5 ft), one of the largest known flying birds (**453**). The large number of water birds, including herons, grebes, ducks, geese, and plovers, may have landed on the asphalt, mistaking its reflective surface for a pond. There are more than 20 species of eagles, hawks, and falcons, with the Golden Eagle being the most common bird. Storks, turkeys, owls, and numerous smaller songbirds complete the bird fauna.

449 Skeleton of ground sloth *Glossotherium harlani* (GCPM). Height 1.8 m (c. 6 ft).

450 Reconstruction of ground sloth.

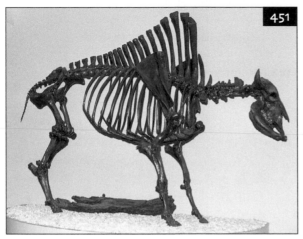

451 Skeleton of bison *Bison antiquus* (GCPM). Height 2.1 m (c. 7 ft).

452 Reconstruction of pronghorn.

Reptiles, amphibians and fish. Seven different lizards (Brattstrom, 1953), nine snakes (La Duke, 1991a), one pond turtle, five amphibians (including toads, frogs, tree frogs, and climbing salamanders) (La Duke, 1991b), and three species of fish (rainbow trout, chub, and stickle-back) (Swift, 1979) are known from La Brea, which suggests that permanent bodies of water existed in this area.

Invertebrates. Freshwater molluscs (five bivalves and 15 gastropods) suggest the presence of ponds or streams at least during part of the year. In addition 11 terrestrial gastropod species, which would have lived on leaf litter, have been recognized. Seven orders of insect include grasshoppers and crickets, termites, true bugs, leafhoppers, beetles, flies, ants, and wasps. Scorpions, millipedes, and several spiders are also known. Many of these were terrestrial and are

453 Skeleton of the teratorn *Teratornis merriami* (GCPM). Height 750 mm (c. 2.5 ft); wingspan about 3.5 m (c. 11.5 ft).

rarely found fossilized. Some of the beetles and flies would have become trapped while feeding on carrion. Others were blown onto the sticky asphalt or became stuck when crawling over it.

Plants. Fossil plants from La Brea include wood, leaves, pine cones, seeds, and microscopic pollen and diatoms.

The fauna and flora from Rancho La Brea are well documented and illustrated by Harris and Jefferson (1985) and Stock and Harris (1992).

PALAEOECOLOGY OF THE RANCHO LA BREA BIOTA

The fauna and flora of the Rancho La Brea 'tar pits' represent a terrestrial ecosystem situated on the western coastal plains of the North American continent, between 30° and 35°N, as at present, but with a cooler, glacial climate. The flora includes many plants that no longer occur in this region and indicates that while glacial winters may have resembled modern winters, the summers were not only cooler, but were also more moist (Johnson, 1977a, b). Annual rainfall was probably twice that of today; freshwater ponds and streams covered the plains, supporting fish, turtles, frogs, toads, molluscs, and aquatic insects, and a rich vegetation included four distinct assemblages.

The slopes of the Santa Monica Mountains were covered with chaparral – tall, densely packed woody bushes of lilac, scrub oak, walnut and elderberry – while the deep, protected canyons were home to redwood, dogwood, and bay. The rivers were lined with sycamore, willow, alder, raspberry, box elder, and live oak, but the plains were covered with coastal sage scrub (drought-tolerant woody bushes) interspersed with wide areas of grass and herbs, with occasional groves of closed-cone pine, valley oak, juniper, and cypress (Harris & Jefferson, 1985).

These wide plains supported large herds of hoofed mammals feeding on the rich vegetation – bison, horse, ground sloths, camels, pronghorns, and mammoths, with occasional visits by peccaries, deer, tapirs, and mastodons. These numerous herbivores in turn supported a diverse population of cursorial carnivores, including the various species of dogs, cats, and bears. There is no need to infer a radically different climate from that of the present to account for the variety of Pleistocene mammals

in this region. In fact the similarity of the smaller mammals to the present-day fauna (rodents, rabbits, shrews), plus the absence of mammals that are elsewhere associated with very cold climates (e.g. musk oxen), suggest that conditions were not significantly colder during the Pleistocene.

In all, more than 600 species have been recorded from La Brea, comprising approximately 160 plants and 440 animals. These include 59 species of mammals (represented by over one million fossils) and 135 different birds (represented by 100,000 specimens). The most significant statistic, however, is the disproportionate percentage of carnivores over herbivores, which does not conform to the normal pyramidal structure of a biological ecosystem in which carnivores form the top of the pyramid, outnumbered by herbivores.

The Rancho La Brea assemblage clearly does not represent an accurate cross section of the ecosystem. The most obvious explanation is that a single herbivore, on becoming trapped in the asphalt, might be pursued by a whole pack of carnivores. When these also became trapped, both would attract additional scavengers to their carcasses. The dogs, cats, and bears, plus the smaller carnivores, such as skunks, weasels, and badgers, constitute 90% of the mammal fauna, while the herbivores make up only 10%. Similarly the carnivorous birds (condors, vultures, teratorns, eagles, hawks, falcons, and owls) comprise approximately 70% of the bird fauna.

A further bias in the La Brea fauna is the preponderance for obvious reasons of young, aged, and maimed individuals. There may also be a sampling bias, as early excavators concentrated on collecting the larger, more impressive, mammals. Given that the preservation potential of a mammal becoming stuck in the asphalt is very high, it is a sobering fact that an entrapment episode of a single herbivore followed by, say, four dire wolves, a sabre-toothed cat, and a coyote need only occur once every decade, over a period of 30,000 years, to account for the number of mammals represented in the collections (Stock & Harris, 1992).

COMPARISON OF RANCHO LA BREA WITH OTHER PLEISTOCENE SITES

Permafrost of Siberia and Alaska

Many animals and plants survived the glaciations by occupying slightly warmer areas to the south of the ice sheets on both the North American and Asian continents. The Russian mammoth steppe was a vast grassy tundra fringing the northern ice sheet; its climate was too dry to support a large build-up of ice, and it was effectively a frozen version of today's hot African grasslands. It was home, 40,000 years ago, to woolly mammoths, woolly rhinos, bison, giant Irish deer, horses, and large predatory cats and although there was a summer thaw, allowing some vegetation, the subsurface was permanently frozen. This gave rise to what must be considered the ultimate Fossil-Lagerstätte – the deep-freeze of the Siberian permafrost.

Woolly mammoths and rhinos in particular were sometimes engulfed in bogs and deep-frozen in the permafrost, where they have remained ever since. Desiccation by freezing, where the moisture is not released to the atmosphere, but forms ice crystals around the mummy, causes the carcass to shrink and shrivel as it dries. (This should not be confused with freeze-drying, where moisture is removed by sublimation and the carcass retains its original form; see Guthrie, 1990.) Carcasses are thus mummified (compare with Baltic amber, Chapter 19), and not only is the coarse outer hair and soft downy inner hair perfectly preserved, the meat is also so fresh that it has been eaten by dogs, and apparently also by humans. One of the first such mammoths to be excavated and examined scientifically was that from Beresovka in Siberia in 1900. According to Kurtén (1986) the excavators attempted to eat the 40,000-year-old meat, but were 'unable to keep it down, in spite of a generous use of spices'. This mammoth is preserved with its final mouthful of food still in its mouth, and is on display at the Zoological Museum in St Petersburg.

Several mammoths, woolly rhinos, bison, horse, and musk ox have been recovered from the permafrost since the 1970s, one of the most well known being the complete baby woolly mammoth (*Mammuthus primigenius*) found at Magadan in Siberia in 1977. The young male (christened 'Dima') was found beneath 2 m (c. 6 ft) of frozen silt and is approximately 40,000 years old. Similar remains have been found in the permafrost of Alaska. In 1976 a partial baby mammoth, with a rabbit, a lynx, and a lemming (or vole), was discovered in the Fairbanks district (Zimmerman & Tedford, 1976) and dated by carbon-14 at 21,300 years BP. The skin, hair, and eyes of the mammoth were well preserved and the liver of the rabbit was recognizable after rehydration, but most of the internal organs were decayed and replaced by bacteria. In 2005, geneticists from McMaster University succeeded in sequencing a portion of the genome of a 27,000-year-old woolly mammoth from Siberia (Poinar *et al.*, 2006), leading to speculation that mammoths might be brought back to life by inserting their DNA into the empty egg cell of an African elephant. One can only speculate as to how far back in time such a technique could be used.

MUSEUMS AND SITE VISITS

Museum

George C. Page Museum of La Brea Discoveries, Hancock Park, Los Angeles, USA.

Sites

The 9 ha (c. 22.5 acre) Hancock Park is situated between Wilshire Boulevard and 6th Street, 11 km (c. 7 miles) west of the civic centre of Los Angeles (**436, 437, 438**). Entrance to the park is free and several 'tar pits' can be observed – relics of sites excavated for asphalt and for fossils. A large lake, surrounded by life-sized reproductions of Pleistocene elephants, represents the flooded site of a former commercial working for asphalt, with methane gas constantly bubbling through the oily water (**437**). A special Observation Pit at the west end of the park surrounds a partially excavated deposit of fossils and asphalt (**440**). Collecting is not possible, but the George C. Page Museum, opened in 1977 at the east end of the park, houses more than two million specimens collected from this site.

REFERENCES

Abdala, F., Cisneros, J. C., Smith, R. M. H. 2006. Faunal aggregation in the early Triassic Karoo Basin: earliest evidence of shelter-sharing behavior among tetrapods? *Palaios* **21**, 507–512.

Abel, O. 1935. *Vorzeitliche Lebensspuren.* Gustav Fischer, Jena, xv + 644 pp.

Agassiz, L. 1841. On the fossil fishes found by Mr. Gardner in the Province of Ceará, in the north of Brazil. *Edinburgh New Philosophical Journal* **30**, 82–84.

Akersten, W. A. 1985. Of dragons and sabertooths. *Terra* **23**, 13–19.

Aldridge, R. J., Theron. J. N. 1993. Conodonts with preserved soft tissue from a new Ordovician Konservat-Lagerstätte. *Journal of Micropalaeontology* **12**, 113–117.

Aldridge, R. J., Theron. J. N., Gabbott, S. E. 1994. The Soom Shale: a unique Ordovician fossil horizon in South Africa. *Geology Today* **10**, 218–221.

Aldridge, R. J., Hou, X-G., Siveter, D. J., Siveter, D. J., Gabbott, S. E. 2007. The systematics and phylogenetic relationships of vetulicolians. *Palaeontology* **50**, 131–168.

Allison, P. A. 1990. Pyrite. 253–255. *In* Briggs, D. E. G., Crowther, P. R. (eds.). *Palaeobiology: a Synthesis.* Blackwell Scientific Publications, Oxford, xiii + 583 pp.

Ander, K. 1941. Die Insektenfauna des Baltischen Bernstein nebst damit verknüpften zoogeographischen Problemen. *Lunds Universitets Årsskrift* N. F. **38**, 3–82.

Anderson, L. I., Dunlop, J. A., Horrocks, C. A., Winkelmann, H. M., Eagar, R. M. C. 1997. Exceptionally preserved fossils from Bickershaw, Lancashire, UK (Upper Carboniferous, Westphalian A (Langsettian)). *Geological Journal* **32**, 197–210.

Anderson, L. I., Trewin, N. H. 2003. An Early Devonian arthropod fauna from the Windyfield cherts, Aberdeenshire, Scotland. *Palaeontology* **46**, 467–509.

Anderson, L. I., Crighton, W. R. B., Hass, H. 2004. A new univalve crustacean from the Early Devonian Rhynie chert hot-spring complex. *Transactions of the Royal Society of Edinburgh: Earth Sciences* **94**, 355–369.

Anderson, O. J., Lucas, S. G. 1998. Redefinition of Morrison Formation (Upper Jurassic) and related San Rafael Group strata, southwestern U. S. *Modern Geology* **22**, 39–69.

Antcliffe, J. B., Brasier, M. D. 2007. *Charnia* and sea pens are poles apart. *Journal of the Geological Society of London* **164**, 49–51.

Arai, M. 2000. Chapadas: relict of mid-Cretaceous interior seas in Brazil. *Revista Brasileira de Geociências* **30**, 436–438.

Ash, S. R., Tidwell, W. D. 1998. Plant megafossils from the Brushy Basin Member of the Morrison Formation near Montezuma Creek Trading Post, southeastern Utah. *Modern Geology* **22**, 321–339.

Bachofen-Echt, A. 1949. *Der Bernstein und Seine Einschlüsse.* Springer-Verlag, Wien, 204 pp.

Baird, G. C., Sroka, S. D., Shabica, C. W., Kuecher, G. J. 1986. Taphonomy of Middle Pennsylvanian Mazon Creek area fossil localities, northeast Illinois: significance of exceptional fossil preservation in syngenetic concretions. *Palaios* **1**, 271–285.

Bartels, C., Briggs, D. E. G., Brassel, G. 1998. *The Fossils of the Hunsrück Slate.* Cambridge University Press, Cambridge, xiv + 309 pp.

Bartels, C., Wuttke, M., Briggs, D. E. G. (eds.). 2002. The *Nahecaris* Project: Releasing the marine life of the Devonian from the Hunsrück Slate of Bundenbach.

Metalla (Bochum) **9.2**, 55–138.

Barthel, K. W. 1964. Zur Entstehung der Solnhofener Plattenkalke (unteres Untertithon). *Mitteilungen der Bayerische Staatssammlung für Paläontologie und Historische Geologie* **4**, 37–69.

Barthel, K. W. 1970. On the deposition of the Solnhofen lithographic limestone (Lower Tithonian, Bavaria, Germany). *Neues Jahrbuch für Geologie und Paläontologie, Abhandlungen* **135**, 1–18.

Barthel, K. W. 1978. *Solnhofen: Ein Blick in die Erdgeschichte.* Ott Verlag, Thun, 393 pp.

Barthel, K. W., Swinburne, N. H. M., Conway Morris, S. 1990. *Solnhofen; a Study in Mesozoic Palaeontology.* Cambridge University Press, Cambridge, x + 236 pp.

Bassett, M. G., Popov, L. E., Aldridge, R. J., Gabbott, S. E., Theron, J. N. 2009. Brachiopoda from the Soom Shale Lagerstätte (Upper Ordovician, South Africa). *Journal of Paleontology* **83**, 614–623.

Bataller, J. R., Masachs, V., De Galvez-Cañero, A. 1953. *Mapa Geológico de España. Hoja 290 ISONA, 1:50,000.* Instituto Geológico y Minero de España y Diputácion de Provincia Lérida., Madrid, 113 pp.

Bauer, A. M., Böhme, W., Weitschat, W. 2005. An Early Eocene gecko from Baltic amber and its implications for the evolution of gecko adhesion. *Journal of Zoology* **265**(4), 237–332.

Bender, P. A., Hancox, P. J. 2004. Newly discovered fish faunas from the Early Triassic, Karoo Basin, South Africa, and their correlative implications. *Gondwana Research* **7**, 185–192.

Benson, R. B. J., Ketchum, H. F., Noè, L. F., Gómez-Pérez, M. 2011. New information on *Hauffiosaurus* (Reptilia, Plesiosauria) based on a new species from the Alum Shale Member (Lower Toarcian: Lower Jurassic) of Yorkshire, UK. *Palaeontology* **54**, 547–571.

Benton, M. J. 2000. *Vertebrate Palaeontology.* Blackwell Science, Oxford, xii + 452 pp.

Benton, M. J., Taylor, M. A. 1984. Marine reptiles from the Upper Lias (Lower Toarcian, Lower Jurassic) of the Yorkshire coast. *Proceedings of the Yorkshire Geological Society* **44**, 399–429.

Bergström, J., Stürmer, W., Winter, G. 1980. *Palaeoisopus, Palaeopantopus* and *Palaeothea,* pycnogonid arthropods from the Lower Devonian Hunsrück Slate, West Germany. *Paläontologische Zeitschrift* **54**, 7–54.

Béthoux, O. 2009. The earliest beetle identified. *Journal of Paleontology* **83**, 931–937.

Beurlen, K. 1962. A geologia da Chapada do Araripe. *Anais da Academia Brasileira de Ciências* **34**, 365–370.

Bilbey, S. A. 1998. Cleveland-Lloyd dinosaur quarry – age, stratigraphy and depositional environments. *Modern Geology* **22**, 87–120.

Bill, P. C. 1914. Über Crustaceen aus dem Voltziensandstein des Elsasses. *Mitteilungen der Geologisches Landesanstalt von Elsass-Lothringen* **8**, 289–338.

Birket-Smith, S. J. R. 1981a. Is *Praecambridium* a juvenile *Spriggina? Zoologische Jahrbuch, Anatomie* **106**, 233–235.

Birket-Smith, S. J. R. 1981b. A reconstruction of the Pre-Cambrian *Spriggina. Zoologische Jahrbuch, Anatomie* **105**, 237–258.

von Bitter, P. H., Purnell, M. A., Tetreault, D. K., Stott, C. A. 2007. Eramosa Lagerstätte – exceptionally preserved soft-bodied biotas with shallow-marine shelly and bioturbating organisms (Silurian, Ontario, Canada).

Geology **35**, 879–882.

Blanc-Louvel, C. 1984. Le genre 'Ranunculus L.' dans le Berriasien (Crétacé inf.) de la province de Lérida (Espagne). *Ilerda* **45**, 83–92.

Blanc-Louvel, C. 1991. Etude complémentaire de *Montsechia vidali* (Zeiller) Teixeira 1954: Nouvelle attribution systématique. *Annales de Paléontologie* **77**, 129–141.

Botha, J., Abdala, F., Smith, R. M. H. 2007. The oldest cynodont: new clues on the origin and early diversification of the Cynodontia. *Zoological Journal of the Linnean Society* **149**, 477–492.

Bottjer, D. J., Elter, W., Hagadorn, J. W., Tang, C. M. (eds.). 2002. *Exceptional Fossil Preservation*. Columbia University Press, New York, xiv + 403 pp.

Bowen, D. Q. 1999. A revised correlation of Quaternary deposits in the British Isles. *Geological Society Special Report* **23**, 1–174.

Braddy, S. J., Aldridge, R. J., Theron, J. N. 1995. A new eurypterid from the late Ordovician Table Mountain Group, South Africa. *Palaeontology* **38**, 563–581.

Braddy, S. J., Aldridge, R. J., Gabbott, S. E., Theron, J. N. 1999. Lamellate book-gills in a late Ordovician eurypterid from the Soom Shale, South Africa: support for a eurypterid–scorpion clade. *Lethaia* **32**, 72–74.

Brattstrom, B. H. 1953. The amphibians and reptiles from Rancho La Brea. *Transactions of the San Diego Society of Natural History* **11**, 365–392.

Breithaupt, B. H. 1998. Railroads, blizzards, and dinosaurs: a history of collecting in the Morrison Formation of Wyoming during the nineteenth century. *Modern Geology* **23**, 441–463.

Brenner, K., Seilacher, A. 1979. New aspects about the origin of the Toarcian Posidonia Shales. *Neues Jahrbuch für Geologie und Paläontologie, Abhandlungen* **157**, 11–18.

Brett, C. E., Seilacher, A. 1991. Fossil Lagerstätten: a taphonomic consequence of event sedimentation. 283–297. *In* Einsele, G., Ricken, W., Seilacher, A. (eds.). *Cycles and Events in Stratigraphy*. Springer-Verlag, Berlin, xix + 955 pp.

Briggs, D. E. G., Clarkson, E. N. K., Aldridge, R. J. 1983. The conodont animal. *Lethaia* **16**, 1–14.

Briggs, D. E. G., Collins, D. 1988. A Middle Cambrian chelicerate from Mount Stephen, British Columbia. *Palaeontology* **31**, 779–798.

Briggs, D. E. G. and Gall, J-C. 1990. The continuum in soft-bodied biotas from transitional environments: a quantitative comparison of Triassic and Carboniferous Konservat-Lagerstätten. *Paleobiology* **16**, 204–218.

Briggs, D. E. G., Erwin, D. H., Collier, F. J. 1994. *The Fossils of the Burgess Shale*. Smithsonian Institution Press, Washington DC, xvii + 238 pp.

Briggs, D. E. G., Raiswell, R., Bottrell, S. H., Hatfield, D., Bartels, C. 1996. Controls on the pyritization of exceptionally preserved fossils: an analysis of the Lower Devonian Hunsrück Slate of Germany. *American Journal of Science* **296**, 633–663.

Briggs, D. E. G., Siveter, D. J., Siveter, D. J. 1996. Soft-bodied fossils from a Silurian volcaniclastic deposit. *Nature* **382**, 248–250.

Briggs, D. E. G., Wilby, P. R., Bernardino, P., Perez-Moreno, B. P., Sanz, J. L., Fregenal-Martínez, M. A. 1997. The mineralisation of dinosaur soft tissue in the Lower Cretaceous of Las Hoyas, Spain. *Journal of the Geological Society* **154**, 587–588.

Briggs, D. E. G., Bartels, C. 2001. New arthropods from the Lower Devonian Hunsrück Slate (Lower Emsian, Rhenish Massif, western Germany). *Palaeontology* **44**, 275–303.

Briggs, D. E. G., Sutton, M. D., Siveter, D. J., Siveter, D. J. 2003. A new phyllocarid (Crustacea: Malacostraca) from the Silurian Fossil-Lagerstätte of Herefordshire, UK. *Proceedings of the Royal Society of London*, Series B **271**, 131–138.

Briggs, D. E. G., Sutton, M. D., Siveter, D. J., Siveter, D. J. 2005. Metamorphosis in a Silurian barnacle. *Proceedings of the Royal Society of London*, Series B **272**, 2365–2369.

Briggs, D. E. G., Siveter, D. J., Siveter, D. J., Sutton, M. D. 2008. Virtual fossils from a 425 million-year-old volcanic ash. *American Scientist* **96**, 474–481.

Briggs, D. E. G., Bartels, C. 2010. Annelids from the Lower Devonian Hunsrück Slate (Lower Emsian, Rhenish Massif, Germany). *Palaeontology* **53**, 215–232.

Brito, P. M., Yabumoto, Y., 2011. An updated review of the fish faunas from the Crato and Santana formations in Brazil, a close relationship to the Tethys fauna. *Bulletin of the Kitakyushu Museum of Natural History and Human History, Series A* **9**, 107–136.

Bromage, T. G., Shermis, S. 1981. The La Brea Woman (HC 1323): descriptive analysis. *Society of California Archaeologists Occasional Papers* **3**, 59–75.

Buatois, L. A., Mangano, M. G., Fregenal-Martínez, M. A., de Gibert, J. M. 2000. Short-term colonization trace-fossil assemblages in a carbonate lacustrine Konservat-Lagerstatte (Las Hoyas fossil site, Lower Cretaceous, Cuenca, central Spain). *Facies* **43**, 145–156.

Buffetaut, E. 1988. The ziphodont mesosuchian crocodile from Messel: a reassessment. *Courier Forschungsinstitut Senckenberg* **107**, 211–221.

Buscalioni, A. D., Fregenal-Martínez, M. A., Bravo, A., Poyato-Ariza, F. J., Sanchíz, B., Baéz, A. M., Cambra-Moo, O., Martín-Closas, C., Evans, S. E., Marugán-Lobón, J. 2008. The vertebrate assemblage of Buenache de la Sierra (Upper Barremian of Serrania de Cuenca, Spain) with insights into its taphonomy and palaeoecology. *Cretaceous Research* **29**, 687–710.

Butterfield, N. J. 1995. Secular distribution of Burgess Shale-type preservation. *Lethaia* **28**, 1–13.

Cadbury, D. 2000. *The Dinosaur Hunters*. Fourth Estate, London, x + 374 pp.

Cairncross, B., Anderson, J. M., Anderson, H. M. 1995. Palaeoecology of the Triassic Molteno Formation, Karoo Basin, South Africa – sedimentological and paleoecological evidence. *South African Journal of Geology* **98**, 452–478.

Calver, M. A. 1968. Distribution of Westphalian marine faunas in northern England and adjoining areas. *Proceedings of the Yorkshire Geological Society* **37**, 1–72.

Cambra-Moo, O., Chamero, B., Marugán-Lobón, J., Martínez-Delclòs, X., Poyato-Ariza, F. J., Buscalioni, A. D. 2006. Estimating the ontogenetic status of an enantiornithine bird from the lower Barremian of El Montsec, Central Pyrenees, Spain. *Estudios Geológicos* **62**, 241–248.

Campos, D., Kellner, A. W. A. 1997. Short note on the first occurrence of Tapejaridae in the Crato Member (Aptian), Santana Formation, Araripe Basin, Northeast Brazil. *Anais da Academia Brasileira de Ciências* **69**, 83–87.

Caron, J-B., Scheltema, A., Schander, C., Rudkin, D. 2006. A soft-bodied mollusc with radula from the Middle Cambrian Burgess Shale. *Nature* **442**, 159–163.

Caron, J-B., Jackson, D. A. 2008. Paleoecology of the Greater Phyllopod Bed community, Burgess Shale. *Palaeogeography, Palaeoclimatology, Palaeoecology* **258**, 222–256.

Caspary, R., Klebs, R. 1907. Die Flora des Bernsteins u. andere fossiler Harze des ostpreussichen Tertiärs. *Abhandlungen der Königlich Preussischen Geologischen Landesanstalt. Berlin* N.F. **4**, 1–182.

Catuneanu, O., Wopfner, H., Eriksson, P. G., Cairncross, B., Rubidge, B. S., Smith, R. M. H., Hancox, P. J. 2005. The Karoo basins of south-central Africa. *Journal of African Earth Sciences* **43**, 211–253.

Chang, M. 2003. *The Jehol Biota: the Emergence of Feathered Dinosaurs, Beaked Birds and Flowering Plants*. Shanghai

Scientific and Technical Publishers, 208 pp.

Channing, A., Edwards, D. 2004. Experimental taphonomy: silicification of plants in Yellowstone hot-spring environments. *Transactions of the Royal Society of Edinburgh: Earth Sciences* **94**, 503–521.

Channing, A., Edwards, D. 2009. Yellowstone hot spring environments and the palaeoecophysiology of Rhynie chert plants: towards a synthesis. *Plant Ecology and Diversity* **2**, 111–143.

Chen, J-Y., Cheng, Y-N., Iten, H. V. (eds). 1997. The Cambrian Explosion and the fossil record. *Bulletin of the National Museum of Natural Science* **10**, 1–318.

Chen, J-Y., Zhou, G-Q. 1997. Biology of the Chengjiang fauna. *Bulletin of the National Museum of Natural Science* **10**, 11–105.

Chen, P-J., Dong, Z-M., Zhen, S-N. 1998. An exceptionally well-preserved theropod dinosaur from the Yixian Formation of China. *Nature* **391**, 147–152.

Chiappe, L. M., Shuan, J., Qiang, J., Norell, M. A. 1999. Anatomy and systematics of the Confuciusornithidae (Theropoda: Aves) from the late Mesozoic of northeastern China. *Bulletin of the American Museum of Natural History* **242**, 1–89.

Claridge, M. F., Lyon, A. G. 1961. Book-lungs in the Devonian Palaeocharinidae (Arachnida). *Nature* **191**, 1190–1191.

Cocks, L. R. M., Brunton, C. H. C., Rowell, A. J., Rust, I. C. 1970. The first Lower Palaeozoic fauna proved from South Africa. *Quarterly Journal of the Geological Society of London* **125**, 583–603.

Cocks, L. R. M., Fortey, R. A. 1986. New evidence on the South African Lower Palaeozoic: age and fossils reviewed. *Geological Magazine* **123**, 437–444.

Conway Morris, S. 1977a. Fossil priapulid worms. *Special Papers in Palaeontology* **20**, 1–95.

Conway Morris, S. 1977b. A new metazoan from the Cambrian Burgess Shale of British Columbia. *Palaeontology* **20**, 623–640.

Conway Morris, S. 1979. Middle Cambrian polychaetes from the Burgess Shale of British Columbia. *Philosophical Transactions of the Royal Society of London*, Series B **285**, 227–274.

Conway Morris, S. 1985. The Middle Cambrian metazoan *Wiwaxia corrugata* (Matthew) from the Burgess Shale and Ogygopsis Shale, British Columbia. *Philosophical Transactions of the Royal Society of London*, Series B **307**, 507–582.

Conway Morris, S. 1986. The community structure of the Middle Cambrian Phyllopod Bed (Burgess Shale). *Palaeontology* **29**, 423–467.

Conway Morris, S. 1990. Late Precambrian and Cambrian soft-bodied faunas. *Annual Reviews of Earth and Planetary Science* **18**, 101–122.

Conway Morris, S. 1993. Ediacaran-like fossils in Cambrian Burgess Shale-type faunas of North America. *Palaeontology* **36**, 593–635.

Conway Morris, S. 1998. *The Crucible of Creation*. Oxford University Press, Oxford, xxiii + 242 pp.

Conway Morris, S. 2006. Darwin's dilemma: the realities of the Cambrian 'explosion'. *Philosophical Transactions of the Royal Society of London*, Series B **361**, 1069–1083.

Conway Morris, S., Peel, J. S. 1990. Articulated halkieriids from the Lower Cambrian of north Grenland. *Nature* **345**, 802–805.

Conwentz, H. 1886. *Die Flora des Bernsteins. 2. Die Angiospermen des Bernsteins*. Danzig, 140 pp.

Conwentz, H. 1890. *Monographie der Baltischen Bernsteinbäume*. Danzig, 203 pp.

Cope, E. D. 1883. A letter to the secretary. *American Philosophical Society Proceedings* **21**, 216–227.

Czeczott, H. 1961. The flora of the Baltic amber and its age. *Prace Muzeum Ziemi. Warszawa* **4**, 119–145.

Daley, A. C., Budd, G. E., Caron, J-B., Edgecombe, G. D.,

Collins, G. 2009. The Burgess Shale anomalocaridid *Hurdia* and its significance for early euarthropod evolution. *Science* **323**, 1597–1600.

Damiani, R. J., Modesto, S. P., Yates, A. M., Neveling, J. 2003. Earliest evidence of cynodont burrowing. *Proceedings of the Royal Society of London*, Series B **270**, 1747–1751.

Davis, S. P., Martill, D. M. 1999. The gonorynchiform fish *Dastilbe* from the Lower Cretaceous of Brazil. *Palaeontology* **42**, 715–740.

Delabroye, A., Vecoli, M. 2010. The end-Ordovician glaciation and the Hirnantian Stage: A global review and questions about Late Ordovician event stratigraphy. *Earth-Science Reviews* **98**, 269–282.

Demko, T. M., Parrish, J. T. 1998. Paleoclimatic setting of the Upper Jurassic Morrison Formation. *Modern Geology* **22**, 283–296.

Denson, N. M., Bachman, G. O., Zeller, H. D. 1959. Uranium-bearing lignite in northwestern South Dakota and adjacent states. *United States Geological Survey Bulletin* **1055-B**, 11–57.

Deprat, J., Mansuy, H. 1912. Etude géologique du Yun-Nan oriental. *Mémoires du Service Géologique de l'Indochine* **1**, 1–370.

Dewell, R. A., Dewell, W. C., McKinney, F. K. 2001. Diversification of the Metazoa: ediacarans, colonies, and the origin of eumetazoan complexity by nested modularity. *Historical Biology* **15**, 193–218.

Dittmar, U. 1996. Profilbilanzierung und verformungsanalyse im südwestlichen Rheinischen Schiefergebirge. Zur konfiguration, deformation und entwicklungsgeschichte eines passiven variskischen kontinenentalrandes. *Beringeria* **17**, 346 pp.

Doberenz, A. R., Wyckoff, R. W. G. 1967. Fine structure in fossil collagen. *Proceedings of the National Academy of Sciences of the United States of America* **57**, 539–541.

Dodson, P., Bakker, R. T., Behrensmeyer, A. K., McIntosh, J. S. 1980. Taphonomy and paleoecology of the dinosaur beds of the Jurassic Morrison Formation. *Paleobiology* **6**, 208–232.

Dubinin, V. B. 1962. Class Acaromorpha: mites or gnatho-somic chelicerate arthropods? 447–473. *In* Rodendorf, B. B. (ed.). *Fundamentals of Palaeontology Volume 9*. Academy of Sciences of the USSR, Moscow, xxxi + 894 pp.

Dunlop, J. A., Anderson, L. I., Kerp, H., Hass, H. 2004. A harvestman (Arachnida: Opiliones) from the Early Devonian Rhynie cherts, Aberdeenshire, Scotland. *Transactions of the Royal Society of Edinburgh: Earth Sciences* **94**, 341–354.

Dunlop, J. A., Fayers, S. R., Hass, H., Kerp, H. 2006. A new arthropod from the early Devonian Rhynie chert, Aberdeenshire (Scotland), with a remarkable filtering device in the mouthparts. *Paläontologische Zeitschrift* **80**, 296–306.

Dzik, J. 2008. Gill structure and relationships of the Triassic cycloid crustaceans. *Journal of Morphology* **269**, 1501–1519.

Edwards, D., Selden, P. A. 1993. The development of early terrestrial ecosystems. *Botanical Journal of Scotland* **46**, 337–366.

Edwards, D., Selden, P. A., Richardson, J. B., Axe, L. 1995. Coprolites as evidence for plant-animal interaction in Siluro-Devonian terrestrial ecosystems. *Nature* **377**, 329–331.

Engel, M. S., Grimaldi, D. A. 2004. New light shed on the oldest insect. *Nature* **427**, 627–630.

Engel, M. S., Grimaldi, D. A., Krishna, K. 2009. Termites (Isoptera): their phylogeny, classification, and rise to eco-logical dominance. *American Museum Novitates* **3650**, 1–27.

Engelmann, G. F., Callison, G. 1998. Mammalian faunas of the Morrison Formation. *Modern Geology* **23**, 343–379.

Erwin, D. H. 2001. Metazoan origins and early evolution. 25–31. *In* Briggs, D. E. G., Crowther, P. R. (eds.). *Palaeobiology II*. Blackwell Scientific Publications, Oxford, xv + 583 pp.

Evanoff, E. 2006. Stratigraphy and climatic significance of the White River Group, Badlands National Park, South Dakota. *Geological Society of America Annual Meeting.* **Paper no. 77-3**.

Evanoff, E., Carpenter, K. 1998. History, sedimentology, and taphonomy of Felch Quarry 1 and associated sandbodies, Morrison Formation, Garden Park, Colorado. *Modern Geology* **22**, 145–169.

Evans, S. E., Milner, A. R. 1996. A metamorphosed salamander from the early Cretaceous of Las Hoyas, Spain. *Philosophical Transactions of The Royal Society of London*, Series B **351**, 627–646.

Fedonkin, M. A. 1990. Precambrian metazoans. 17–24. *In* Briggs, D. E. G., Crowther, P. R. (eds.). *Palaeobiology: a Synthesis*. Blackwell Scientific Publications, Oxford, xiii + 583 pp.

Fedonkin, M. A., Gehling, J. G., Grey, K., Narbonne, G. M., Vickers-Rich, P. 2007. *The Rise of Animals – Evolution and Diversification of the Kingdom Animalia*. The Johns Hopkins University Press, xvi + 327 pp.

Feduccia, A., Tordoff, H. B. 1979. Feathers of *Archaeopteryx*: asymmetric vanes indicate aerodynamic function. *Science* **203**, 1021–1022.

Fisher, D. C. 1979. Evidence of subaerial activity of *Euproops danae* (Merostomata, Xiphosurida). 379–477. *In* Nitecki, M. H. (ed.). *Mazon Creek Fossils*. Academic Press, New York, 581 pp.

Fletcher, T. P., Collins, D. H. 1998. The Middle Cambrian Burgess Shale and its relationship to the Stephen Formation in the southern Canadian Rocky Mountains. *Canadian Journal of Earth Science* **35**, 413–436.

Ford, T. D. 1958. Pre-Cambrian fossils from Charnwood Forest. *Proceedings of the Yorkshire Geological Society* **31**, 211–217.

Fortey, R. 1997. *Life: an Unauthorised Biography*. Flamingo, London, xiv + 399 pp.

Fortey, R. A., Theron, J. N. 1995. A new Ordovician arthropod, *Soomaspis*, and the agnostid problem. *Palaeontology* **37**, 841–861.

Franzen, J. L. 1985. Exceptional preservation of Eocene vertebrates in the lake deposits of Grube Messel (West Germany). *Philosophical Transactions of the Royal Society of London*, Series B **311**, 181–186.

Franzen, J. L., Gingerich, P. D., Habersetzer, J., Hurum, J. H., von Koenigswald, W., Smith, B. H. 2009. Complete primate skeleton from the middle Eocene of Messel in Germany: morphology and paleobiology. *PLoS ONE* **4**, e5273 (27 pp.).

Fraser, N. C., Grimaldi, D. A., Olsen, P. E. and Axsmith, B. 1996. A Triassic Lagerstätte from eastern North America. *Nature* **380**, 615–619.

Fregenal-Martínez, M. A., Meléndez, N. 2000. The lacustrine fossiliferous deposits of the las Hoyas subbasin (Lower Cretaceous, Serranía de Cuenca, Iberian Ranges, Spain). *In* Gierlowski-Kordesch, E. H., Kelts, K. R. (eds.). *Lake basins through space and time. AAPG Studies in Geology* **46**, 303–314.

Fregenal-Martínez, M. A., Buscalioni, A. D. 2009. Las Hoyas Konservat-Lagerstätte: a field trip to a Barremian subtropical continental (wetland) ecosystem. 131–152. *In* Alcalà, L., Royo-Torres, R. (coord.). 2009. Mesozoic terrestrial ecosystems in eastern Spain. *¡Fundamental!* **14**, 1–153.

Frey, E., Martill, D. M. 1994. A new pterosaur from the Crato Formation (Lower Cretaceous, Aptian) of Brazil. *Neues Jahrbuch für Geologie und Paläontologie, Abhandlungen* **1994**, 379–412.

Frey, E., Martill, D. M. 1995. A possible oviraptosaurid theropod from the Santana Formation (Lower Cretaceous, Albian) of Brazil. *Neues Jahrbuch für Geologie und Paläontologie, Monatshefte* **1995**, 397–412.

Frey, E., Martill, D. M., Buchy, M-C. 2003. A new crested ornithocheirid from the Lower Cretaceous of northeast Brazil and the unusual death of an unusual pterosaur. 55–64. *In* Buffetaut, E., Mazin, J-M. (eds.). *Evolution and Palaeobiology of Pterosaurs*. Geological Society of London Special Publication 217.

Frickhinger, K. A. 1994. *The Fossils of Solnhofen*. Goldschneck-Verlag, Weidert, 336 pp.

Gabbott, S. E. 1998. Taphonomy of the Ordovician Soom Shale Lagerstätte: an example of soft tissue preservation in clay minerals. *Palaeontology* **41**, 631–667.

Gabbott, S. E. 1999. Orthoconic cephalopods and associated fauna from the late Ordovician Soom Shale Lagerstätte, South Africa. *Palaeontology* **42**, 123–148.

Gabbott, S. E., Aldridge, R. J., Theron, J. N. 1995. A giant conodont with preserved muscle tissue from the Upper Ordovician of South Africa. *Nature* **374**, 800–803.

Gabbott, S. E., Aldridge, R. J., Theron, J. N. 1998. Chitinozoan chains and cocoons from the Upper Ordovician Soom Shale Lagerstätte, South Africa: implications for affinity. *Journal of the Geological Society of London* **155**, 447–452.

Gabbott, S. E., Norry, M. J., Aldridge, R. J., Theron, J. N. 2001. Preservation of fossils in clay minerals; a unique example from the Upper Ordovician Soom Shale, South Africa. *Proceedings of the Yorkshire Geological Society* **53**, 237–244.

Gabbott, S. E., Hou, X-G., Norry, M. J., Siveter, D. J. 2004. Preservation of early Cambrian animals of the Chengjiang biota. *Geology* **32**, 901–904.

Gaines, R. R., Briggs, D. E. G., Zhao, Y-L. 2008. Cambrian Burgess Shale-type deposits share a common mode of fossilization. *Geology* **36**, 755–758.

Gall, J-C. 1971. Faunes et paysages du Grès à Voltzia du nord des Vosges. Essai paléoécologique sur le Buntsandstein supérieur. *Mémoires du Service de la Carte Géologique d'Alsace et de Lorraine* **34**, 1–318.

Gall, J-C. 1972. Fossil-Lagerstätten aus dem Buntsandstein der Vogesen (Frankreich) und ihre ökologische Deutung. *Neues Jahrbuch für Geologie und Paläontologie, Monatshefte* **1972**, 285–293.

Gall, J-C. 1983. The Grès à Voltzia delta. 134–148. *In* Gall, J.-C. (ed.). *Ancient Sedimentary Environments and the Habitats of Living Organisms*. Springer-Verlag, Berlin, xxii + 219 pp.

Gall, J-C. 1985. Fluvial depositional environment evolving into deltaic setting with marine influences in the Buntsandstein of northern Vosges. 449–477. *In* Mader, D. (ed.). *Aspects of Fluvial Sedimentation in the Lower Triassic Buntsandstein of Europe*. Lecture Notes in Earth Sciences 4. Springer-Verlag, Berlin, viii + 626 pp.

Gall, J-C. 1990. Les voiles microbiens. Leur contribution à la fossilisation des organismes au corps mou. *Lethaia* **23**, 21–28.

Gall, J-C., Grauvogel, L. 1966. Ponts d'invertébrés du Buntsandstein supérieur. *Annales de Paléontologie (Invertébrés)* **52**, 155–161.

Gall, J-C., Grauvogel, L. 1967. Faune du Buntsandstein. III. Quelques annélides du Grès à Voltzia des Vosges. *Annales de Paléontologie (Invertébrés)* **53**, 105–110.

Gall, J-C., Grauvogel, L. 1971. Faune du Buntsandstein. IV. Palaega pumila sp. nov., un isopode (Crustacé Eumalacostracé) du Buntsandstein des Vosges (France). *Annales de Paléontologie (Invertébrés)* **57**, 77–89.

Gall, J-C., Grauvogel-Stamm, L. 1984. Genèse des gisements fossilifères du Grès à Voltzia (Anisien) du nord du Vosges (France). *Géobios, Mémoire Special* **8**, 293–297.

Gall, J-C., Grauvogel-Stamm, L. 2005. The early Middle Triassic 'Grès à Voltzia' Formation of eastern France: a model of environmental refugium. *Comptes Rendus Palevol* **4**, 637–652.

Gardner, G. 1841. Geological notes made during a journey from the coast into the interior of the Province of Ceará, in the north of Brazil, embracing an account of a deposit of fossil fishes. *Edinburgh New Philosophical Journal* **30**, 75–82.

Gardner, G. 1846. *Travels in the Interior of Brazil, Principally Through the Northern Provinces.* Reeve, Benham and Reeve, London, xvi + 562 pp. (Reprinted AMS Press, New York, 1970.)

Gastaldo, R. A., Rolerson, M. W. 2008. *Katbergia* gen. nov., a new trace fossil from the Late Permian and Early Triassic of the Karoo Basin: Implications for paleoenvironmental conditions at the P/Tr extinction event. *Palaeontology* **51**, 215–229.

Gehling, J. G. 1987. Earliest known echinoderm – a new Ediacaran fossil from the Pound Supergroup of South Australia. *Alcheringa* **11**, 337–345.

Gehling, J. G. 1988. A cnidarian of actinian-grade from the Ediacaran Pound Supergroup, South Australia. *Alcheringa* **12**, 299–314.

Gehling, J. G. 1991. The case for the Ediacaran fossil roots to the metazoan tree. *Geological Society of India Memoir* **20**, 181–224.

Gehling, J. G. 1999. Microbial mats in terminal Proterozoic siliciclastics: Ediacaran death masks. *Palaios* **14**, 40–57.

Gibert, J. M. de, Fregenal-Martínez, M. A., Buatois, L. A., Mángano M. G. 2000. Trace fossils and their palaeoecological significance in Lower Cretaceous lacustrine conservation deposits, El Montsec, Spain. *Palaeogeography, Palaeoclimatology, Palaeoecology* **156**, 89–101.

Glaessner, M. F. 1961. Pre-Cambrian animals. *Scientific American* **204**, 72–78.

Glaessner, M. F. 1984. *The Dawn of Animal Life. A Biohistorical Study.* Cambridge University Press, Cambridge, 244 pp.

Glaessner, M. F., Wade, M. 1966. The Late Precambrian fossils from Ediacara, South Australia. *Palaeontology* **9**, 599–628.

Gomez, B., Martín-Closas, C., Barale, G., Solé de Porta, N., Thévenard, F., Guignard, G. 2002. *Frenelopsis* (Coniferales: Cheirolepidiaceae) and related male organ genera from the Lower Cretaceous of Spain. *Palaeontology* **45**, 997–1036.

Gomez, B., Daviero-Gomez, V., Martin-Closas, C., de la Fuente, M. 2006. *Montsechia vidalii*, an early aquatic angiosperm from the Barremian of Spain (Abstract). *7th European Palaeobotany and Palynology Conference*, Prague 49.

Göppert, H. R., Berendt, G. C. 1845. *Der Bernstein und die in ihm Befindlichen Pflanzenreste der Vorwelt.* Vol. 1. Berlin.

Göppert, H. R., Menge, A. 1883. *Die Flora des Bernsteins unde ihre Beziehungen zur Flora der Tertiarformation und der Oegenwart.* Volume 1. Danzig.

Gould, S. J. 1989. *Wonderful Life: the Burgess Shale and the Nature of History.* Norton, New York, 323 pp.

Grabau, A. W. 1923. Cretaceous mollusca from north China. *Bulletin of the Geological Survey of China* **5**, 183–197.

Grabau, A. W. 1928. *Stratigraphy of China, Part 2.* Geological Survey of China, Peking, 774 pp.

Graham, L. E., Cook, M. E., Hanson, D. T., Pigg, K. E., Graham, J. M. 2010. Structural, physiological, and stable carbon isotopic evidence that the enigmatic Paleozoic fossil *Prototaxites* formed from rolled liverwort mats. *American Journal of Botany* **97**, 268–275.

Grauvogel, L., Gall, J-C. 1962. *Progonionemus vogesiacus* nov. gen, nov. sp., une méduse du Grès à Voltzia des Vosges septentrionales. *Bulletin du Service de la Carte Géologique d'Alsace et de Lorraine* **15**, 17–27.

Grauvogel-Stamm, L. 1978. La flore du Grès à Voltzia (Buntsandstein supérieur) des Vosges du Nord (France): morphologie, anatomie, interprétations phylogénique et paléogéographique. *Mémoires des Sciences Géologiques*, n. 50. Institut de Géologie de l'Université Louis Pasteur, Strasbourg.

Grauvogel-Stamm, L., Ash, S. R. 2005. Recovery of the

Triassic land flora from the end-Permian life crisis. *Comptes Rendus Palevol* **4**, 593–608.

Gregory, H. E. 1938. The San Juan Country. *United States Geological Survey Professional Paper* **188**.

Grimaldi, D. A. 1990. Insects from the Santana Formation, Lower Cretaceous of Brazil. *Bulletin of the American Museum of Natural History* **195**, 1–191.

Grimaldi, D. A. 1996. *Amber: Window to the Past.* Harry N. Abrams Inc. and American Museum of Natural History, New York, 216 pp.

Grimaldi, D. A., Shedrinsky, A., Ross, A., Baer, N. S. 1994. Forgeries of fossils in 'amber': history, identification and case studies. *Curator* **37**, 251–274.

Grimaldi, D. A., Engel, M. S. 2005. *Evolution of the Insects.* Cambridge University Press, New York, 755 pp.

Gu, Z-W. 1962. *The Jurassic and Cretaceous of China.* Science Press, Beijing, 84 pp.

Gupta, N. S., Cambra-Moo, O., Briggs, D. E. G., Love, G. D., Fregenal-Martínez, M. A., Summons, R. E. 2008. Molecular taphonomy of macrofossils from the Cretaceous Las Hoyas Formation, Spain. *Cretaceous Research* **29**, 1–8.

Guthrie, R. D. 1990. *Frozen Fauna of the Mammoth Steppe.* University of Chicago Press, Chicago, 338 pp.

Habersetzer, J., Storch, G. 1990. Ecology and echolocation of the Eocene Messel bats. 213–233. *In* Hanak, V., Horacek, T., Gaisler, J. (eds.). *European Bat Research 1987.* Charles University Press, Prague, 719 pp.

Habersetzer, J., Richter, G., Storch, G. 1992. Palaeoecology of the Middle Eocene Messel bats. *Historical Biology* **8**, 235–260.

Habgood, K. S., Hass, H., Kerp, H. 2004. Evidence for an early terrestrial food web: coprolites from the Early Devonian Rhynie chert. *Transactions of the Royal Society of Edinburgh: Earth Sciences* **94**, 371–389.

Hancock, N. J., Hurst, J. M., Fürsich, F. T. 1974. The depths inhabited by Silurian brachiopod communities. *Journal of the Geological Society of London* **130**, 151–156.

Hancox, P. J., Rubidge, B. S. 2001. Breakthroughs in the biodiversity, biogeography, biostratigraphy, and basin analysis of the Beaufort group. *Journal of African Earth Sciences* **33**, 563–577.

Harmon, D. M., Engelmann, G. F., Shuster, R. D. 2003. *Roadside Geology of Nebraska.* Mountain Press, Missoula, Montana, 264 pp.

Harris, J. M., Jefferson, G. T. 1985. Rancho La Brea: treasures of the tar pits. *Natural History Museum of Los Angeles County, Science Series* **31**, 1–87.

Hauff, B. 1921. Untersuchung der Fossilfundstätten von Holzmaden im Posidonienschiefer des oberen Lias Württembergs. *Palaeontographica* **64**, 1–42.

Hauff, B., Hauff, R. B. 1981. *Das Holzmadenbuch.* Repro-Druck, Fellbach, 136 pp.

Heads, S. W., Martill, D. M., Loveridge, R. F. 2005. An exceptionally preserved antlion (Insecta, Neuroptera) with colour pattern preservation from the Cretaceous of Brazil. *Palaeontology* **48**, 1409–1417.

Heads, S. W., Martill, D. M., Loveridge, R. F. 2008. Palaeoentomological paradise: the Cretaceous Crato Formation of Brazil. *Bulletin of the Royal Entomological Society* **32**, 91–98.

Heer, O. 1859. *Flora Tertiaria Helvetiae: die Tertiäre Flora der Schweiz.* Volume 3. Wintherthur, 377 pp.

Hirst, S. 1923. On some arachnid remains from the Old Red Sandstone (Rhynie Chert Bed, Aberdeenshire). *Annals and Magazine of Natural History*, 9th Series **70**, 455–474.

Hirst, S., Maulik, S. 1926. On some arthropod remains from the Rhynie Chert (Old Red Sandstone). *Geological Magazine* **63**, 69–71.

Ho, T. Y. 1965. The amino acid composition of bone and tooth proteins in late Pleistocene mammals. *Proceedings of* the *National Academy of Sciences of the United States of America* **54**, 26–31.

Hou, L-H., Zhou, Z., Martin, L., Feduccia, A. 1995. A beaked bird from the Jurassic of China. *Nature* **377**, 616–618.

Hou, X-G., Bergström, J. 1997. Arthropods of the Lower Cambrian Chengjiang fauna, southwest China. *Fossils and Strata* **45**, 1–116.

Hou, X-G., Aldridge, R. J., Bergström, J., Siveter, D. J., Siveter, D. J., Feng, X-H. 2004. *The Cambrian Fossils of Chengjiang, China: the Flowering of Early Animal Life*. Blackwell, Oxford, xii + 233 pp.

Hu, Y-M., Meng, J., Wang, Y-Q., Li, C-K. 2005. Large Mesozoic mammals fed on young dinosaurs. *Nature* **433**, 149–152.

Hu, Y-M., Meng, J., Chuankui, L., Wang, Y-Q. 2009. New basal eutherian mammal from the Early Cretaceous Jehol biota, Liaoning, China. *Proceedings of the Royal Society, Series B* **277**, 229–236.

Hueber, F. M. 2001. Rotted wood-alga-fungus: The history and life of *Prototaxites* Dawson 1859: *Review of Palaeobotany and Palynology* **116**, 123–158.

Jeram, A. J., Selden, P. A., Edwards, D. 1990. Land animals in the Silurian: arachnids and myriapods from Shropshire, England. *Science* **250**, 658–661.

Ji, Q., Ji, S-A. 1996. On discovery of the earliest bird fossil in China and the origin of birds. *Chinese Geology* **233**, 30–33.

Ji, Q., Ji, S-A. 1997. Protarchaeopterygid bird (*Protarchaeopteryx* gen.nov.) – fossil remains of archaeopterygids from China. *Chinese Geology* **238**, 38–41.

Ji, Q., Currie, P. J., Norell, M. A., Ji, S-A. 1998. Two feathered dinosaurs from northeastern China. *Nature* **393**, 753–761.

Ji, Q., Luo, Z-X., Yuan, C-X., Wible, J. R., Zhang, J-P., Georgi, J. A. 2002. The earliest known eutherian mammal. *Nature* **416**, 816–822.

Johnson, D. L. 1977a. The Californian Ice-Age refugium and the Rancholabrean extinction problem. *Quaternary Research* **8**, 149–153.

Johnson, D. L. 1977b. The late Quaternary climate of coastal California: evidence for an Ice Age refugium. *Quaternary Research* **8**, 154–179.

Johnson, G. A. L., Nudds, J. R. 1975. Carboniferous coral geochronometers. 27–42. *In* Rosenberg, G. D., Runcorn, S. K. (eds.). *Growth Rhythms and the History of the Earth's Rotation*. John Wiley, London, 559 pp.

Jordan, D. S., Branner, J. C. 1908. The Cretaceous fishes of Ceará, Brazil. *Smithsonian Miscellaneous Collections* **25**, 1–29.

Jubb, R. A., Gardiner, B. G. 1975. A preliminary catalogue of identifiable fossil fish material from southern Africa. *Annals of the South African Museum* **67**, 381–440.

Kamenz, C., Dunlop, J. A., Scholtz, G., Kerp, H., Hass, H. 2008. Microanatomy of Early Devonian book lungs. *Biology Letters* **4**, 212–215.

Kauffman, E. G. 1979. Benthic environments and paleo-ecology of the Posidonienschiefer (Toarcian). *Neues Jahrbuch für Geologie und Paläontologie, Abhandlungen* **157**, 18–36.

Keilbach, R. 1982. Bibliographie und liste der Arten tierischer Einschlüsse in fossilen Harzen sowie ihrer Aufbewahrungsorte. *Deutsche Entomologische Zeitschrift* N. F. **29**, 129–286, 301–491.

Keiser, D., Weitschat, W. 2005. First record of ostracods (Crustacea) in Baltic amber. *Hydrobiologia* **538**, 107–114.

Kellner, A. W. A. 1996. Remarks on Brazilian dinosaurs. *Memoirs of the Queensland Museum* **39**, 611–626.

Kellner, A. W. A. 1999. Short note on a new dinosaur (Theropoda, Coelurosauria) from the Santana Formation (Romualdo Member, Albian), Northeastern Brazil. *Boletim do Museu Nacional, Geologia* **49**, 1–8.

Kellner, A. W. A. 2002. A review of avian Mesozoic fossil feathers. 389–404. *In* Chiappe, L. M., Witmer, L. M.

(eds.). *Mesozoic Birds: Above the Heads of the Dinosaurs*. University of California Press, Los Angeles, xii + 520 pp.

Kelman, R., Feist, M., Trewin, N. H., Hass, H. 2004. Charophyte algae from the Rhynie chert. *Transactions of the Royal Society of Edinburgh: Earth Sciences* **94**, 445–455.

Kennedy, G. E. 1989. A note on the ontogenetic age of the Rancho La Brea hominid, Los Angeles, California. *Bulletin of the Southern California Academy of Sciences* **88**, 123–126.

Kerp, H., Trewin, N. H., Hass, H. 2004. New gametophytes from the Early Devonian Rhynie chert. *Transactions of the Royal Society of Edinburgh: Earth Sciences* **94**, 411–428.

Keupp, H. 1977a. Ultrafazies und Genese der Solnhofener Plattenkalke (Oberer Malm, Südliche Frankenalb). *Abhandlung der Naturhistorischen Gesellschaft Nürnberg* **37** 1–128.

Keupp, H. 1977b. Der Solnhofener Plattenkalk – ein Blau-grünalgen-Laminit. *Paläontologische Zeitschrift* **51**, 102–116.

Kevan, P. G., Chaloner, W. G., Savile, D. B. O. 1975. Interrelationships of early terrestrial arthropods and plants. *Palaeontology* **18**, 391–417.

Kidston, R., Lang, W. H. 1917. On Old Red Sandstone plants showing structure, from the Rhynie Chert bed, Aberdeen-shire. Part I. *Rhynia gwynne-vaughani* Kidston and Lang. *Transactions of the Royal Society of Edinburgh* **51**, 761–784.

Kidston, R., Lang, W. H. 1920a. On Old Red Sandstone plants showing structure, from the Rhynie Chert bed, Aberdeenshire. Part II. Additional notes on *Rhynia gwynne-vaughani*, Kidston and Lang; with descriptions of *Rhynia major*, n.sp., and *Hornia lignieri*, n. g., n. sp. *Transactions of the Royal Society of Edinburgh* **52**, 603–627.

Kidston, R., Lang, W. H. 1920b. On Old Red Sandstone plants showing structure, from the Rhynie Chert bed, Aberdeenshire. Part III. *Asteroxlon mackiei*, Kidston and Lang. *Transactions of the Royal Society of Edinburgh* **52**, 643–680.

Kidston, R., Lang, W. H. 1921a. On Old Red Sandstone plants showing structure, from the Rhynie Chert bed, Aberdeenshire. Part IV. Restorations of the vascular cryptogams, and discussion of their bearing on the general morphology of the Pteridophyta and the origin of the organisation of land-plants. *Transactions of the Royal Society of Edinburgh* **52**, 831–854.

Kidston, R., Lang, W. H. 1921b. On Old Red Sandstone plants showing structure, from the Rhynie Chert bed, Aberdeenshire. Part V. The Thallophyta occuring in the peat-bed; the succession of the plants throughout a vertical section of the bed, and the conditions of accumulation and preservation of the deposit. *Transactions of the Royal Society of Edinburgh* **52**, 855–902.

Kirkland, J. I. 1998. Morrison fishes. *Modern Geology* **22**, 503–533.

Klass, K-D., Zompro, O., Kristensen, N. P., Adis, J. 2002. Mantophasmatodea: a new insect order with extant members in the Afrotropics. *Science* **296**, 1456–1459.

Klebs, R. 1886. Gastropoden im Bernstein. *Jahrbuch der Königlich Preussischen Geologischen Landesanstalt und Bergakademie zu Berlin* **188**, 366–394.

Koch C. L., Berendt, G. C. 1854. Die im Bernstein befindlichen Crustaceen, Myriapoden, Arachniden und Apteren der Vorwelt. *In* Berendt, G. C. (Menge, A., ed.). *Die im Bernstein Befindlichen Organischen Reste der Vorwelt*. Berlin, **1** (2), pp. 1–124, pl. I-XVIII.

Koenigswald, R. von. 1930. Die Fauna des Bundenbacher Schiefers in ihren Beziehungen zum Sediment. *Zentralblatt für Mineralogie, Geologie und Paläontologie* B, 241–247.

Kornilovich, N. 1903. Has the structure of striated muscle of insects in amber been preserved? *Protokol Obshchestva Estestvoispytatele pri Imperatorskom Yur'evskom Universitete. Yur'ev.* (Dorpat) **13**, 198–206.

Kovács-Endrödy, E. 1987. The earliest known vascular plant, or a possible ancestor of vascular plants, in the flora of the

Lower Silurian Cedarberg Formation, Table Mountain Group, South Africa. *Annals of the Geological Survey of South Africa* **20**, 893–906.

Krings, M., Kerp, H., Hass, H., Taylor, T. N., Dotzler, N. 2007a. A filamentous cyanobacterium showing structured colonial growth from the Early Devonian Rhynie chert. *Review of Palaeobotany and Palynology* **146**, 265–276.

Krings, M., Taylor, T. N., Hass, H., Kerp, H., Dotzler, N., Hermsen, E. J. 2007b. Fungal endophytes in a 400-million-yr-old land plant: infection pathways, spatial distribution, and host responses. *New Phytologist* **174**, 648–657.

Kuecher, G. J., Woodland, B. G., Broadhurst, F. M. 1990. Evidence of deposition from individual tides and of tidal cycles from the Francis Creek Shale (host rock to the Mazon Creek Biota), Westphalian D (Pennsylvanian), northeastern Illinois. *Sedimentary Geology* **68**, 211–221.

Kühl, G., Bergström, J., Rust, J. 2008. Morphology, paleobiology and phylogenetic position of *Vachonisia rogeri* (Arthropoda) from the Lower Devonian Hunsrück Slate (Germany). *Palaeontographica* A **286**, 123–157.

Kühl, G., Briggs, D. E. G., Rust, J. 2009. A great-appendage arthropod with a radial mouth from the Lower Devonian Hunsrück Slate, Germany. *Science* **323**, 771–773.

Kühl, G., Rust, J. 2010. Re-investigation of *Mimetaster hexagonalis*: a marrellomorph arthropod from the Lower Devonian Hunsrück Slate (Germany). *Paläontologische Zeitschrift* **84**, 397–411.

Kühne, W. G. 1961. Präparation von flachen Wirbeltierfossilien auf künstlicher Matrix. *Paläontologische Zeitschrift* **35**, 251–252.

Kundrat, M., Cruickshank, A. R. I., Manning, T. W., Nudds, J. R. 2008. Embryos of therizinosauroid theropods from the Upper Cretaceous of China: diagnosis and analysis of ossification patterns. *Acta Zoologica* **88**, 231–251.

Kurtén, B. 1986. *How to Deep-Freeze a Mammoth*. Columbia University Press, New York, vii + 121 pp.

La Duke, T. C. 1991a. The fossil snakes of Pit 91, Rancho La Brea, California. *Contributions in Science* **424**, 1–28.

La Duke, T. C. 1991b. First record of salamander remains from Rancho La Brea. *Abstract of the Annual Meeting of the California Academy of Science* **7**.

Lacasa-Ruiz, A. 1985. Nota sobre las plumas fósiles del yacimiento eocretácico de 'La Pedrera – La Cabrua' en la Sierra del Montsec (Prov. Lleida, España). *Ilerda* **46**, 227–238.

Lacasa-Ruiz, A. 1986. Nota preliminar sobre el hallazgo de restos óseos de un ave fósil en el yacimiento neocomiense del Montsec. Prov. de Lérida, España. *Ilerda* **47**, 203–206.

Lacasa-Ruiz, A. 1989. An Early Cretaceous fossil bird from Montsec Mountain (Lleida, Spain). *Terra Nova* **1**, 45–46.

Lacasa-Ruiz, A., Martínez-Delclòs, X. 1986. *Meiatermes, Nuevo Género Fósil de Insecto Isóptero (Hodotermitidae) de las Calizas Necomienses del Montsec (Provincia de Lérida, España)*. Lleida, Institut d'Estudis Ilerdencs, 65 pp.

Larsson, S. G. 1978. *Baltic amber – a Palaeobiological Study*. Entomonograph Volume 1. Scandinavian Science Press Ltd., Klampenborg, Denmark, 192 pp.

Leidy, J. 1869. The extinct mammalian fauna of Dakota and Nebraska. *Journal of the Academy of Natural Science, Philadelphia* **7**, 23–472.

Liu, Y., Liu, Y., Ji, S., Yang, Z. 2006. U–Pb zircon age for the Daohugou Biota at Ningcheng of Inner Mongolia and comments on related issues. *Chinese Science Bulletin* **51**, 2634–2644.

Liu, H. P., McKay, R. M., Young, J. N., Witzke, B. J., McVey, K. J., Liu, X. 2006. A new Lagerstätte from the Middle Ordovician St. Peter Formation in northeast Iowa, USA. *Geology* **34**, 969–972.

Liu, J., Rubidge, B., Li, J. 2009. New basal synapsid supports Laurasian origin for therapsids. *Acta Palaeontologica Polonica* **54**, 393–400.

Long, J. A. 1988. The extraordinary fishes of Gogo. *New Scientist* **120** (1639), 40–44.

Long, J. A., Trinajstic, K. 2010. The Late Devonian Gogo Formation Lagerstätten of Western Australia: Exceptional early vertebrate preservation and diversity. *Annual Review of Earth and Planetary Sciences*, **38**, 255–279.

Lourenço, W. R., Weitschat, W. 2000. New fossil scorpions from the Baltic amber – implications for Cenozoic biodiversity. *Mitteilungen aus dem Geologisch-Paläontologisches Institut der Universität Hamburg* **84**, 247–259.

Lourenço, W. R., Gall, J-C. 2004. Fossil scorpions from the Buntsandstein (Early Triassic) of France. *Comptes Rendus Palevol* **3**, 369–378.

Lutz, H. 1987. Die Insekten-Thanatocoenose aus dem Mittel-Eozän der 'Grube Messel' bei Darmstadt: Erste Ergebnisse. *Courier Forschungsinstitut Senckenberg* **91**, 189–201.

McMenamin, A. S. 1998. *The Garden of Ediacara*. Columbia University Press, New York, xvi + 295 pp.

Maier, W., Richter, G., Storch, G. 1986. *Leptictidium nasutum* – ein archaisches Säugetier aus Messel mit aussergewöhnlichen biologischen Anpassungen. *Natur und Museum* **116**, 1–19.

Maisey, J. G. 1991. *Santana fossils: an Illustrated Atlas*. T. F. H. Publications, New Jersey, 459 pp.

Martill, D. M. 1988. Preservation of fish in the Cretaceous Santana Formation of Brazil. *Palaeontology* **31**, 1–18.

Martill, D. M. 1989. The Medusa effect: instantaneous fossilization. *Geology Today* **5**, 201–205.

Martill, D. M. 1993. *Fossils of the Santana and Crato Formations, Brazil*. (Field Guide to Fossils No.5). The Palaeontological Association, London, 159 pp.

Martill, D. M. 2007. The age of the Cretaceous Santana Formation fossil Konservat Lagerstätte of north-east Brazil: a historical review and an appraisal of the biochronostratigraphic utility of its palaeobiota. *Cretaceous Research* **28**, 895–920.

Martill, D. M., Unwin, D. M. 1989. Exceptionally well preserved pterosaur wing membrane from the Cretaceous of Brazil. *Nature* **340**, 138–140.

Martill, D. M., Filgueira, J. B. M. 1994. A new feather from the Lower Cretaceous of Brazil. *Palaeontology* **37**, 483–487.

Martill, D. M., Frey, E. 1995. Colour patterning preserved in Lower Cretaceous birds and insects: the Crato Formation of N. E. Brazil. *Neues Jahrbuch für Geologie und Paläontologie, Monatshefte* **1995**, 118–128.

Martill, D. M., Cruickshank, A. R. I., Frey, E., Small, P. G., Clarke, M. 1996. A new crested maniraptoran dinosaur from the Santana Formation (Lower Cretaceous) of Brazil. *Journal of the Geological Society of London* **153**, 5–8.

Martill, D. M., Frey, E., Sues, H. D., Cruickshank, A. R. I. 2000. Skeletal remains of a small theropod dinosaur with associated soft structures from the Lower Cretaceous Santana Formation of northeastern Brazil. *Canadian Journal of Earth Sciences* **37**, 891–900.

Martill, D. M., Bechly, G., Loveridge, R. F. (eds.). 2007. *The Crato Fossil Beds of Brazil: Window into an Ancient World*. Cambridge University Press, Cambridge, xvi + 625 pp.

Martill, D. M., Brito, P. M., Washington-Evans, J. 2008. Mass mortality of fishes in the Santana Formation (Lower Cretaceous, ?Albian) of northeast Brazil. *Cretaceous Research* **29**, 649–658.

Martin, L. D. 1985. The relationship of *Archaeopteryx* to other birds. 177–183. *In* Hecht, M. K., Ostrom, J. H., Viohl, G., Wellnhofer, P. (eds.). *The Beginnings of Birds*. Proceedings of the International *Archaeopteryx* Conference, Eichstätt, 1984, 382 pp.

Martínez-Delclòs, X., Martinell, J. 1995. The oldest known record of social insects. *Journal of Paleontology* **69**, 594–599.

Melott, A. L., Lieberman, B. S., Laird, C. M., Martin, L. D.,

Medvedev, M. V., Thomas, B. C., Cannizzo, J. K., Gehrels, N. and Jackman, C. H. 2004. Did a gamma-ray burst initiate the late Ordovician mass extinction? *International Journal of Astrobiology* **3**, 55–61.

Meyer, H. von. 1861. *Archaeopteryx lithographica* (Vogel-Feder) und *Pterodactylus* von Solnhofen. *Neues Jahrbuch für Mineralogie, Geologie und Paläontologie* **1861**, 678–679.

Mierzejewski, P. 1976a. Scanning electron microscope studies on the fossilization of Baltic amber spiders (preliminary note). *Annals of the Medical Section of the Polish Academy of Sciences* **21**, 81–82.

Mierzejewski, P. 1976b. On application of scanning electron microscope to the study of organic inclusions from Baltic amber. *Rocznik Polskiego Towarzystwa Geologicznego (w Krakowie)* **46**, 291–295.

Mierzejewski, P. 1978. Electron microscope study on the milky impurities covering arthropod inclusions in Baltic amber. *Prace Muzeum Ziemi. Warszawa* **28**, 79–84.

Mikulic, D. G., Briggs, D. E. G., Kluessendorf, J. 1985. A Silurian soft-bodied biota. *Science* **228**, 715–717.

Modesto, S. P. 2000. *Eunotosaurus africanus* and the Gondwanan ancestry of anapsid reptiles. *Palaeontologia Africana* **36**, 15–20.

Modesto, S. P., Smith, R. M. H. 2001. A new Late Permian captorhinid reptile: a first record from the South African Karoo. *Journal of Vertebrate Paleontology* **21**, 405–409.

Mohr, B. A. R., Friis, E. M. 2000. Early angiosperms from the Lower Cretaceous Crato Formation (Brazil), a preliminary report. *International Journal of Plant Science* **161**, 155–167.

Mohr, B. A. R., Bernades-de-Oliveira, M. E. C., Loveridge, R. F. 2007. The macrophyte flora of the Crato Formation. 537–565. *In* Martill, D. M., Bechly, G., Loveridge, R. F. (eds.). *The Crato Fossil Beds of Brazil: Window into an Ancient World*. Cambridge University Press, Cambridge, xvi + 625 pp.

Moore, E. C., Mehlin, R. E., Kepferie, R. C. 1959. Uranium-bearing lignite in southwestern North Dakota. *United States Geological Survey Bulletin* **1055-E**, 147–166.

Murphy, E. C., Hoganson, J. W., Forsman, N. F. 1993. The Chadron, Brule and Aikaree formations in North Dakota. *North Dakota Geological Survey Report of Investigation* **96**, 1–144.

Naish, D., Martill, D. M., Frey, E. 2004. Ecology, systematics and biogeographical relationships of dinosaurs, including a new theropod, from the Santana Formation (?Albian, Early Cretaceous) of Brazil. *Historical Biology* **16**, 57–70.

Naish, D., Martill, D. M., Merrick, I. 2007. Birds of the Crato Formation. 525–533. *In* Martill, D. M., Bechly, G., Loveridge, R. F. (eds.). *The Crato Fossil Beds of Brazil: Window into an Ancient World*. Cambridge University Press, Cambridge, xvi + 625 pp.

Nicolas, M., Rubidge, B. S. 2010. Changes in Permo-Triassic terrestrial tetrapod ecological representation in the Beaufort Group (Karoo Supergroup) of South Africa. *Lethaia* **43**, 45–59.

Nitecki, M. H. (ed.). 1979. *Mazon Creek Fossils*. Academic Press, New York, 581 pp.

Novacek, M. 1985. Evidence for echolocation in the oldest known bats. *Nature* **315**, 140–141.

Nudds, J. R., Brito, P. M., Washington Evans, J. 2005. The original syntypes of *Vinctifer comptoni* and *Notelops brama* from the Santana Formation (Cretaceous) of northeast Brazil. *Journal of Vertebrate Paleontology* **25**, 716–719.

Nudds, J. R., Selden, P. A. 2008. *Fossil Ecosystems of North America*. Manson Publishing, London, 288 pp.

O'Harra, C. C. 1910. The badland formations of the Black Hills Region. *South Dakota School of Mines Bulletin* **9**, 1–152.

O'Harra, C. C. 1920. The White River Badlands. *South Dakota School of Mines Bulletin* **13**, 1–181.

Orr, P. J., Briggs, D. E. G., Siveter, D. J., Siveter, D. J. 2000. Three-dimensional preservation of a non-biomineralized arthropod in concretions in Silurian volcaniclastic rocks from Herefordshire, England. *Journal of the Geological Society of London* **157**, 173–186.

Ortega, F., Escaso, F., Sanz, J. L. 2010. A bizarre, humped Carcharodontosauria (Theropoda) from the Lower Cretaceous of Spain. *Nature* **467**, 203–206.

Ostrom, J. H. 1974. *Archaeopteryx* and the origin of flight. *Quarterly Review of Biology* **49**, 27–47.

Ostrom, J. H. 1985. The meaning of *Archaeopteryx*. 161–176. *In* Hecht, M. K., Ostrom, J. H., Viohl, G., Wellnhofer, P. (eds.). *The Beginnings of Birds*. Proceedings of the International *Archaeopteryx* Conference, Eichstätt, 1984, 382 pp.

Padian, K. 1998. Pterosaurians and Avians from the Morrison Formation (Upper Jurassic, western U.S.). *Modern Geology* **23**, 57–68.

Paterson, J. R., Edgecombe, G. D., García-Bellido, D. C., Jago, J. B., Gehling, J. G. 2010. Nektaspid arthropods from the Lower Cambrian Emu Bay Shale Lagerstätte, South Australia, with a reassessment of lamellipedian relationships. *Palaeontology* **53**, Part 2, 377–402.

Penney, D. 2010. *Amber*. Siri Scientific Press, Manchester, 30 pp.

Peters, D. S. 1989. Ein vollständiges Exemplar von Palaeotis weigelti. *Courier Forschungsinstitut Senckenberg* **107**, 223–233.

Petrunkevitch, A. 1942. A study of amber spiders. *Transactions of the Connecticut Academy of Arts and Sciences* **34**, 119–464.

Petrunkevitch, A. 1950. Baltic amber spiders in the collections of the Museum of Comparative Zoology. *Bulletin of the Museum of Comparative Zoology, Harvard University* **103**, 259–337.

Petrunkevitch, A. 1958. Amber spiders in European collections. *Transactions of the Connecticut Academy of Arts and Sciences* **41**, 97–400.

Poinar, G. O. 1992. *Life in Amber*. Stanford University Press, Stanford, California, xiii + 350 pp.

Poinar, G. O., Hess, R. 1982. Ultrastructure of 40-million-year-old insect tissue. *Science* **215**, 1241–1242.

Poinar, G. O., Poinar, R. 1999. *The Amber Forest: a Reconstruction of a Vanished World*. Princeton University Press, Princeton, New Jersey, xviii + 239 pp.

Poinar, H.N., Schwarz, C., Qi, J., Shapiro, B., MacPhee, R.D.E., Buigues, B., Tikhonov, A., Huson, D.H., Tomsho, L.P., Auch, A., *et al.* 2006. Metagenomics to paleogenomics: large-scale sequencing of mammoth DNA. *Science* **311**, 392–394.

Poplin, C., Heyler, D. (eds.). 1994. *Quand le Massif Central était sous l'Équateur. Un Écosystème Carbonifère à Montceau-les-Mines*. Comité des Travaux Historiques et Scientifiques, Paris, 341 pp.

Powers, W. E. 1945. White River Formation of North Dakota. *Geological Society of America Bulletin* **56**, 1192.

Poyato-Ariza, F. J. 1997. A new assemblage of Spanish Early Cretaceous teleostean fishes, formerly considered 'leptolepids': phylogenetic relevance. *Comptes Rendus de l'Academie des Sciences de Paris* **325**, 373–379.

Quirke, T. T. 1918. The geology of the Kildeer Mountains, North Dakota. *Journal of Geology* **26**, 255–271.

Rabadà, D. 1993. Crustáceos decápodos lacustres de las calizas litográficas del Cretácico inferior de España: Las Hoyas (Cuenca) y el Montsec de Rúbies (Lleida). *Cuadernos de Geología Ibérica* **17**, 345–370.

Rasnitsyn, A. P., Martínez-Delclòs, X. 2000. Wasps (Insecta: Vespida = Hymenoptera) from the Early Cretaceous of Spain. *Acta Geologica Hispanica* **35**, 65–95.

Rayner, R. J. 1986. *Promissum pulchrum*: the unfulfilled promise? *South African Journal of Science* **82**, 106–107.

Remy, W., Selden, P. A., Trewin, N. H. 1999. Gli strati di Rhynie. 28–35. *In* Pinna, G. (ed.). *Alle Radici della Storia Naturale d'Europa.* Jaca Book, Milan, 254 pp.

Remy, W., Selden, P. A., Trewin, N. H. 2000. Der Rhynie Chert, Unter-Devon, Schottland. 28–35. *In* Pinna, G., Meischner, D. (eds.). *Europäische Fossillagerstätten.* Springer, Berlin, 264 pp.

Ren, D., Shih, C-K., Gao, Y-Z., Zhao, Y-Y. 2010. *Silent Stories: Insect Fossil Treasures from the Dinosaur Era of Northeastern China.* Science Press, Beijing, 324pp.

Retallack, G. J. 1994. Were the Ediacaran fossils lichens? *Paleobiology* **20**, 523–544.

Retallack, G. J., Smith, R. M. H., Ward, P. D. 2003. Vertebrate extinction across the Permian-Triassic boundary in the Karoo Basin, South Africa. *Geological Society of America Bulletin* **115**, 1133–1152.

Retallack, G. J., Metzger, C. A., Greaver, T., Jahren, A. H., Smith, R. M. H., Sheldon, N. D. 2006. Middle-Late Permian mass extinction on land. *Geological Society of America Bulletin* **118**, 1398–1411.

Reynolds, R. L. 1985. Domestic dog associated with human remains at Rancho La Brea. *Bulletin of the Southern California Academy of Sciences* **84**, 76–85.

Rice, P. C. 1993. *Amber: the Golden Gem of the Ages.* 1993 Revision. The Kosciuszko Foundation Inc., NY, x + 289 pp.

Rice, C. M., Trewin, N. H., Anderson, L. I. 2002. Geological setting of the Early Devonian Rhynie cherts, Aberdeenshire, Scotland: an early terrestrial hot spring system. *Journal of the Geological Society of London* **159**, 203–214.

Richardson, E. S., Johnson, R. G. 1971. The Mazon Creek faunas. *Proceedings of the North American Paleontological Convention,* 5–7 September 1969, Field Museum of Natural History **1**, 1222–1235.

Richmond, D. R., Morris, T. H. 1998. Stratigraphy and cataclysmic deposition of the Dry Mesa Dinosaur Quarry, Mesa County, Colorado. *Modern Geology* **22**, 121–143.

Richter, G., Wedmann, S. 2005. Ecology of the Eocene Lake Messel revealed by analysis of small fish coprolites and sediments from a drilling core. *Palaeogeography, Palaeoclimatology, Palaeoecology* **223**, 147–161.

Riegraf, W. 1977. *Goniomya rhombifera* (Goldfuss) in the Posidonia Shales (Lias epsilon). *Neues Jahrbuch für Geologie und Palaontologie, Monatshefte* **1977**, 446–448.

Roemer, C. F. 1862. Asteriden und Crinoiden von Bundenbach. *Verhandlungen der Naturhistorischen Vereins der Preussischens Rheinland und Westfalens.* Bonn **20**, 109.

Rolfe, W. D. I. 1980. Early invertebrate terrestrial faunas. 117–157. *In* Panchen, A. L. (ed.). *The Terrestrial Environment and the Origin of Land Vertebrates.* Systematics Association Special Volume **15**. Academic Press, London and New York, 633 pp.

Ross, A. 1998. *Amber: the Natural Time Capsule.* The Natural History Museum, London, 73 pp.

Rubidge, S. 1956. The origin of the Rubidge collection of Karoo fossils. *South African Museums Association Bulletin* **6**, 107–113.

Rubidge, B. S. 2005. Re-uniting lost continents – Fossil reptiles from the ancient Karoo and their wanderlust. *South African Journal of Geology* **108**, 135–172.

Runnegar, B. 1992. Evolution of the earliest animals. 65–93. *In* Schopf, J. W. (ed.). *Major Events in the History of Life.* Jones and Bartlett, Boston, MA, xv + 190 pp.

Russell, D., Béland, P., McIntosh, J. S. 1980. Paleoecology of the dinosaurs of Tendaguru (Tanzania). *Mémoires de la Société géologique de France* **139**, 169–175.

Sanz, J. L., Bonaparte, J. F., Lacasa, A. 1988. Unusual Early Cretaceous birds from Spain. *Nature* **331**, 433–435.

Sanz, J. L., Chiappe, L. M., Buscalioni, A. D. 1995. The osteology of *Concornis lacustris* (Aves: Enantiornithes) from the Lower Cretaceous of Spain and a reexamination of its phylogenetic significance. *American Museum Novitates* **3133**, 1–23.

Sanz, J. L., Chiappe, L. M., Perez-Moreno, B. P., Moratalla, J. J., Hernandez-Carrasquilla, F., Buscalioni, A. D., Ortega, F., Poyato-Ariza, F. J., Rasskin-Gutman, D. and Martinez-Delclos, X. 1997. A nestling bird from the Early Cretaceous of Spain: implications for avian skull and neck evolution. *Science* **276**, 1543–1546.

Sanz, J. L., Chiappe, L. M., Fernandez-Jalvo, Y., Ortega, F., Sanchez-Chillon, B., Poyato-Ariza, F. J., Pérez-Moreno, B. P. 2001a. An Early Cretaceous pellet. *Nature* **409**, 998–1000.

Sanz, J. L., Fregenal-Martínez, M. A., Melendez, N., Ortega, F. 2001b. Las Hoyas. 356–359. *In:* Briggs, D. E. G., Crowther, P. R. (eds.). *Paleobiology II.* Blackwell, Oxford, 580 pp.

Sanz, J. L., Pérez-Moreno, B. P., Chiappe, L.M., Buscalioni, A. D. 2002. The birds from the Lower Cretaceous of Las Hoyas (province of Cuenca, Spain). 209–229. *In* Chiappe, L. M., Witmer, L. M. (eds.). *Mesozoic Birds: Above the Heads of the Dinosaurs.* University of California Press, Los Angeles, xii + 520 pp.

Savage, D. E. 1951. Late Cenozoic vertebrates of the San Francisco Bay region. *University of California Publications in Geological Sciences* **28**, 215–314.

Schaal, S., Ziegler, W. (eds.). 1992. *Messel. An Insight into the History of Life and of the Earth.* Clarendon Press, Oxford, 322 pp.

Schawaller, W. 1978. Neue Pseudoskorpione aus dem Baltischen Bernstein der Stuttgarter Bernsteinsammlung (Arachnida: Pseudoscorpionidea). *Stuttgarter Beiträge zur Naturkunde,* Series B **42**, 1–22.

Schlüter, T. 1990. Baltic Amber. 294–297. *In* Briggs, D. E. G., Crowther, P. R. (eds.). *Palaeobiology: a Synthesis.* Blackwell Scientific Publications, Oxford, xiii + 583 pp.

Schopf, K. M., Baumiller, T. K. 1998. A biomechanical approach to Ediacaran hypotheses: how to weed the Garden of Ediacara. *Lethaia* **31**, 89–97.

Schram, F. R. 1979. The Mazon Creek biotas in the context of a Carboniferous faunal continuum. 159–190. *In* Nitecki, M. H. (ed.). *Mazon Creek Fossils.* Academic Press, New York, 581 pp.

Schubert, K. 1961. Neue Untersuchungen uuber Bau und Leiben der Bernsteinkiefern [*Pinus succunifera* (Conw.) emend.]. *Beihefte zum Geologischen Jahrbuch* **45**, 1–149.

Schulz, R., Buness, H., Gabriel, G., Pucher, R., Rolf, C., Wiederhold, H., Wonik, T. 2005. Detailed investigation of preserved maar structures by combined geophysical surveys. *Bulletin of Volcanology* **68**, 95–106.

Scott, A. C., Anderson, J. M., Anderson, H. M. 2004. Evidence of plant–insect interactions in the Upper Triassic Molteno Formation of South Africa. *Journal of the Geological Society of London* **161**, 401–410.

Scourfield, D. J. 1926. On a new type of crustacean from the old Red Sandstone (Rhynie Chert Bed, Aberdeenshire) – *Lepidocaris rhyniensis,* gen. et sp. nov. *Philosophical Transactions of the Royal Society of London,* Series B **214**, 153–187.

Scourfield, D. J. 1940. Two new and nearly complete specimens of young stages of the Devonian fossil crustacean *Lepidocaris rhyniensis. Proceedings of the Linnean Society* **152**, 290–298.

Seilacher, A. 1982. Ammonite shells as habitats in the Posidonia Shales of Holzmaden – floats or benthic islands? *Neues Jahrbuch für Geologie und Paläontologie, Monatshefte* **1982**, 98–114.

Seilacher, A. 1989. Vendozoa: organismic construction in the Proterozoic biosphere. *Lethaia* **22**, 229–239.

Seilacher, A. 1992. Vendobionta and Psammocorallia: lost constructions of Precambrian evolution. *Journal of the Geological Society of London* **149**, 607–613.

Seilacher, A., Reif, W-E., Westphal, F. 1985. Sedimentological, ecological and temporal patterns of fossil Lagerstätten. *Philosophical Transactions of the Royal*

Society of London, Series B **311**, 5–23.

Selden, P. A. 1985. Eurypterid respiration. *Philosophical Transactions of the Royal Society of London*, Series B **309**, 219–226.

Selden, P. A. 2002. First British Mesozoic spider, from Cretaceous amber of the Isle of Wight, southern England. *Palaeontology* **45**, 973–983.

Selden, P. A., Edwards, D. 1989. Colonisation of the land. 122–152. *In* Allen, K. C., Briggs, D. E. G. (eds.). *Evolution and the Fossil Record*. Belhaven Press, London, xiii + 265 pp.

Selden, P. A., Shear, W. A., Bonamo, P. M. 1991. A spider and other arachnids from the Devonian of New York, and reinterpretations of Devonian Araneae. *Palaeontology* **34**, 241–281.

Selden, P. A., Gall, J-C. 1992. A Triassic mygalomorph spider from the northern Vosges, France. *Palaeontology* **35**, 211–235.

Selden, P. A., Penney, D. 2003. Lower Cretaceous spiders (Arthropoda: Arachnida: Araneae) from Spain. *Neues Jahrbuch für Geologie und Paläontologie, Monatshefte* **2003**, 175–192.

Selden, P. A., Anderson, H. M., Anderson, J. M. 2009. A review of the fossil record of spiders (Araneae) with special reference to Africa, and description of a new specimen from the Triassic Molteno Formation of South Africa. *African Invertebrates* **50**, 105–116.

Sepkoski, J. J. 1984. A kinetic model of Phanerozoic taxonomic diversity. III. Post-Paleozoic families and mass extinctions. *Paleobiology* **10**, 246–267.

Sereno, P. C., Tan, L., Brusatte, S. L., Kriegstein, H. J., Zhao, X-J., Cloward, K. 2009. Tyrannosaurid skeletal design first evolved at small body size. *Science* **326**, 418–422.

Shabica, C. W., Hay, A. A. (eds.). 1997. *Richardson's Guide to the Fossil Fauna of Mazon Creek*. Northeastern Illinois University, Chicago, xvii + 308 pp.

Shaw, C. A., Quinn, J. P. 1986. Rancho La Brea: a look at coastal southern California's past. *California Geology* **39**, 123–133.

Shcherbakov, D. E. 2008. Madygen, Triassic Lagerstätte number one, before and after Sharov. *Alavesia* **2**, 113–124.

Shear, W. A. 1991. The early development of terrestrial ecosystems. *Nature* **351**, 283–289.

Shear, W. A., Selden, P. A. 2001. Rustling in the undergrowth: animals in early terrestrial ecosystems. 29–51. *In* Gensel, P. G., Edwards, D. (eds.). *Plants Invade the Land: Evolutionary and Environmental Perspectives*. Columbia University Press, New York, x + 304 pp.

Shear, W. A., Selden, P. A., Gall, J-C. 2009. Millipedes from the Grès à Voltzia, Triassic of France, with comments on Mesozoic millipedes (Diplopoda: Helminthomorpha: Eugnatha). *International Journal of Myriapodology* **2**, 1–13.

Shu, D-G., Luo, H-L., Conway Morris, S. 1999. Lower Cambrian vertebrates from south China. *Nature* **402**, 42–46.

Shu, D-G., Conway Morris, S., Han, J. 2001. Primitive deuterostomes from the Chengjiang Lagerstätte (Lower Cambrian, China). *Nature* **414**, 419–424.

Shu, D-G., Conway Morris, S., Zhang, Z-F., Han, J. 2009. The earliest history of the deuterostomes: the importance of the Chengjiang Fossil-Lagerstätte. *Proceedings of the Royal Society of London*, Series B **277**, 165–174.

Sidor, C. A., Smith, R. M. H. 2004. A new galesaurid (Therapsida: Cynodontia) from the Lower Triassic of South Africa. *Palaeontology* **47**, 535–556.

Sidor, C. A., Smith, R. M. H. 2007. A second burnetiamorph therapsid from the Permian Teekloof Formation of South Africa and its associated fauna. *Journal of Vertebrate Paleontology* **10**, 420–430.

Siveter, D. J., Sutton, M. D., Briggs, D. E. G., Siveter, D. J. 2003. An ostracod crustacean with soft parts from the Lower Silurian. *Science* **302**, 1749–1751.

Siveter, D. J., Sutton, M. D., Briggs, D. E. G., Siveter, D. J.

2004. A Silurian sea spider. *Nature* **431**, 978–980.

Siveter, D. J., Aitchison, J. C., Siveter, D. J., Sutton, M. D. 2007a. The Radiolaria of the Herefordshire Konservat-Lagerstätte (Silurian), England. *Journal of Micropalaeontology* **26**, 1–8.

Siveter, D. J., Fortey, R. A., Sutton, M. D., Briggs, D. E. G., Siveter, D. J. 2007b. A Silurian 'marrellomorph' arthropod. *Proceedings of the Royal Society of London*, Series B **274**, 2223–2229.

Siveter, D. J., Siveter, D. J., Sutton, M. D., Briggs, D. E. G. 2007c. Brood care in a Silurian ostracod. *Proceedings of the Royal Society of London*, Series B **274**, 465–469.

Siveter, D. J., Sutton, M. D., Briggs, D. E. G., Siveter, D. J. 2007d. A new probable stem lineage crustacean with three-dimensionally preserved soft parts from the Herefordshire (Silurian) Lagerstätte, UK. *Proceedings of the Royal Society of London*, Series B **274**, 2099–2107.

Siveter, D. J., Briggs, D. E. G., Siveter, D. J., Sutton, M. D. 2010. An exceptionally preserved myodocopid ostracod from the Silurian of Herefordshire, UK. *Proceedings of the Royal Society of London*, Series B **277**, 1539–1544.

Small, H. 1913. Geologia e suprimento de água subterrânea no Ceará e parte do Piaui. *Inspectorat Obras contra Secas, Series Geologia* **25**, 1–180.

Smith, R. M. H. 1987. Helical burrow casts of therapsid origin from the Beaufort Group (Permian) of South Africa. *Palaeogeography, Palaeoclimatology, Palaeoecology* **60**, 155–170.

Smith, R. M. H. 1993a. Vertebrate taphonomy of Late Permian floodplain deposits in the southwestern Karoo Basin of South Africa. *Palaios* **8**, 45–67.

Smith, R. M. H. 1993b. Sedimentology and ichnology of floodplain paleosurfaces in the Beaufort Group (Late Permian), Karoo Sequence, South Africa. *Palaios* **8**, 339–357.

Smith, R. M. H. 1995. Changing fluvial environments across the Permian–Triassic boundary in the Karoo Basin, South Africa and possible causes of tetrapod extinctions. *Palaeogeography, Palaeoclimatology, Palaeoecology* **117**, 81–104.

Smith, R. M. H., Evans, S. E. 1995. An aggregation of juvenile *Youngina* from the Beaufort Group, Karoo Basin, South Africa. *Palaeontologia Africana* **32**, 45–49.

Smith, R. M. H., Ward, P. D. 2001. Pattern of vertebrate extinctions across an event bed at the Permian–Triassic boundary in the Karoo Basin of South Africa. *Geology* **28**, 227–230.

Smith, R. M. H., Botha, J. 2005. The recovery of terrestrial vertebrate diversity in the South African Karoo Basin after the end-Permian extinction. *Comptes Rendus Palevol* **4**, 623–636.

Soriano, C., Delclòs, X., Ponomarenko, A. G. 2007. Beetle associations (Insecta: Coleoptera) from the Barremian (Lower Cretaceous) of Spain. *Alavesia* **1**, 81–88.

Sperling, E. A., Vinther, J. 2010. A placozoan affinity for *Dickinsonia* and the evolution of late Proterozoic metazoan feeding modes. *Evolution & Development* **12**, 201–209.

Spix, J. B. von, Martius, C. F. P. 1823–1831. *Reise in Brasilien*. Munich, 1388 pp.

Sprigg, R. C. 1947. Early Cambrian (?) jellyfishes from the Flinders Ranges, South Australia. *Transactions of the Royal Society of South Australia* **71**, 212–224.

Sprigg, R. C. 1949. Early Cambrian 'jellyfishes' of Ediacara, South Australia and Mount John, Kimberley District, Western Australia. *Transactions of the Royal Society of South Australia* **73**, 72–99.

Steiner, G., Salvini-Plawen, L. 2001. *Acaenoplax* – polychaete or mollusc? *Nature* **414**, 601–602.

Steiner, M. B., Eshet, Y., Rampino, M. R., Schwindt, D. M. 2003. Fungal abundance spike and the Permian–Triassic boundary in the Karoo Supergroup (South Africa). *Palaeogeography, Palaeoclimatology, Palaeoecology* **194**, 405–414.

Stock, C. 1930. Rancho La Brea: a record of Pleistocene life in California. *Natural History Museum of Los Angeles County, Science Series* **1**, 1–84.

Stock, C., Harris, J. M. 1992. Rancho La Brea: a record of Pleistocene life in California. *Natural History Museum of Los Angeles County, Science Series* **37**, 1–113.

Størmer, L. 1976. Arthropods from the Lower Devonian (Lower Emsian) of Alken-an-der-Mosel, Germany. Part 5. Myriapoda and additional forms, with general remarks on fauna and problems regarding invasion of land by arthropods. *Senckenbergiana Lethaea* **57**, 87–183.

Sturm, M. 1978. Maw contents of an Eocene horse (*Propalaeotherium*) out of the oil shale of Messel near Darmstadt. *Courier Forschungsinstitut Senckenberg* **30**, 120–122.

Stürmer, W. 1970. Soft parts of cephalopods and trilobites: some surprising results of x-ray examination of Devonian slates. *Science* **170**, 1300–1302.

Stürmer, W., Bergström, J. 1973. New discoveries on trilobites by x-rays. *Paläontologische Zeitschrift* **47**, 104–141.

Stürmer, W., Bergström, J. 1976. The arthropods *Mimetaster* and *Vachonisia* from the Devonian Hunsrück Shale. *Paläontologische Zeitschrift* **50**, 78–111.

Südkamp, W. H. 1997. Discovery of soft parts of a fossil brachiopod in the 'Hunsrückschiefer' (Lower Devonian, Germany). *Paläontologische Zeitschrift* **71**, 91–95.

Südkamp, W. H. 2007. An atypical fauna in the Lower Devonian Hunsrück Slate of Germany. *Paläontologische Zeitschrift* **81**, 181–204.

Südkamp, W. H., Burrow, C. J. 2007. The acanthodian *Machaeracanthus* from the Lower Devonian Hunsrück Slate of the Hunsrück region (Germany). *Paläontologische Zeitschrift* **81**, 97–104.

Sullivan, C., Reisz, R. R., Smith, R. M. H. 2003. The Permian mammal-like herbivore *Diictodon*, the oldest known example of sexually dimorphic armament. *Proceedings of the Royal Society of London*, Series B **270**, 173–178.

Sundell, K. A. 1997. Oreodonts: extinct large burrowing mammals of the Oligocene. *Tate Museum Publication* **2**, 31–43.

Sutcliffe, O. E., Briggs, D. E. G., Bartels, C. 1999. Ichnological evidence for the environmental setting of the Fossil-Lagerstätten in the Devonian Hunsrück Slate, Germany. *Geology* **27**, 275–278.

Sutton, M. D., Briggs, D. E. G., Siveter, D. J., Siveter, D. J. 2001a. Methodologies for the visualization and reconstruction of three-dimensional fossils from the Silurian Herefordshire Lagerstätte. *Palaeontologia Electronica* **4**(1), 1–17. (http://palaeo-electronica.org/2001_1/s2/issue1_01.htm)

Sutton, M. D., Briggs, D. E. G., Siveter, D. J., Siveter, D. J. 2001b. A three-dimensionally preserved fossil polychaete worm from the Silurian of Herefordshire, England. *Proceedings of the Royal Society of London*, Series B **268**, 2355–2363.

Sutton, M. D., Briggs, D. E. G., Siveter, D. J., Siveter, D. J., Orr, P. J. 2002. The arthropod *Offacolus kingi* (Chelicerata) from the Silurian of Herefordshire, England: computer based morphological reconstructions and phylogenetic affinities. *Proceedings of the Royal Society of London*, Series B **269**, 1195–1203.

Sutton, M. D., Briggs, D. E. G., Siveter, D. J., Siveter, D. J., Gladwell, D. J. 2005. A starfish with three-dimensionally preserved soft parts from the Silurian of England. *Proceedings of the Royal Society of London*, Series B **272**, 1001–1006.

Sutton, M. D., Briggs, D. E. G., Siveter, D. J., Siveter, D. J. 2006. Fossilized soft tissues in a Silurian platyceratid gastropod. *Proceedings of the Royal Society of London*, Series B **273**, 1039–1044.

Sutton, M. D., Briggs, D. E. G., Siveter, D. J., Siveter, D. J. 2010. A soft-bodied lophophorate from the Silurian of England. *Biology Letters* **7**, 146–149 (published online August 2010).

Swift, C. C. 1979. Freshwater fish of the Rancho La Brea

deposit. *Abstracts, Annual Meeting Southern California Academy of Science* **88**, 44.

Swisher III, C. C., Wang, Y-Q., Wang, X-L., Xu, X., Wang, Y. 1999. Cretaceous age for the feathered dinosaurs of Liaoning, China. *Nature* **400**, 58–61.

Taylor, T. N., Hass, H., Kerp, H. 1997. A cyanolichen from the Lower Devonian Rhynie Chert. *American Journal of Botany* **84**, 992–1004.

Taylor, T. N., Klavins, S. D., Krings, M., Taylor, E. L., Kerp, H., Hass, H. 2004. Fungi from the Rhynie chert: a view from the dark side. *Transactions of the Royal Society of Edinburgh: Earth Sciences* **94**, 457–473.

Taylor, T. N., Kerp, H., Hass, H. 2005. Life history biology of early land plants: Deciphering the gametophyte phase. *Proceedings of the National Academy of Sciences of the USA* **102**, 5892–5897.

Taylor, W. A., Wellman, C. H. 2009. Ultrastructure of enigmatic phytoclasts (banded tubes) from the Silurian–Lower Devonian: evidence for affinities and role in early terrestrial ecosystems. *Palaios* **24**, 167–180.

Theron, J. N., Rickards, R. B., Aldridge, R. J. 1990. Bedding plane assemblages of *Promissum pulchrum*, a new giant Ashgill conodont from the Table Mountain Group, South Africa. *Palaeontology* **33**, 577–594.

Tillyard, R. J. 1928. Some remarks on the Devonian fossil insects from the Rhynie chert beds, Old Red Sandstone. *Transactions of the Entomological Society of London* **76**, 65–71.

du Toit, A. L. 1921. The Carboniferous glaciation of South Africa. *Transactions of the Geological Society of South Africa* **24**, 188–227.

Trewin, N. H. 1985. Mass mortalities of Devonian fish – the Achanarras Fish Bed, Caithness. *Geology Today* **2**, 45–49.

Trewin, N. H. 1986. Palaeoecology and sedimentology of the Achanarras fish bed of the Middle Old Red Sandstone, Scotland. *Transactions of the Royal Society of Edinburgh: Earth Sciences* **77**, 21–46.

Trewin, N. H. 1994. Depositional environment and preservation of biota in the Lower Devonian hot-springs of Rhynie, Aberdeenshire, Scotland. *Transactions of the Royal Society of Edinburgh: Earth Sciences* **84**, 433–442.

Ulmer, G. 1912. Die Trichopteren des baltischen Bernsteins. *Beiträge zur Naturkund Preussens, Königsberg* **10**, 1–380.

Unwin, D. M., Martill, D. M. 2007. Pterosaurs of the Crato Formation. 475–524. *In* Martill, D. M., Bechly, G., Loveridge, R. F. (eds.). *The Crato Fossil Beds of Brazil: Window into an Ancient World*. Cambridge University Press, Cambridge, xvi + 625 pp.

Van Roy, P., Orr, P. J., Botting, J. P., Muir, L. A., Vinther, J., Lefebvre, B., el Hariri, K., Briggs, D. E. G. 2010. Ordovician faunas of Burgess Shale type. *Nature* **465**, 215–218.

Vidal, L. M. 1899. Compte-rendu des excursions dans la province de Lérida du 11 au 15 Octobre. *Bulletin de la Société Géologique de France, Sér.* **3**, 20, 884–900.

Vidal, L. M. 1902. Nota sobre la presencia del tramo Kimeridgense en el Montsech (Lerida) y hallazgo de un batracio en sus hiladas. *Memorias de la Real Academia de Ciencias y Artes de Barcelona* **4**, 263–267.

Viohl, G. 1985. Geology of the Solnhofen lithographic limestone and the habitat of *Archaeopteryx*. 31–44. *In* Hecht, M. K., Ostrom, J. H., Viohl, G., Wellnhofer, P. (eds.). *The Beginnings of Birds*. Proceedings of the International *Archaeopteryx* Conference, Eichstätt, 1984, 382 pp.

Viohl, G. 1996. The paleoenvironment of the Late Jurassic fishes from the southern Franconian Alb (Bavaria, Germany). 513–528. *In* Arratia, G., Viohl, G. (eds.). *Mesozoic fishes – Systematics and Paleoecology*. Proceedings of the International meeting, Eichstätt, 1993. Dr Friedrich Pfeil Verlag, Munich, 576 pp.

Vullo, R., Buscalioni, A. D., Marugán-Lobón, J., Moratalia, J. J. 2009. First pterosaur remains from the Early Cretaceous Lagerstätte of Las Hoyas, Spain: palaeoecological significance. *Geological Magazine* **146**, 931–936.

Wang, S-S., Hu, H-G., Li, P-X., Wang, Y-Q. 2001. Further discussion on geologic age of Sihetun vertebrate assemblage in western Liaoning, China: evidence from Ar-Ar dating. *Acta Petrologica Sinica* **17**, 663–668.

Wang, X-L., Kellner, A. W. A., Zhou, Z., Campos, D. 2005. Pterosaur diversity and faunal turnover in Cretaceous terrestrial ecosystems in China. *Nature* **437**, 875–879.

Ward, P. D., Montgomery, D. R., Smith, R. M. H. 2000. Altered river morphology in South Africa related to the Permian–Triassic extinction. *Science* **289**, 1741–1743.

Ward, P. D., Botha, J., Buick, R., De Kock, M. O., Erwin, D. H., Garrison, G., Kirschvink, J. L., Smith, R. H. M. 2005. Abrupt and gradual extinction among Late Permian land vertebrates in the Karoo Basin, South Africa. *Science* **307**, 709–714.

Weitschat, W. 1980. *Leben im Bernstein*. Geologisch-Paläontologisches Institut der Universität Hamburg, Hamburg, 48 pp.

Weitschat, W., Wichard, W. 2002. *Atlas of Plants and Animals in Baltic Amber*. Verlag Dr Friedrich Pfeil, Munich, 256 pp.

Wellnhofer, P., Buffetaut, E. 1999. Pterosaur remains from the Cretaceous of Morocco. *Paläontologische Zeitschrift* **73**, 133–142.

Westphal, F. 1980. *Chelotriton robustus*, n. sp., ein Salamandride aus dem Eozän der Grube Messel bei Darmstadt. *Senckenbergiana Lethaea* **60**, 475–487.

Whalley, P. E., Jarzembowski, E. A. 1981. A new assessment of *Rhyniella*, the earliest known insect, from the Devonian of Rhynie, Scotland. *Nature* **291**, 317.

Wheeler, W. M. 1915. The ants of the Baltic amber. *Schriften der (Königlichen) Physikalischen-Ökonomischen Gesellschaft zu Königsberg* **55**, 1–11.

Whittington, H. B. 1971. Redescription of *Marrella splendens* (Trilobitoidea) from the Burgess Shale, Middle Cambrian, British Columbia. *Bulletin of the Geological Survey of Canada* **209**, 1–24.

Whittington, H. B. 1975. The enigmatic animal *Opabinia regalis*, Middle Cambrian, Burgess Shale, British Columbia. *Philosophical Transactions of the Royal Society of London*, Series B **271**, 1–43.

Whittington, H. B., Briggs, D. E. G. 1985. The largest Cambrian animal, *Anomalocaris*, Burgess Shale, British Columbia. *Philosophical Transactions of the Royal Society of London*, Series B **309**, 569–609.

Whittle, R. J., Gabbott, S. E., Aldridge, R. J., Theron, J. N. 2009. An Ordovician lobopodian from the Soom Shale Lagerstätte, South Africa. *Palaeontology* **52**, 561–567.

Wild, R. 1990. Holzmaden. 282–285. *In* Briggs, D.E.G., Crowther, P.R. (eds). *Palaeobiology: a Synthesis*. Blackwell Scientific Publications, Oxford, xiii + 583 pp.

Wilson, H. M., Almond, J. E. 2001. New euthycarcinoids and an enigmatic arthropod from the British Coal Measures. *Palaeontology* **44**, 143–156.

Wilson, H. M., Martill, D. M. 2001. A new japygid dipluran from the Lower Cretaceous of Brazil. *Palaeontology* **44**, 1025–1031.

Wilson, H. M., Anderson, L. I. 2004. Morphology and taxonomy of Paleozoic millipedes (Diplopoda: Chilognatha: Archipolypoda) from Scotland. *Journal of Paleontology* **78**, 169–184.

Wolfe, A. P., Tappert, R., Muehlenbachs, K., Boudreau, M., McKellar, R. C., Basinger, J. F. 2009. A new proposal concerning the botanical origin of Baltic amber. *Proceedings of the Royal Society of London*, Series B **276**, 3403–3412.

Wunderlich, J. 1982. Sex im Bernstein: ein fossiles Spinnenpaar. *Neue Entomologische Nachträge* **2**, 9–11.

Wunderlich, J. 1986. *Spinnen Gestern und Heute. Fossil Spinnen in Bernstein und Ihre Heute Lebenden Verwandten*. Erich Bauer Verlag, Wiesbaden, 283 pp.

Wunderlich, J. 1988. Die fossilen Spinnen (Araneae) im Baltischen Bernstein. *Beiträge zur Araneologie* **3**, 1–280.

Xu, X., Tang, Z-L., Wang, X-L. 1999a. A therizinosauroid dinosaur with integumentary structures from China. *Nature* **399**, 350–354.

Xu, X., Wang, X-L., Wu, X-C. 1999b. A dromaeosaurid dinosaur with a filamentous integument from the Yixian Formation of China. *Nature* **401**, 262–266.

Xu, X., Zhou, Z-H., Wang, X-L. 2000. The smallest known non-avian theropod dinosaur. *Nature* **408**, 705–708.

Xu, X., Cheng, Y-N., Wang, X-L., Chang, C-H. 2002a. An unusual oviraptorosaurian dinosaur from China. *Nature* **419**, 291–293.

Xu, X., Norell, M. A., Wang, X-L., Makovicky, P. J., Wu, X-C. 2002b. A basal troodontid from the Early Cretaceous of China. *Nature* **415**, 780–784.

Xu, X., Zhou, Z-H., Wang, X-L., Kuang, X-W. Zhang, F-C., Du, X-K. 2003. Four-winged dinosaurs from China. *Nature* **421**, 335–340.

Xu, X., Norell, M. A. 2004. A new troodontid dinosaur from China with avian-like sleeping posture. *Nature* **431**, 838–841.

Xu, X., Norell, M. A., Kuang, X., Wang, X-L., Zhao, Q., Jia, C-K. 2004. Basal tyrannosauroids from China and evidence for protofeathers in tyrannosaurids. *Nature* **431**, 680–684.

Yalden, D. W. 1985. Forelimb function in *Archaeopteryx*. 91–97. *In* Hecht, M. K., Ostrom, J. H., Viohl, G., Wellnhofer, P. (eds.). *The Beginnings of Birds*. Proceedings of the International *Archaeopteryx* Conference, Eichstätt, 1984, 382 pp.

Zangerl, R., Richardson, E. S. 1963. The paleoecological history of two Pennsylvanian black shales. *Fieldiana Geology Memoir* **4**, 1–352.

Zhang, F-C., Kearns, S. L., Orr, P. J., Benton, M. J., Zhou, Z., Johnson, D., Xu, X., Wang, X.-L. 2010. Fossilized melanosomes and the colour of Cretaceous dinosaurs and birds. *Nature* **463**, 1075–1078.

Zhang, X., Hua, H., Reitner, J. 2006. A new type of Precambrian megascopic fossils: the Jinxian biota from northeastern China. *Facies* **52**, 169–181.

Zhao, F-C., Caron, J-B., Hu, S-X., Zhu, M-Y. 2009. Quantitative analysis of taphofacies and paleocommunities in the early Cambrian Changjiang Lagerstätte. *Palaios* **24**, 826–839.

Zheng, X-T., You, H-L., Xu, X., Dong, Z-M. 2009. An Early Cretaceous heterodontosaurid dinosaur with filamentous integumentary structures. *Nature* **458**, 333–336.

Zhou, Z-H., Barrett, P. M., Hilton, J. 2003. An exceptionally preserved Lower Cretaceous ecosystem. *Nature* **421**, 807–814.

Zhu, M-Y., Zhang, J-M., Li, G-X. 2001. Sedimentary environments of the early Cambrian Chengjiang biota: sedimentology of the Yu'anshan Formation in Chengjiang County, eastern Yunnan. *In*: Zhu, M-Y, Van Iten, H., Peng, S-C., Li, G-X. (eds). The Cambrian of South China. *Acta Palaeontologica Sinica* **40**, 80–105.

Zimmerman, M. R., Tedford, R. H. 1976. Histologic structures preserved for 21,300 years. *Science* **194**, 183–184.

Zompro, O. 2001. The Phasmatodea and Raptophasma n. gen., Orthoptera incertae sedis, in Baltic amber (Insecta: Orthoptera). *Mitteilungen der Geologisch-Paläontologisches Institut der Universität Hamburg* **85**, 229–261.

Zompro, O. Adis, J., Weitschat, W. 2002. A review of the order Mantophasmatodea (Insecta). *Zoologischer Anzeiger* **241**, 269–279.

INDEX

T - #0939 - 101024 - C288 - 261/194/13 - PB - 9781840761603 - Matt Lamination